MATHEMATICS
&Climate

MATHEMATICS
&Climate

HANS KAPER

Mathematics and Climate Research Network (MCRN)

Argonne National Laboratory
Argonne, Illinois

Georgetown University
Washington, District of Columbia

HANS ENGLER

Georgetown University
Washington, District of Columbia

Society for Industrial and Applied Mathematics
Philadelphia

Library of Congress Cataloging-in-Publication Data

Kaper, H. G.
  Mathematics and climate / Hans Kaper, Mathematics and Climate Research Network (MCRN), Argonne National Laboratory, Argonne, Illinois, Georgetown University, Washington, District of Columbia, Hans Engler, Georgetown University, Washington, District of Columbia.
       pages cm
  Includes bibliographical references and index.
  ISBN 978-1-611972-60-3
  1. Climatology--Mathematical models. I. Engler, Hans, 1953- II. Title.
  QC981.K37 2013
  551.601'51--dc23

                          2013019839

Partial royalties from the sale of this book are donated to the Jürgen Moser Lecture Fund.

 is a registered trademark.

We dedicate this book
to the next generation
of students and researchers.

Contents

List of Figures

Preface

Understanding the Earth's climate system and predicting its behavior under a range of "what if" scenarios are among the greatest challenges for science today. The stakes are high, and decision makers have more questions than science can answer.

The year 2013 is devoted to "Mathematics of Planet Earth" (MPE2013). This book is part of the effort to realize one of the goals of MPE2013, "to encourage research in identifying and solving fundamental questions about planet Earth" [69]. Mathematics has an important role to play; controlled physical experiments are out of the question, and the only way we can study Earth's climate system is through mathematical models, computational experiments, and data analysis.

The purpose of this book is to introduce students to mathematically interesting topics from climate science. The book grew out of a course on *Mathematics and Climate* for first-year Master's level students in the Mathematics and Statistics Department of Georgetown University. Likewise, our target audience for the book consists of advanced undergraduate students and beginning graduate students in mathematics.

Students today appear to be well attuned to the need for scientific arguments in the debate about climate and climate change, probably because their lives will be influenced directly by the uncontrolled experiments that have been going on since the early days of industrialization. The course was designed to respond to this interest. It differed from a general introduction to applied mathematics; instead of a course where the emphasis is on common techniques that can be used in many application areas, the emphasis here was on mathematical and statistical techniques in the context of climate.

In writing this book, we have tried to make the discussion of climate issues understandable to readers coming from fields other than geophysics. We assume that the student reader is familiar with calculus, linear algebra, ordinary differential equations, and statistics, and is comfortable with mathematical models of natural phenomena.

While the mathematical and statistical material mostly comes from standard courses in applied mathematics and multivariate statistics, many of the applications in climate science were found in original research papers from the past decades. Throughout, we emphasize the use of conceptual models—mathematical descriptions that retain only some essential aspects of the climate system. Through inspired model reduction and sometimes just clever guessing, it is often possible to come up with such models that are still capable of reproducing faithfully quite complex phenomena. They often lead to new scientific questions and have at times given rise to new terms and concepts in climate science. We also discuss some statistical topics in depth. This is ordinarily not done in a course in applied mathematics but is essential in climate science, which has been revolutionized by enhanced capabilities for data collection and data analysis.

The book consists of twenty chapters and three appendices. Every chapter concludes with a set of exercises. Some exercises serve to work out details, others serve to extend the theory presented in the text, and some exercises are meant to stimulate critical thinking and discussion in class. Also, in some exercises students are asked to explain technical results in simple terms, to emphasize the importance of good communication skills in the public arena.

Although the book touches on many aspects of the Earth's climate system, it is not meant to give a general introduction to climate science, and we certainly do not claim that it will take the reader to the frontiers of climate science research. But students who are looking for an exciting area of applied mathematics or statistics may find that climate offers ample opportunities for applications that are relevant well beyond the classroom.

The introductory Chapter 1 addresses some general questions. What is climate? What do we mean by climate change? Is it possible to construct a model of Earth's climate system? What would such a model look like? What are the challenges, both mathematical and statistical? What can we learn from data?

Chapter 2 introduces the concept of a global energy balance model—a conceptual model of the state of the climate system formulated in terms of a single variable, namely the global mean surface temperature. The current climate corresponds to an equilibrium state, but the model also points to the possibility of other equilibrium states. Here the concept of a dynamical system and the phenomenon of bifurcation make their first appearance.

Chapter 3 is concerned with the Earth's oceans. Ocean circulation is driven by temperature and salinity differences, and a simple way to study it at the conceptual level is through box models. Box models give rise to low-dimensional dynamical systems. We discuss a very simple one-dimensional model and demonstrate the phenomenon of bifurcation and hysteresis.

Conceptual models of the climate system most often involve systems of ordinary differential equations. Dynamical systems theory provides a framework for the study of the qualitative behavior of their solutions. In Chapter 4 we therefore introduce the basic elements of this theory.

In a mathematical model, forcing scenarios are realized by changing parameter values. In many instances, the interest is on critical parameter values where the qualitative behavior of the solution changes. Such phenomena fall under the rubric of bifurcation theory, which is introduced in Chapter 5.

The next two chapters are devoted to applications of the theory of dynamical systems and bifurcations. Chapter 6 continues the discussion of box models describing ocean circulation, with a focus on the two-box model first introduced by Stommel. This model gives rise to a two-dimensional dynamical system, as well as some interesting bifurcation phenomena. Chapter 7 discusses the famous Lorenz model, introduced in 1963 as a caricature of the Earth's atmosphere, and the notion of chaotic dynamics.

Chapter 8 shifts attention toward statistics. The chapter describes some of the specific challenges for statistics that arise in climate science and introduces basic terminology.

Chapter 9 presents regression analysis—the statistical technique for estimating functional relationships between observed variables and some of its applications in climate science.

The best known indicator of anthropogenic effects on the composition of the Earth's atmosphere is the Keeling curve—a record of CO_2 data collected since 1958 on the Mauna Loa Volcano in Hawaii Island. Chapter 10 describes the statistical techniques that are used to obtain and analyze this curve.

The following Chapter 11 is devoted to the Fourier transform as a tool to extract information from time series. We introduce the fast Fourier transform (FFT), which is an indispensable tool in many areas of science. We illustrate its use in climate science with applications in paleoclimatology—the study of climate in prehistoric times. A theory due to Milankovitch suggests that past glacial cycles are correlated to cyclical changes in the amount of solar energy reaching the Earth from the Sun. We show how Fourier analysis is used to test this theory.

In Chapter 12, we generalize the theory of Chapter 2 to the case of a climate system where the mean temperature varies with latitude and thermal energy is exchanged among latitudes by diffusion. The resulting zonal energy balance model gives rise to a partial differential equation. We use it to introduce the concept of a spectral method, which allows us to reduce the partial differential equation to an infinite set of ordinary differential equations. With a proper choice of the parameter values, the model reproduces the current climate with ice caps near the North and South Pole surprisingly well.

The short Chapter 13 discusses some general characteristics of the Earth's atmosphere and serves as an introduction to Chapter 14, which deals with the fundamental equations of hydrodynamics—the continuity equation and Navier–Stokes equation. These are partial differential equations for the density and velocity, respectively, which are at the core of every general circulation model (GCM). They are often approximated by the shallow water equations and the Boussinesq approximation, which we also introduce in this chapter.

Chapter 15 discusses climate models as abstract infinite-dimensional dynamical systems. An abstract climate model is reduced to a finite-dimensional dynamical system by a dimension-reduction and truncation procedure. We illustrate this approach by deriving the Lorenz equations from the system of partial differential equations for fluid motion and heat conduction between horizontal planes, which is itself a highly simplified model of the atmosphere.

Chapter 16 is devoted to El Niño–Southern Oscillation (ENSO), one of the fundamental drivers of the planetary climate system. We present two conceptual models of ENSO, a harmonic recharge-oscillator model and a nonlinear delayed-oscillator model. The latter is formulated as a delay differential equation, and we introduce some basic theory for such equations.

The next two chapters address the cryosphere and the biosphere. In Chapter 17, we pay particular attention to sea ice, where we find an application for scaling laws. In Chapter 18, the focus is on the biogeochemistry of the carbon cycle and on the phenomenon of algal blooms. Algal blooms are described by a variant of the nutrient–phytoplankton–zooplankton (NPZ) model, coupled with an equation for vertical diffusion. The model involves nonlocal dynamics and gives rise to interesting bifurcation phenomena.

In the final two chapters we return to the domain of statistics. Chapter 19 is devoted to the statistics of extreme events. Extreme weather events capture the attention of the public and can have huge societal impacts. The last Chapter 20 is concerned with data assimilation, a technique to incorporate observational data into a mathematical model. This technique is finding increasing uses in climate science. We present a Bayesian approach, including ensemble filtering techniques.

Three appendices contain auxiliary material: a list of units and symbols, a glossary of common terms, and several MATLAB® programs. An extensive bibliography and index are included at the end. A website for this book may be accessed at http://www.siam.org/books/ot131.

Resources

The following texts are recommended for background reading:

1. R. T. PIERREHUMBERT, *Principles of Planetary Climate*, Cambridge University Press, Cambridge, UK, and New York, NY, USA, 2010.

2. F. W. TAYLOR, *Elementary Climate Physics*, Oxford University Press, New York, NY, USA, 2005.

3. D. ARCHER AND R. PIERREHUMBERT, EDS., *The Warming Papers: The Scientific Foundation for the Climate Change Forecast*, Wiley, Chichester, UK, 2011.

Several data repositories are accessible through the Web:

1. U.S. National Oceanic and Atmospheric Administration (NOAA), National Climatic Data Center (NCDC), *Climate Data Online*, http://www.ncdc.noaa.gov/cdo-web/

2. National Snow and Ice Data Center (NSIDC): data on snow, ice, cryosphere, and climate. http://nsidc.org/

3. National Center for Atmospheric Research (NCAR), *Climate Data Guide*, https://climatedataguide.ucar.edu/

Acknowledgments

In preparing the following text, we have benefited from discussions with our colleagues in the "Mathematics and Climate Research Network" (MCRN), a network funded by the National Science Foundation linking researchers across the United States to develop the mathematics needed to better understand the Earth's climate [66]. The publication of this text is consistent with one of the objectives of the MCRN, namely "to prepare and disseminate educational material for the undergraduate- and graduate-level curriculum." The first author wishes to thank his immediate colleagues, Chris Jones (University of North Carolina), director of MCRN, and Mary Lou Zeeman (Bowdoin College), co-director of MCRN, for their constructive comments and continued interest in this project. Thanks are also due to the Georgetown students who took the course on which this book is based, for their persistent comments, questions, and curiosity.

The authors are indebted to their friends and colleagues Ali Arab, Arjen Doelman, Ken Golden, Nikki Lovenduski, Dick McGehee, Esther Widiasih, and Antonios Zagaris for useful comments, suggestions of additional resources, and permission to use some of their instructional material.

The authors appreciate the feedback they received from colleagues and anonymous reviewers of an earlier draft of the book and thank the staff of SIAM for their patience and technical support.

We thank our spouses, Helen Kaper and Susan Wexler, for their patience and understanding during the period of planning, writing, and rewriting of the book.

Washington, DC   Hans G. Kaper
February, 2013   Hans Engler
                                                                   Georgetown University

Chapter 1

Climate and Mathematics

In this chapter, we introduce the Earth's climate system and explain what we mean by "climate." We emphasize a system-level approach to modeling and use a simple energy balance model to illustrate the significance of conceptual models. We show the decisive role of statistics in discussing climate change and stress the distinction between climate variability and climate change. We conclude with a discussion of a data-driven approach to mathematical and statistical modeling that is unique to climate science.

Keywords: Earth's climate system, climate, weather, climate models, climate variability, climate change, climate and data.

1.1 ▪ Earth's Climate System

What is climate? What do we mean by climate change? Is it possible to construct a model of Earth's climate system? What would such a model look like? What can we learn from such a model? What can we learn from data?

Let us begin at the beginning: "What is climate?" As a wit once put it, "Climate is what we expect, weather is what we get." Weather and climate are indeed closely connected. They are measured by the same physical variables—temperature, precipitation, humidity, atmospheric pressure, and so on—but they are not the same. In a colloquial sense, climate is "averaged weather." When we talk about climate, we mean weather averaged over space and time, so local variations and diurnal (day-night) and random fluctuations have been eliminated. Thus, *climate* refers to the mean state of the *climate system* or, put differently, to the statistics of weather. Accordingly, the term *climate change* refers to changes in the statistics of weather over time.

So what is the Earth's "climate system" and how do we approach it mathematically? Figure 1.1 is a cartoon taken from the most recent Assessment Report (AR4) issued by the Intergovernmental Panel on Climate Change (IPCC) [103]. The IPCC is an international scientific body involving thousands of scientists from all over the world. It was established in 1988 jointly by the United Nations Environment Programme and the World Meteorological Organization, "to provide the world with a clear scientific view on the current state of knowledge in climate change and its potential environmental and socio-economic impacts." Since its establishment, the IPCC has issued four Assessment Reports, the last one (IPCC AR4) in 2007, for which it was awarded the Nobel Peace Prize. The fifth Assessment Report is currently in preparation and expected to be completed in 2014.

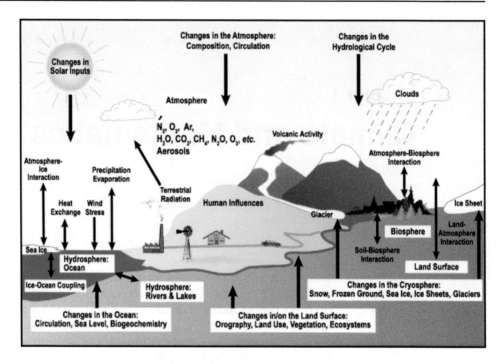

Figure 1.1. *Schematic view of the Earth's climate system. Reprinted with permission from IPCC.*

Looking at the cartoon, we see a climate system that is schematically made up of five components: the *atmosphere*; the *hydrosphere* (oceans, lakes, and other bodies of water); the *cryosphere* (snow and ice); the *lithosphere* (land surface); and the *biosphere* (all living things). The components do not exist in isolation; they are interconnected and interact at several levels, either directly or indirectly. The system as a whole is powered by solar radiation and evolves under the influence of its own internal dynamics through ocean currents and atmospheric circulation. In addition, there are external factors which drive the system; these are called *forcings* and include both natural phenomena such as cyclical changes in the Earth's orbit around the Sun, volcanic eruptions, and variations in the solar output, and human-induced (*anthropogenic*) factors like changes in atmospheric composition, land use, and so on. Let us agree that this is a *complex system*, and let us also agree that it is not obvious how to approach it mathematically.

1.2 ▪ Modeling Earth's Climate

As mathematicians, we are used to setting up models for physical phenomena, usually in the form of equations. For example, we recognize the second-order differential equation

$$L\ddot{x}(t) + g \sin x(t) = 0$$

as a model for the motion of a physical pendulum under the influence of gravity. Every symbol in the equation has its counterpart in the physical world: $x(t)$ stands for the angle between the arm of the pendulum and its rest position (straight down) at the time t; L is the length of the pendulum arm; and g is gravitational acceleration. The mass of the bob turns out to be unimportant and therefore does not appear in the equation. The model is understood by all to be an approximation, and part of the modeling effort consists in outlining the assumptions that went into its formulation. For example, it is assumed that there is no friction in the pendulum joint, there is no air resistance, the arm of the pendulum is massless, and the pendulum bob is idealized to be a single point. Understanding these assumptions and the resulting limitations of the model is an essential part of the modeling effort. Note that the modeling assumptions can all be assessed by an expert who is not a mathematician: a clockmaker can estimate the effect of friction in the joint, the difficulty of making a slender pendulum arm, and the effort in making a bob that offers little air resistance. As mathematicians, we take the differential equation and apply the tools of the trade to extract information about the behavior of the physical pendulum. For example, we can find its period—which is important in the design of pendulum clocks—in terms of measurable quantities.

Would it be possible to develop a "mathematical model" of the Earth's climate system in a similar fashion? Such a model should stay close to physical reality, climate scientists should be able to assess the assumptions, and mathematicians should be able to extract information from it.

Taking the point of view reflected in the cartoon of Figure 1.1, we would need to select variables that describe the state of the climate system (air temperature, humidity, fractions of aerosols and trace gases in the atmosphere, strength of ocean currents, rate of evaporation from vegetation cover, change in land use due to natural cycles and human activity, and many many more), take the rules that govern their evolution (laws of motion for gases and fluids, chemical reaction laws, land use and vegetation patterns, and so on), and translate all this into the language of mathematics. It is not at all clear that this can be done equally well for all components of the system. The laws for airflow over a mountain range may be well known, but it is much harder, for example, to model crop use and changes in vegetation. The ranges and limitations of any such model would remain subject to debate, much more so than in the case of the pendulum equation. The resulting equations would likely cover many pages and be far too unwieldy for a mathematical analysis. The only way to deal with a model of this type would be through numerical simulations, which is indeed the approach that prevails in the climate science research community and that is reflected in the IPCC reports. Such an approach has, of course, become possible thanks to the development of sophisticated algorithms and ever more powerful supercomputers. In fact, climate research has been a major driver for the emergence of the new discipline of *computational science*.

In this book we take a different approach, one that is more attuned to complex systems. One might call it the *system-level approach*. Here, the emphasis is not so much on the processes that go on inside the system as on the qualitative behavior of the climate system (or some of its subsystems) as a whole. Surprisingly often, mathematics can offer perspectives that complement or provide insight into the results of observations and large-scale computational experiments. Through inspired model reduction and sometimes just clever guessing, it is often possible to come up with relatively simple conceptual models that retain some essential features observed in the physical world and reproduce complex phenomena quite faithfully.

1.3 ▪ Conceptual Models

What do we mean by a system-level approach? Here is an example, which we will pursue in more detail in Chapter 2. At the conceptual level, the Earth's climate system is a heat engine that is driven by the Sun. The Earth receives energy from the Sun and maintains a balance by radiating energy back into space, as sketched in Figure 1.2. If too much energy is radiated back, the climate system will cool off; if not enough is radiated back it will heat

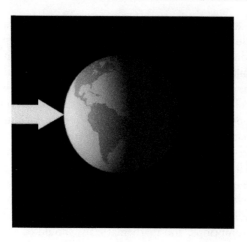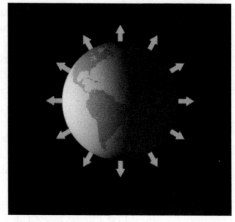

Figure 1.2. *Conceptual model of Earth's climate system: incoming sunlight (shortwave radiation) and outgoing heat (longwave radiation).*

up. This principle is formalized in an *energy balance model*, a simple ordinary differential equation, even simpler than the pendulum equation, for the global mean temperature. The model overlooks all the details of Figure 1.1 and admittedly reflects an extreme point of view, but it already leads to some interesting observations.

The equilibrium state of the climate system can be summed up in the picture shown in Figure 1.3. It shows incoming energy (solid line) and outgoing energy (dotted line) as a function of the global mean surface temperature. The solar output, Earth's motion about the Sun, Earth's topography, and so on are all as today. The dotted line is increasing because a warmer planet radiates more energy. The solid line increases from a low to a high level because at low temperatures (below freezing) the planet is covered by snow and ice,

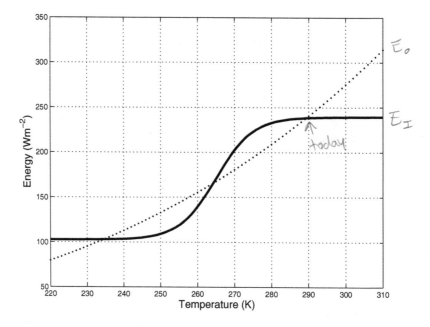

Figure 1.3. *An energy balance model showing that the Earth's climate system can have multiple equilibrium states.*

which reflect back more solar energy and leave less to reach the planet's surface, while at higher temperatures there is less snow and ice, and oceans and land masses absorb more energy. Just as in economics the price for a commodity is expected to settle where supply and demand curves meet, here the equilibrium temperature is expected to be where the two curves intersect. In this case, there are three intersection points. The one furthest to the right corresponds to today's climate. But there are two more, which correspond to much colder climates. Why is the planet at today's climate when much colder climates are also possible? Has the planet ever been in one of the much colder climate states in the past? (The answer is yes.) Is there any danger that Earth could again revert to a much colder climate in the future? How would this happen? Mathematics can raise these questions from a very simple climate model and also support or rule out certain answers using an analysis of the model. The very fact that the Earth's climate system can have more than one equilibrium state, as suggested by this elementary model, has been confirmed subsequently by climate researchers using much more complicated models. Thus, even a simple energy balance model such as this one can lead to new concepts and advance the scientific discussion.

1.4 ▪ **Climate and Statistics**

As stated earlier, climate is the statistics of weather. Here is an example to convince you that understanding statistics is important for the discussion of climate and climate change.

Global mean surface temperatures have been recorded since about 1880. Figure 1.4 shows the global mean temperature anomalies—that is, the difference between the ob-

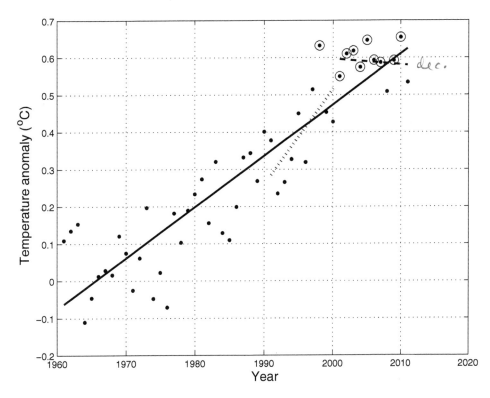

Figure 1.4. *Global mean temperature anomalies for the period 1960–2011 relative to the average global mean temperature since 1880, with various trend lines and records.*

served global mean temperature and the overall average of these temperatures since 1880. Here are some observations and conclusions that have been made based on the data in the plot.

- Most temperature anomalies for the last 50 years are positive. Therefore, the last 50 years have been warmer than the previous half-century, and global warming is a reality.

- A trend line for the years 1960–2011 (the solid line in the plot) shows a mean increase of the global temperature of about 0.014°C per year. Therefore, global warming is a reality.

- A trend line for the years 2001–2011 (the dashed line in the plot) shows a mean decrease of the global mean temperature of about 0.001°C per year. Therefore, global warming, if it ever happened, has stopped, and perhaps the planet is now getting cooler.

- A trend line for the years 1990–2000 (the dotted line in the plot) shows a mean increase of the global temperature of about 0.025°C per year. Together with the cooling of the following decade (dashed line), this shows that global warming did happen, but it has now come to a halt. Since human activity has not changed during this time, the observed temperature variations cannot be anthropogenic.

- Of the 10 warmest years since 1962 (plotted with circles), nine occurred since 2001. This cannot be due to chance, and therefore global warming is a reality.

These conclusions are all based on the same data, and they have been obtained with valid statistical computations. Clearly, they cannot all be correct. How can these conclusions be reconciled?

1.5 ▪ Climate Variability and Climate Change

One of the major challenges for climate science is to identify *climate change* as distinct from *climate variability*. Direct and indirect records of past climates show that the Earth's climate has changed considerably in the past. Figure 1.5 shows two paleoclimate records, one going back 65 million years and the other 3 million years. Both show a significant and complex evolution, a mix of variability and change, with gradual trends of warming and cooling. Of course, these records are not the result of direct temperature measurements; rather, they have been reconstructed from "proxy data"—measurements of quantities that are known to be closely related to the planetary temperature. In this case, the proxy data came from oxygen isotope ratios in deep-sea sediment cores. How reliable are these reconstructions? Which of these changes are likely due to chance and natural variability, and which are actual changes?

The graph in Figure 1.6 suggests more questions. It shows temperatures reconstructed from isotope ratios in ice core samples from the Antarctic. Even though it was reconstructed from very different proxy data, the graph shows a close resemblance to the graph for the last 400,000 years in the bottom panel of Figure 1.5. The periodic coming and going of ice ages is clearly visible, with maybe a longer period for the past one million years or so than before. It is also clear that warming and cooling have occurred on different time scales: a period of rapid warming is followed by a period of slow cooling. Assuming that changes have been separated from variability, can we explain the changes? Can we identify changes in related records that might shed light on the mechanisms for the observed evolution? Clearly, we have more questions than answers, and every answer generates new questions.

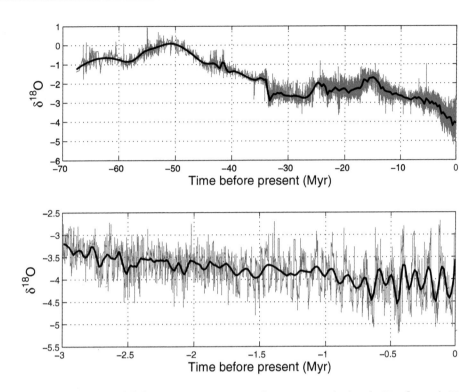

Figure 1.5. *Global mean temperature over the past 65 Myr (top) and 3 Myr (bottom). Time increases from left to right.*

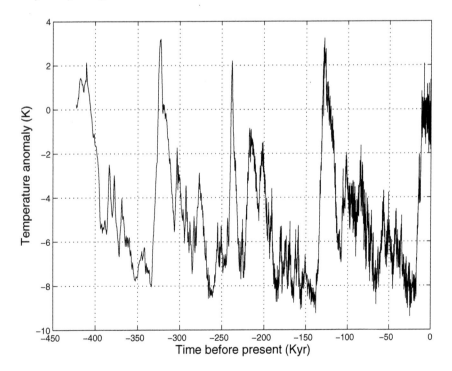

Figure 1.6. *Global mean temperature over the past 420 Kyr. Time increases from left to right. Reprinted courtesy of the U.S. Geological Survey.*

1.6 ▪ Models from Data

In 1904, the British meteorologist SIR GILBERT WALKER (1868–1958) was asked to predict monsoon events in southern and southeast Asia. During his studies, he observed that in years of reduced rainfall over Australia and Indonesia, characterized by unusually high air pressure readings in the western Pacific (Darwin, Australia), barometer readings in the central and eastern Pacific (Tahiti, French Polynesia) tended to be low, and vice versa. He coined the term "Southern Oscillation" for this phenomenon and observed that the resulting droughts over Australia were associated with mild winters in western Canada. Initially criticized for suggesting that these weather patterns might be connected halfway around the globe, he was vindicated when his observations were connected to ocean temperature oscillations off the coast of South America, which occur every three to seven years. These oscillations had been noticed by fishermen in that region, who gave them the name "El Niño," since their adverse effect on local fishing yields tended to occur around Christmas. A comprehensive explanation connecting the two phenomena was first proposed by the Norwegian-American meteorologist JACOB BJERKNES (1897–1975) in the 1960s. The El Niño–Southern Oscillation (ENSO) phenomenon is now considered to be the dominant mode in global climate variability.

Climate scientists have identified several other spatially related or temporally recurring patterns in today's climate system, both in the atmosphere and in the ocean. For example, the North Atlantic Oscillation (NAO) is a North-South oscillatory pattern in atmospheric pressure over the North Atlantic, characterized by the pressure difference between the "Icelandic low" and the "Azores high." This is familiar to people in western and central Europe, where it influences much of the daily weather. The NAO was also discovered by Sir Gilbert.

The Atlantic Multidecadal Oscillation (AMO) is a temporal pattern in sea surface temperatures in the North Atlantic, with a period of about half a century. It has been observed since the mid-1800s. A graph is shown in Figure 1.7. Recent work has suggested that positive anomalies in the AMO (unusually high ocean surface temperatures) may be associated with strong snowfalls on the East Coast of the United States, as happened during the extremely snowy winter of 2009–2010.

In climate science, pairs of related patterns are called *dipole structures*—pairs of regions such that certain variables of interest move together within each region but move in op-

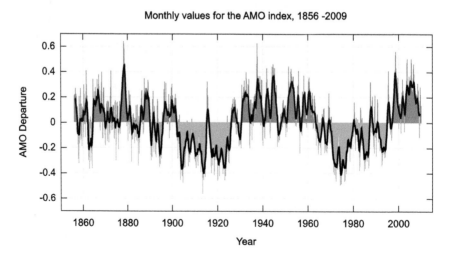

Figure 1.7. *Ocean temperature anomalies for AMO, 1856–2009* [51].

posite ways when compared between regions. Thus, ENSO is characterized by a dipole with centers over the western Pacific and over the central/eastern Pacific. These dipoles interact, and there is strong evidence that their influence can extend over thousands of kilometers, a phenomenon known as *teleconnection*. We emphasize that teleconnections are initially purely observational phenomena. They may manifest themselves differently in different dipole pairs and may require very different physical explanations. Many of the structures are poorly understood, in the sense that not even conceptual models exist for them, although they contribute to global weather variability over time scales of decades. Indeed, a better understanding of interdecadal climate variability is one of the main current goals of the IPCC.

A systematic search for dipole structures and teleconnections is now possible. By applying data mining techniques to the vast amounts of data from satellite observations and computer simulations and data that predate the satellite age, one can reconstruct recent climate states.[1] Figure 1.8 shows some results for sea-level pressure data for the period 1948–1967 generated from the NCEP/NCAR Reanalysis project [75]. The color codes correspond to aggregates of various physical observations. Black lines correspond to teleconnections, which have been identified from 20 years of processed ("reanalyzed") data. Some of these teleconnections are well known. Others may represent new climatological phenomena, they may be consequences of known teleconnections, or they may just be the result of spurious observations.

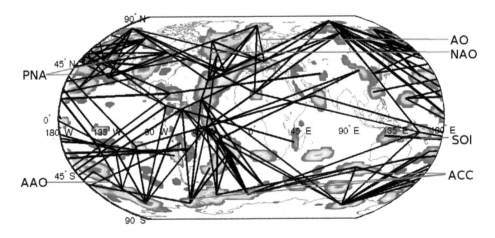

Figure 1.8. *Dipoles in NCEP sea-level data for the period 1948–1967. The color background shows the regions of high activity. The edges represent dipole connections between regions.*

What can we infer from these empirical observations? Do these dipole structures represent oscillatory modes of the climate system? What are possible coupling mechanisms that cause the teleconnections? Is it too far-fetched to think of the climate system as a system of coupled oscillators?

---

[1]This approach was already articulated by Sir Gilbert, who wrote in 1932: "I think that the relationships of world weather are so complex that our only chance of explaining them is to accumulate the facts empirically; we know that it was impossible to explain cyclones (lows) until data of the upper air conditions were available, and there is a strong presumption that when we have data of pressure and temperature at 10 and 20 km, we shall find a number of new relations that are of vital importance."

1.7 ▪ Exercises

1. Like any complex system, the Earth's climate system has many built-in feedback mechanisms. They can be positive (self-reinforcing), negative (self-inhibiting), or sometimes both. For example, if the global average temperature increases, more water evaporates; the atmosphere becomes more humid, and the cloud cover increases; clouds reflect more of the incoming sunlight than soil and water, so less energy reaches the Earth's surface; as a result, the global average temperature decreases (negative feedback). On the other hand, water vapor is a greenhouse gas; greenhouse gases increase the opacity of the atmosphere for infrared radiation, so more heat that is radiated out by the Earth's surface in the form of infrared radiation is trapped by the atmosphere; as a result, the global average temperature increases (positive feedback). Use Figure 1.1 to identify other positive or negative feedback loops in the Earth's climate system.

2. Arctic sea ice extent averaged over January 2011 was 13.55 million square km. This was the lowest January ice extent recorded since satellite records began in 1979. It was 50,000 square km below the record low of 13.60 million square km set in 2006, and 1.27 million square km below the 1979 to 2000 average. What do these data suggest about the Earth's climate?

3. The extent of the Northern Hemisphere snow cover during January 2011 was 1.76 million square km above the long-term average of 46.7 million square km, marking the sixth largest January snow cover extent and the tenth largest monthly snow cover extent on record for the hemisphere. It also marked the fourth consecutive January with above-average snow cover extent for the Northern Hemisphere. What do these data suggest about the Earth's climate?

4. The strength of a preferred climate pattern is quantified by its *index*. For example, the Southern Oscillation Index (SOI) is the mean sea-level pressure (MSLP) anomaly at Tahiti (eastern Pacific) minus the MSLP anomaly at Darwin, Australia (western Pacific), normalized by the long-term mean and standard deviation of the MSLP difference. The SOI is available from the 1860s [70]. Similar indices are associated with the other preferred patterns. Explain how these indices can be used to detect teleconnections.

5. Regional climate in different locations may obviously vary out of phase just because of their position on the globe. For example, it is winter in the Northern Hemisphere when it is summer in the Southern Hemisphere. Explain how regional climates in different locations in the same hemisphere may vary out of phase, and give an example.

6. Figure 1.8 shows many possible dipole connections. Choose two such connections which are identified in the figure and not discussed in the text and write a brief summary for each, using publicly available resources.

7. There are several dipole connections in Figure 1.8 that have been identified and named in the climatological literature, but are not labeled in that figure. Identify one such connection and write a brief summary of it, using publicly available resources.

8. Consider again Figure 1.8. If there is a dipole connection between regions A and B and another one between regions A and C, then it is possible that also a con-

nection between regions B and C is found. Identify two such "triangles" of dipole connections in the figure.

9. Consider Figure 1.8. Some of the dipole connections identified there might come from two regions in opposite hemispheres of the globe that have a similar climate which varies six months out of phase. Are there any obvious candidates for this effect in the figure?

10. Consider Figure 1.8. Some of the dipole connections identified there might be spurious (due to chance). How would you go about identifying such spurious connections? Are there any obvious candidates in the figure?

Chapter 2

Earth's Energy Budget

In this chapter we introduce the concept of a global energy balance model (EBM). A global EBM summarizes the state of the Earth's climate system in a single variable, namely the temperature at the Earth's surface averaged over the entire globe. The time evolution of this variable is governed by an ordinary differential equation, which reduces to an algebraic equation at equilibrium. We show that, with reasonable values of the physical parameters, the current state of the Earth's climate corresponds to a stable equilibrium, but as the solar input decreases, a transition to a second stable equilibrium state, some 50 degrees colder, may occur.

Keywords: Solar spectrum, insolation, conservation law, energy balance, albedo, greenhouse effect, Snowball Earth, bifurcation, hysteresis.

2.1 ▪ Solar Radiation

The Earth's climate system receives virtually all its energy from the Sun. This energy comes in the form of electromagnetic radiation, which originates from different depths in the Sun's interior. Some of the energy is removed by absorption in the solar photosphere, but a good approximation of the *solar energy spectrum* as it reaches the top of the Earth's atmosphere is that of a black body at a temperature of 5,780 K. Further absorption in the Earth's atmosphere gives the solar spectrum at the Earth's surface its more ragged appearance (Figure 2.1).

The *solar constant* or *total solar irradiance* is the solar energy flux through a unit area of an imaginary sphere at the mean distance of the Earth from the Sun; its measured value is 1,368 Wm^{-2}. The solar constant is not really constant; it varies by about 0.1% with the sunspot cycle and has changed more over the lifetime of the Sun.

The more general term *insolation* (*in*cident *sol*ar rad*iation*) refers to the actual energy per unit area that reaches a specific location, usually on the Earth's surface, in a given situation. Unlike the solar constant, which is a constant for all practical purposes, the surface insolation varies substantially with latitude, time of day, season, and cloud cover, and goes to zero at night. It follows from the definition that the insolation at the top of the atmosphere equals the solar constant when the Earth is at its mean distance from the Sun. Figure 2.2 shows the annual average surface insolation in Europe.

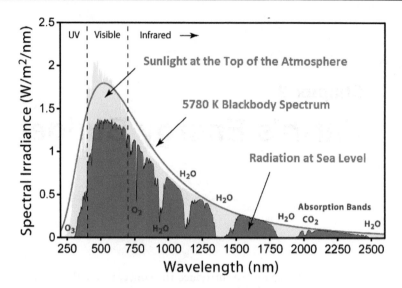

Figure 2.1. *The observed solar spectrum at the top of the atmosphere (yellow) and at sea level on a clear day (red). The solid line (black) is the spectrum the Sun would have if it were a black body at the temperature of 5,780 K. Image courtesy of Global Warming Art.*

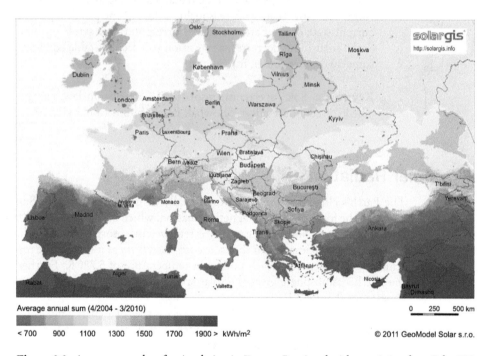

Figure 2.2. *Average annual surface insolation in Europe. Reprinted with permission from SolarGIS.*

2.2 ▪ Energy Balance Models

Although the details of the radiative energy balance of the Earth's atmosphere are complicated and involve many pathways and inherent uncertainties (see Figure 2.3), it is possible to gain some insight from a very simple observation, namely that the global average temperature at the Earth's surface increases if the amount of energy reaching the Earth

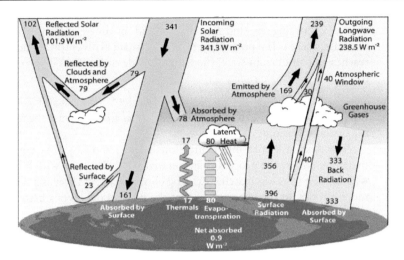

Figure 2.3. *Detailed radiative energy balance* [112].

from the Sun exceeds the amount of energy emitted by the Earth and released into the stratosphere, and decreases if the converse is true. To translate this observation into a mathematical equation, recall the sketch in Figure 1.2 and think of the Earth as a homogeneous solid sphere. We ignore differences in the atmosphere's composition, as well as differences among continents and oceans, topography, and all other local features, and characterize the state of the entire system by a single variable, namely the global mean surface temperature T. We are interested in the evolution of T over time (t).

Consider all components of the climate system that can exchange heat with outer space—all oceans, the entire atmosphere, and the soil of all land masses to a depth of several meters. At time t, the average temperature throughout this entire system is $T(t)$. The energy needed to raise this temperature by one degree Celsius (one kelvin) is known as the *heat capacity* of the system. The heat capacity is commonly measured in watt-years per square meter (W yr m^{-2}), since its actual value depends on the medium under consideration. Estimates vary from approximately 0.55 W yr m^{-2} for soil/atmosphere to approximately 90 W yr m^{-2} for the ocean/atmosphere [36, Supporting online material]. However, we assume for simplicity that the heat capacity is constant over the entire globe and equal to its average value C. If the average temperature, after a time interval Δt, is $T(t + \Delta t) = T(t) + \Delta T$, then the amount of energy needed to reach this temperature is $AC\Delta T$, where A is the surface area of the planet. Since the Sun is the only source of energy, this quantity must equal the net gain of energy due to solar radiation. If E_{in} is the average amount of solar energy reaching one square meter of the Earth's surface per unit time, and E_{out} is the average amount of energy emitted by one square meter of the Earth's surface and released into the stratosphere per unit time, it must be the case that

$$AC\Delta T = A(E_{\text{in}} - E_{\text{out}})\Delta t.$$

Canceling A and dividing both sides of the equation by Δt and letting Δt tend to zero, we see that the evolution of T is described by the global *energy balance model* (EBM),

$$C\frac{dT}{dt} = E_{\text{in}} - E_{\text{out}}. \qquad (2.1)$$

In general, both E_{in} and E_{out} depend on T. If the climate system is forced, E_{in} and/or E_{out} depend explicitly on time, and Eq. (2.1) is a nonautonomous differential equation. In the absence of forcing, E_{in} and E_{out} still depend on time but only implicitly through T. In that case, there is the possibility that the climate evolves toward an equilibrium. Once it reaches an equilibrium state, the left-hand side of Eq. (2.1) is zero, and the EBM reduces to

$$E_{in} = E_{out}. \qquad (2.2)$$

Eq. (2.1) is also referred to as a *zero-dimensional* EBM because the model does not account for spatial variations.

The next task is to find expressions for E_{in} and E_{out}. As with every mathematical modeling effort, the process is an iterative one. We start with a basic model and check whether the solution matches existing climate data. If there is a discrepancy, we bring in more physics, in the expectation that the new model gives a better match with observations.

2.3 ▪ Basic Model

Viewed from the Sun, the Earth is a flat disc whose radius is equal to the radius of the Earth, R, and whose surface area is πR^2. Hence, the amount of solar energy intercepted by the Earth per unit time is $\pi R^2 S_0$, where S_0 is the solar constant. A fraction of this energy is reflected back into space before it reaches the Earth's surface; this fraction is called the (planetary) *albedo* and denoted by α_p (or simply α, if no confusion is possible). The planetary albedo is an average of the local albedo, averaged over the entire globe. The local albedo is a highly variable quantity which depends on many local factors, such as the cloud cover, the composition of the Earth's atmosphere (in particular, the presence of aerosols), the presence of ice on the Earth's surface, etc. The remaining fraction $1 - \alpha$ (called the *co-albedo*) of the incoming solar radiation reaches the Earth's surface. Hence, the amount of solar energy reaching the Earth's surface per unit time is $(1 - \alpha)\pi R^2 S_0$.

By assumption, this energy is distributed uniformly over the Earth's surface area, which is $4\pi R^2$, so the amount of solar energy flowing into a square meter of the Earth's surface per unit time is $(1 - \alpha)\pi R^2 S_0 / (4\pi R^2) = \frac{1}{4}(1 - \alpha)S_0$. We use the abbreviation $Q = \frac{1}{4}S_0$, so the amount of solar energy flowing into the Earth's surface per unit area and per unit time is

$$E_{in} = (1 - \alpha)Q, \quad Q = \tfrac{1}{4}S_0. \qquad (2.3)$$

Next, we turn our attention to E_{out}. Like the Sun, the Earth emits electromagnetic radiation. But unlike the Sun's, the Earth's energy spectrum is almost entirely in the infrared regime (very long wavelengths). The amount of energy radiated out depends on the temperature. As a first approximation, we assume that the Earth radiates as a black body with an effective surface temperature T. Then the average amount of energy radiated out per unit area per unit time follows the Stefan–Boltzmann law,

$$E_{out} : T \mapsto E_{out}(T) = \sigma T^4. \qquad (2.4)$$

Here, σ is Stefan's constant, $\sigma = 5.67 \cdot 10^{-8}$ Wm^{-2}K^{-4}.

With these expressions for E_{in} and E_{out}, the differential equation (2.1) becomes

$$C\frac{dT}{dt} = (1 - \alpha)Q - \sigma T^4, \qquad (2.5)$$

which reduces at equilibrium to the energy balance equation

when $\dfrac{dT}{dt} = 0$
$$(1 - \alpha)Q = \sigma T^4. \qquad (2.6)$$

This equation can be solved for the temperature T; the solution is

$$T^* = \left(\frac{(1-\alpha)Q}{\sigma} \right)^{1/4}. \tag{2.7}$$

A typical value for the Earth's albedo is $\alpha = 0.30$. Recall that $S_0 = 1,368\,\mathrm{Wm^{-2}}$, so $Q = 342\,\mathrm{Wm^{-2}}$. Thus, the basic model gives the equilibrium temperature $T^* = 254.8\,\mathrm{K}$.

The actual value of the surface temperature is $287.7\,\mathrm{K}$. A good part of the difference can be attributed to the *greenhouse effect* of the Earth's atmosphere—the effect of gases like carbon dioxide (CO_2), water vapor, and methane, and aerosols (water droplets, dust particles, etc.).

2.4 ▪ Greenhouse Effect

Since the Sun and the Earth have emitting temperatures that differ by an order of magnitude, their spectra barely overlap (see Figure 2.4). When measured above the atmosphere, about 9% of the solar energy is in the form of ultraviolet radiation (wavelength less than $0.4\,\mu\mathrm{m}$), approximately 38% is visible (wavelength between $0.4\,\mu\mathrm{m}$ and $0.7\,\mu\mathrm{m}$), and about 53% lies in the near infrared regime (wavelength between $0.7\,\mu\mathrm{m}$ and $4\,\mu\mathrm{m}$) [109]. On the other hand, the emission spectrum of the Earth lies entirely in the middle and far infrared regime (wavelengths greater than $5\,\mu\mathrm{m}$). Due to their chemical properties, greenhouse gases increase the opacity of the atmosphere in the infrared (long-wavelength)

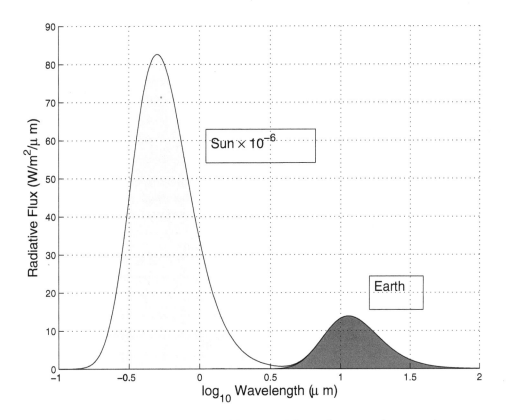

Figure 2.4. *Black-body emission curves of the Sun and the Earth. Reprinted courtesy of NASA.*

regime, so they affect the quantity E_{out} but not E_{in}. As a consequence, the global average temperature increases. To maintain an energy balance, we must somehow compensate this increase in temperature. The simplest way to do this is by reducing the expression for E_{out} by a factor ε, where $0 < \varepsilon < 1$. (Here we see an example of what in climate science is known as *parameterization*.) Thus, the energy balance equation is modified to

$$(1-\alpha)Q = \varepsilon\sigma T^4, \tag{2.8}$$

with a unique solution,

$$T^* = \left(\frac{(1-\alpha)Q}{\varepsilon\sigma}\right)^{1/4}. \tag{2.9}$$

With $\alpha = 0.30$ and $S_0 = 1,368\,\mathrm{Wm^{-2}}$, we recover the value $T = 287.7\,\mathrm{K}$ if we choose $\varepsilon = 0.62$, so with a 38% reduction we can recover the global mean surface temperature in 2010. Of course, we have no idea how realistic this choice of ε is, so in a sense the agreement is artificial.

2.5 ▪ Multiple Equilibria

So far, we have assumed that the planetary albedo is constant and independent of the surface temperature. This assumption overlooks the fact that water turns into snow and ice when the temperature drops below a certain value. Snow and ice have a much higher albedo than open water, so the value $\alpha = 0.30$ may not be appropriate at very low temperatures. We therefore generalize the previous model and introduce a temperature-dependent albedo, $\alpha : T \mapsto \alpha(T)$, such that

$$\alpha(T) \approx \begin{cases} 0.7 & \text{if } T < 250\,\mathrm{K}, \\ 0.3 & \text{if } T > 280\,\mathrm{K}. \end{cases}$$

These constraints guarantee that more energy is reflected if the global mean temperature is sufficiently low and the Earth is covered by snow and ice. The generalization can be accomplished, for example, by setting

$$\alpha(T) = 0.5 - 0.2 \cdot \tanh\left(\frac{T-265}{10}\right).$$

The graph of the co-albedo $1 - \alpha(T)$ vs. T is S-shaped, monotonically increasing from a value just above 0.3 at 250 K to a value just below 0.7 at 280 K, with an inflection point at 265 K, midway between 250 K and 280 K. The factor 10 is more or less arbitrary; it serves to give the graph a reasonable slope at the inflection point.

The incoming energy is now a function of the mean surface temperature, $E_{\text{in}} : T \mapsto E_{\text{in}}(T) = (1-\alpha(T))Q$. With $E_{\text{out}} : T \mapsto E_{\text{out}}(T) = \varepsilon\sigma T^4$ as in Section 2.4, the governing equation for the global mean surface temperature becomes

$$C\frac{dT}{dt} = (1-\alpha(T))Q - \varepsilon\sigma T^4, \tag{2.10}$$

which gives the energy balance equation

$$(1-\alpha(T))Q = \varepsilon\sigma T^4. \tag{2.11}$$

This equation can no longer be solved analytically for the equilibrium temperature. Of course, it is easy to find solutions numerically; however, that is not the point of this section's discussion. Something more interesting is happening here.

Consider Figure 2.5. (This is the same as Figure 1.3, which we used to illustrate a system-level approach to climate modeling.) The S-shaped solid curve is the graph of $(1-\alpha(T))Q$ vs. T; the dotted curve is the graph of $\varepsilon\sigma T^4$ vs. T for $\varepsilon = 0.6$, which increases monotonically with a monotonically increasing slope. The two graphs intersect in three points, at $T_1^* \approx 288\,\text{K}$, $T_2^* \approx 265\,\text{K}$, and $T_3^* \approx 233\,\text{K}$. Each of these points corresponds to an equilibrium state, so this model indicates that the climate system can have *multiple equilibria*.

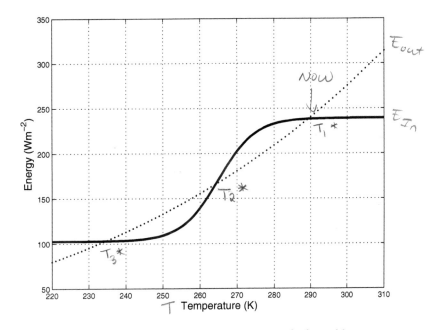

Figure 2.5. *The Earth's climate system can have multiple equilibria.*

Suppose the climate system is in the intermediate equilibrium state at T_2^*. Then we claim that any perturbation, no matter how small, will drive the system away from T_2^*. Indeed, if T is just above T_2^*, the right-hand side of Eq. (2.10) is positive so T is increasing with time, and if T is just below T_2^*, the right-hand side of Eq. (2.10) is negative so T is decreasing with time. In either case, the temperature will move away from T_2^*, and the intermediate equilibrium state at T_2^* is *unstable*. An unstable equilibrium cannot persist and will not be observed in practice.

On the other hand, T_1^* and T_3^* correspond to *stable* equilibrium states. The former corresponds to the current climate, the latter to a climate which is more than 50 degrees colder than the current climate, where the Earth is entirely covered with snow and ice.

2.6 ▪ Budyko's Model

So far, we have assumed that the Earth radiates like a black body, so the outgoing radiation follows the Stefan–Boltzmann law, Eq. (2.4), or a modification thereof. But satellites have been collecting data about the energy radiated out by the Earth, and these satellite data can suggest other, possibly more realistic, models for the outgoing radiation. The approach to use observational data was, in fact, first suggested in the 1960s by Budyko [10], who proposed the simple linear model

$$E_{\text{out}} : T \mapsto E_{\text{out}}(T) = a + bT - (a_1 + b_1 T)n,$$

where a, b, a_1, and b_1 are to be determined from observational data. The factor n is a "cloudiness" coefficient (a number between 0 and 1), so the model actually accounts for average cloudiness conditions, the effects of infrared absorbing gases, and the variability of water vapor, and comes closer to a realistic description of the Earth's atmosphere. Budyko's work appeared before the era of satellites and relied on data from 260 Earth-based observation stations.

Once satellite data became available, Budyko's model was validated, and today it is usually taken in the form

$$E_{\text{out}} : T \mapsto E_{\text{out}}(T) = A + BT. \tag{2.12}$$

The actual values of A and B vary with temperature. The best fit with observational data for the current climate on the Northern Hemisphere is obtained with $A = 203.3\,\text{Wm}^{-2}$ and $B = 2.09\,\text{Wm}^{-2}\text{deg}^{-1}$ [79]. (Here, temperatures are measured in degrees C.) For comparison, a linear expansion of $E_{\text{out}}(T) = \sigma(273.15 + T)^4$ about $T = 0$ would lead to coefficients $A^* = 315.6\,\text{Wm}^{-2}$ and $B^* = 4.62\,\text{Wm}^{-2}\text{deg}^{-1}$.

Keeping a temperature-dependent co-albedo, the governing equation (2.1) becomes

$$C\frac{dT}{dt} = (1 - \alpha(T))Q - (A + BT), \tag{2.13}$$

which leads to the energy balance equation

$$(1 - \alpha(T))Q = A + BT. \tag{2.14}$$

The graphics are almost the same as in Figure 2.5. The graph of $(1 - \alpha(T))Q$ vs. T (solid curve) is unchanged, and instead of the Stefan–Boltzmann curve (dotted curve) we now have a straight line, which can intersect the S-shaped solid curve in three points, indicating three possible equilibrium states. The intermediate state is unstable, as before. The two extreme states are stable; the state with the higher global mean temperature corresponds to the current climate, and the state with the lower temperature to a deep-freeze climate, where the Earth is completely covered with snow and ice.

2.7 ▪ Snowball Earth

The discussion in the preceding sections suggests that the Earth's climate system can have multiple equilibrium states, including a stable state in which temperatures are similar to those seen today, with global averages well above the freezing temperature of water, and another stable state which is some 50 degrees colder. Recall that the time scale for an EBM ranges over millions of years, so the first equilibrium state encompasses all recent ice ages and interglacial periods, including the most severe glaciations of the last million years. The other, much colder state corresponds to a complete glaciation of the Earth, with all oceans frozen to a depth of several kilometers and almost the entire planet covered with ice—a *Snowball Earth* state. The possibility of such a scenario, although hard to imagine, raises serious questions for climate science. For example, has the Earth ever been in a snowball state? If so, what caused its transition to such a state, how did it transition out of that state, and could it return to a snowball state in the future?

There is indeed fairly strong geological evidence that the Earth's climate may have been in the Snowball Earth state up to four times during the Neoproterozoic age, between 750 and 580 million years ago [40]. The evidence comes from geological deposits that can form only during glaciations and that have been found in tropical areas around the globe at what was then sea level. In addition, there are related deposits which point to large build-ups of CO_2 in the atmosphere during these same periods, which were subsequently brought down rapidly to normal levels, and there are geological indications of very little biological activity during these times.

These observations have puzzled geologists for a long time but can be neatly explained by periods of complete global glaciation lasting millions of years and terminated by rapid warming events. During a period of global glaciation, there would have been very little biological productivity, with life surviving near thermal vents at the floor of the ocean. There would still have been a build-up of CO_2 in the atmosphere due to volcanic activity. Ordinarily, this build-up is checked by the chemical erosion of rocks, even if there were no plant life to take up CO_2, but most continents would have been covered by ice and the CO_2 would have continued to increase, leading to a huge greenhouse effect. In our very simple EBM, the greenhouse effect decreases the value of ε in Eq. (2.11). As ε decreases, the dotted curve in Figure 2.5 ($\varepsilon = 0.6$) moves down and the stable equilibrium corresponding to complete glaciation disappears. The climate would then suddenly transition to a warm equilibrium (warmer than today), and the snowball would turn into a hothouse.

Less E_{out} so ε dec. [handwritten annotation in left margin]

There is indeed evidence in the geological record that such rapid transitions occurred at the end of these global glaciations. There is also speculation that these glaciation periods contributed to what is known as the Cambrian Explosion, a rapid diversification of life on the planet about 550 million years ago, during which many early versions of the current multicellular life forms first appeared.

If these snowball periods appeared in the distant past of the planet, why did they not happen again during the past 500 million years? And could they happen again in the future? A mathematical model can also shed light on these questions. It is known from other geological evidence that, during the Neoproterozoic age, the continents were closer to the equator than they are today. Since chemical erosion of rocks is a powerful mechanism to remove CO_2 from the atmosphere, the continents in the lower latitudes would have remained ice free, and CO_2 levels in the atmosphere would have remained low even while polar caps expanded and the planet became colder during naturally occurring ice ages. Greenhouse effects would have been shut off, leading to a runaway glaciation. This is less likely to happen with today's distribution of continents, since the continents are closer to the poles and chemical erosion would diminish earlier during an ice age, leading to an earlier onset of a CO_2 increase and a greenhouse effect that would check further glaciation.

2.8 • Bifurcation

We now take another look at the model of Section 2.5 and ask what happens to the equilibrium solutions if the solar constant changes. Recall that $Q = \frac{1}{4}S_0$, where S_0 is the solar constant.

If the solar constant decreases, the solid curve in Figure 2.5 moves down, and the equilibrium states at T_1^* and T_2^* merge and then disappear, leaving only the deep-freeze state at T_3^*. If the solar constant increases, the equilibrium states at T_2^* and T_3^* coalesce, and we are left with an ice-free Earth in the equilibrium state at T_1^*. The disappearance and emergence of solutions indicate *qualitative changes* in the climate system. When such a

change is the result of a small smooth variation of a parameter of the problem, we say that
a *bifurcation* occurs. The parameter that triggers the change is identified as the *bifurcation*
parameter. In the case under consideration, the parameter is the solar constant or, rather,
the ratio Q/Q_0, where $Q_0 = 342\,\mathrm{W\,m^{-2}}$ is the value of Q for the current climate. (It is
generally recommended to choose a bifurcation parameter that is dimensionless.)

Figure 2.6 gives the graphs of the equilibrium temperatures T_1^* (upper branch), T_2^*
(middle branch), and T_3^* (lower branch) as functions of the bifurcation parameter Q/Q_0.
A diagram of this type is called a *bifurcation diagram*. It is common practice to represent
stable solutions (T_1^* and T_3^*) by solid curves and unstable solutions (T_2^*) by dashed curves.
The bifurcation diagram shows that, as the solar input decreases quasi-statically from its
current value ($Q = Q_0$), the mean surface temperature T_1^* decreases until it reaches a *tip-*
ping point at the critical value T_{fi}. The climate system transits to the lower branch, the
planet turns white, and its temperature equilibrates at T_3^*. The diagram also shows the
possibility of a reverse scenario. If the planet today were in the deep-freeze state and the
solar input were to increase quasi-statically, the mean surface temperature T_3^* would increase until it reached another tipping point at the critical value T_{if}. All the snow and ice
would melt, and the climate would settle onto an equilibrium state with temperature T_1^*.
Since the paths for increasing and decreasing values of the bifurcation parameter are distinct, we see that *hysteresis* is built into the current climate model. Bifurcation phenomena
play an important role in nonlinear systems and will be discussed in Chapter 5.

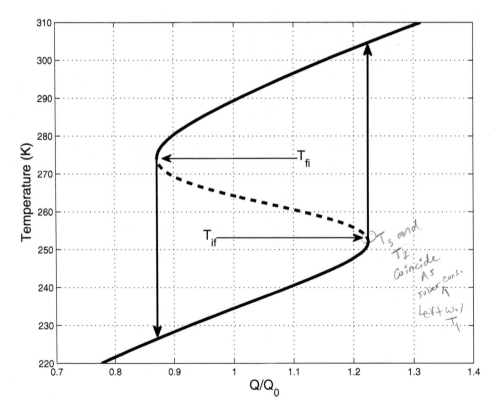

Figure 2.6. *Mean surface temperatures at equilibrium as a function of the solar constant (in*
units of its present value).

2.9 ▪ Exercises

1. Planck's law describes the *radiance*—that is, the amount of radiant energy leaving a unit area—of a perfect black body of a given temperature per unit time, per unit spectral interval, and per unit solid angle. Radiance is therefore expressed in units of $J\,s^{-1}m^{-2}$(wavelength)$^{-1}$(steradian)$^{-1}$. Recall that joule (J) is a measure of energy and watt (W) a unit of power, and that power measures the amount of energy delivered per unit time; hence, if the wavelength is measured in meters, the unit of radiance is $W\,m^{-3}sr^{-1}$. The formula is

$$B(\lambda, T) = \frac{2hc^2}{\lambda^5} \frac{1}{e^{hc/\lambda kT} - 1},$$

where λ is the wavelength and T the temperature. The constants in this formula are Planck's constant, $h = 6.625 \cdot 10^{-34}\,J\,s$; Boltzmann's constant, $k = 1.38 \cdot 10^{-23}\,J\,K^{-1}$; and the speed of light, $c = 3 \cdot 10^8\,m\,s^{-1}$. Check that the quantity $hc/\lambda kT$ is dimensionless and that B has the dimension of radiance.

2. The *intensity* is the integral of the radiance over all solid angles spanning a hemisphere. To do the integration, choose a spherical coordinate system with the polar axis along the normal to the surface directed toward the exterior of the black body. A black body is a perfect Lambertian scatterer, so the radiance observed along the hemisphere follows Lambert's cosine law and is the actual radiance times the cosine of the polar angle ϕ. Furthermore, the infinitesimal solid angle is $d\Omega = \sin \phi \, d\phi \, d\theta$, where θ is the azimuthal angle. Thus, the intensity is

$$I(\lambda, T) = \int_0^{2\pi} \int_0^{\pi/2} B(\lambda, T) \cos \phi \sin \phi \, d\phi \, d\theta = \pi B(\lambda, T).$$

Plot the intensity $I(\lambda, T)$ as a function of λ for the temperature range from 4,000 to 6,000 K. Find the maximum of the function $\lambda \mapsto I(\lambda, T)$ and verify that the value of λ where the maximum occurs increases as the temperature decreases (Wien's law).

3. The Stefan–Boltzmann law (or Stefan's law) gives the total energy radiated per unit surface area of a perfect black body per unit time as a function of its temperature T,

$$F = \sigma T^4;$$

cf. Eq. (2.4). Stefan's law is derived by integrating the intensity found in Exercise 2 over all wavelengths. Use the identity $\int_0^\infty x^3 (e^x - 1)^{-1} dx = \frac{1}{15}\pi^4$ to obtain the expression $\sigma = 2\pi^5 k^4 / (15h^3 c^2)$ for Stefan's constant and verify the numerical value $\sigma = 5.670 \cdot 10^{-8} J\,m^{-2}K^{-4}s^{-1}$.

4. Assume that the temperature of the Sun is T and that the Sun radiates as a perfect black body, so the total energy radiated by the Sun per unit surface area and per unit time is $F_S = \sigma T^4$. Assume, furthermore, that this energy is radiated uniformly in all directions. The amount of energy passing per unit time through a unit area at the Earth's surface is found by multiplying F_S by the ratio of the surface area of the Sun (πR_S^2) and the surface area of a sphere at the mean distance from the Sun to the Earth (πR_{SE}^2), $S = (R_S/R_{SE})^2 F_S$. Taking $R_S = 695.5 \cdot 10^6\,m$ and $R_{SE} = 149.6 \cdot 10^9\,m$, and assuming $T = 5,780\,K$, find the value of S. How close is this value to the solar constant? What could explain the difference?

5. Figure 2.4 shows that the maximum radiative flux per unit area and per wavelength unit at the surface of the Sun is about $6 \cdot 10^6$ times as large as the corresponding flux at the top of the Earth's atmosphere. Explain why this must be the case, using Planck's Law from Exercise 1 and the black-body temperatures of the Sun and the Earth given in the text.

6. Reconcile the maximum value of the spectral irradiance from the Sun at the top of the Earth's atmosphere shown in Figure 2.1 with the maximum radiative flux per unit area at the surface of the Sun shown in Figure 2.4, using the inverse square law of radiation intensity.

7. The amount of energy captured by the Earth as a whole at the top of the atmosphere is found by multiplying the quantity S obtained in Exercise 4 by the area of a disc with radius equal to the mean radius of the Earth (πR_E^2). Taking $R_E = 6,378.1 \cdot 10^3$ m, determine how much energy is captured by the Earth per unit time. Not all this energy reaches the Earth's surface. A fraction α is reflected back into space, and a fraction β of the remaining energy is transmitted to the Earth's surface. If the Sun is at an angle ϕ with respect to the zenith, an additional factor $\cos \phi$ is needed to allow for the increase in area illuminated by one solar constant. Convince yourself that the amount of energy received at the Earth's surface per unit time is given by an expression of the form

$$F = (1-\alpha)\beta S \cos \phi \ \mathrm{Wm}^{-2}.$$

The parameters α and β are major unknowns, since they vary significantly with time and depend on the topography, cloud cover, and many other factors. How could you measure them?

8. Satellite data of the Sun's radiation indicate that the solar constant S_0 varies approximately between $1,365.5 \ \mathrm{Wm}^{-2}$ and $1,367 \ \mathrm{Wm}^{-2}$, with a period of about 11 years.

   (i) Use the balance equation (2.8) with $\alpha = 0.3$ and $\varepsilon = 0.6$ to estimate the resulting variation (difference of max and min) in the Earth's global mean surface temperature T. Use a suitable linearization if possible.

   (ii) Use the balance equation (2.14),

   $$(1-\alpha)Q = A + BT, \tag{2.15}$$

   with $A = 203.3$ and $B = 2.09$ to estimate the resulting variation (difference of max and min) in the surface temperature T. Use $\alpha = 0.3$.

   (iii) Explain (conceptually, without formulas) why the actual variation in surface temperature would be less than what you computed above. Hint: Think of the heat capacity of the oceans.

9. Which of the following effects represent forcing? Positive, negative, unclear? Are there possible feedbacks present? Explain your answers.

   (i) Increased air traffic results in the release of more CO_2 into the atmosphere.

   (ii) Increased air traffic results in more contrails (artificial clouds) in the atmosphere.

   (iii) Deforestation leads to higher albedo and less evaporation cooling from vegetation.

(iv) Deforestation leads to the release of more aerosols into the atmosphere because of the burning of vegetation.

10. The function $f : x \mapsto f(x) = \tanh x$ connects the value -1 smoothly with the value 1. Explain why.

   (i) Find a function f that connects the value $a - \frac{1}{2}b$ with the value $a + \frac{1}{2}b$.

   (ii) Show by introducing a scaling parameter that one can achieve the transition from a value just above $a - \frac{1}{2}b$ to a value just below $a + \frac{1}{2}b$ over an arbitrarily small interval.

   (iii) The *logistic function*, $g : x \mapsto g(x) = (1 + e^{-x})^{-1}$, is another function that connects the value -1 smoothly with the value 1. Explain why. Find the relation between the functions f and g.

11. This exercise is about the energy balance equation

$$C \frac{dT}{dt} = (1 - \alpha(T))Q - \varepsilon \sigma T^4.$$

Figure 2.7 gives the graph of the right-hand side for $\varepsilon = 0.6$.

   (i) Sketch the solution $T(t)$ of this equation for $t > 0$ if $T(0) = 230, 240, 260, 270,$ and 300.

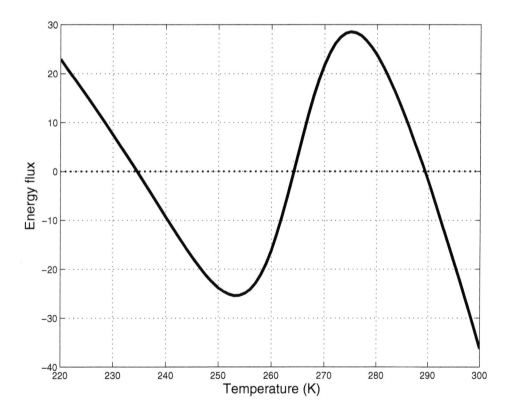

Figure 2.7. *Graph of the function $T \mapsto Q(1 - \alpha(T)) - \varepsilon \sigma T^4$ for $\varepsilon = 0.6$.*

(ii) Sketch the solution $T(t)$ of this equation for $t > 0$ if $T(0) = 285$. Then sketch the solution of this equation with the same initial data in the same coordinate system if C is twice as large. Explain your answer.

(iii) If ε is decreased due to an increased greenhouse effect, the entire curve is shifted upwards (roughly speaking). Sketch the solution $T(t)$ of the equation for $t > 0$ if $T(0) = 280$. Then sketch the solution of this equation with the same initial data in the same coordinate system if ε is decreased. Explain your answer.

12. Consider again the energy balance equation $C(dT/dt) = (1 - \alpha(T))Q - \varepsilon\sigma T^4$. Let T^* be an equilibrium solution.

(i) Find the (nonlinear) differential equation for the deviation $x(t) = T(t) - T^*$.

(ii) Show that, as long as $T(t)$ is sufficiently close to T^*, x essentially satisfies the *linearized equation* $C\dot{x} = -Dx$, where $D = \alpha'(T^*)Q + 4\varepsilon\sigma T^{*3}$. Find the general solution of this equation.

(iii) Find the condition(s) for T^* to be *linearly stable* ("linearly" because we consider only the linearized differential equation for x).

(iv) Assuming that T^* is linearly stable, find the typical response time to a perturbation—for example, the time it takes for the deviation to reach the value $x(t) = 0.1$ if $x(0) = 1$. Describe how the response time changes with C and give a physical interpretation.

13. Consider the energy balance equation

$$(1 - \alpha)Q = A + BT.$$

(i) Find the equilibrium temperature T_0 that goes with the current value Q_0 of Q. (This is very easy.)

(ii) Find the change δT in the temperature if Q changes by a small amount δQ, assuming that α, A, and B do not change. (This is also very easy.)

(iii) In reality, α, A, and B change with temperature. Explain why. Hint: Think of feedback mechanisms that could be responsible for such changes.

(iv) Assume that the temperature dependence of α, A, and B is of the form

$$\alpha : T \mapsto \alpha(T) = \alpha_0 + \alpha_1\,\delta T + \cdots,$$
$$A : T \mapsto A(T) = A_0 + A_1\,\delta T + \cdots,$$
$$B : T \mapsto B(T) = B_0 + B_1\,\delta T + \cdots$$

for $T = T_0 + \delta T$. Here, δT is small and the dots indicate higher-order terms in δT. Show that, up to higher-order terms in δT, δT and δQ must satisfy the relation

$$(1 - \alpha_0)\delta Q = (\alpha_1 Q_0 + B_0 + A_1 + B_1 T_0)\delta T.$$

(v) Show that the last relation can be written in the form

$$B_0\,\delta T = (1 - \alpha_0) f\,\delta Q,$$

where $f = 1/(1 - g)$, $g = g_1 + g_2$, and $g_1 = -(A_1 + B_1 T_0)/B_0$ and $g_2 = -\alpha_1 Q_0/B_0$. Compare this relation to the one found under (ii) above and explain why the factor f is called the *climate gain*. Give plausible arguments why the factor g_1 represents the water-vapor feedback mechanisms and g_2 the ice and snow feedback mechanisms.

(vi) Where does the cloud feedback come in? Where do you think the greatest uncertainty in any of these factors is?

14. Tung [113] introduced a zonal EBM, which generalizes the global equation (2.13) as follows. Let (θ, ϕ) denote the spherical coordinates on the globe, θ being latitude $(-\frac{1}{2}\pi < \theta < \frac{1}{2}\pi)$ and ϕ being longitude $(-\pi < \phi < \pi)$. Define $y = \sin\theta$, so $-1 < y < 1$; $y = -1$ corresponds to the South Pole and $y = 1$ to the North Pole. Define the *zonal average temperature* $T : (t, y) \mapsto T(t, y)$ as the surface temperature at latitude y averaged over longitude. This function is assumed to satisfy the energy equation

$$C\frac{\partial T}{\partial t} = (1 - \alpha(y))s(y)Q - (A + BT) + D(\overline{T} - T).$$

Here, $s(y)$ is the fraction of the incident solar radiation at latitude y, averaged over time; $\alpha(y)$ is the albedo, which depends on latitude; D is a positive constant; Q, A, and B have the same meaning as before.

(i) Draw a plausible graph of s for a planet with a tilted axis of rotation and explain why $s(0) > 1$ and $s(\pm 1) < 1$.

(ii) The last term in the energy equation represents energy transfer across latitudes, where $\overline{T}$ is the global average temperature, $\overline{T}(t) = \frac{1}{2}\int_{-1}^{1} T(t, y)\,dy$. Explain the effect of this term.

(iii) Verify that $\overline{T}$ satisfies a global energy equation of the form

$$C\frac{d\overline{T}}{dt} = (1 - \alpha_p)Q - (A + B\overline{T}).$$

Find the expression for the planetary albedo α_p.

Chapter 3

Oceans and Climate

In this chapter we turn our attention to the world's oceans and their role in maintaining the Earth's energy balance and supporting the carbon cycle. The ability to transport energy around the globe is due to the ocean circulation, which typically operates on a decadal time scale and is driven by temperature and salinity contrasts. We present a simple mathematical model to describe the thermohaline circulation and show how the circulation pattern can vary with temperature and salinity.

Keywords: Ocean circulation, conveyor belt, thermohaline circulation, salinity, box models, bifurcation.

3.1 ▪ Ocean Circulation

The ocean is a major player in the Earth's climate system. Its importance rests both in its capacity to transport significant amounts of energy around the globe and in its capability to take up huge amounts of CO_2 from the atmosphere. CO_2 is somewhat soluble in water, so it tends to get washed out of the atmosphere and into the ocean by the action of rainfall and wave breaking. This CO_2 is consumed by tiny single-cell organisms (*phytoplankton*) through photosynthesis and converted to more complex molecules, which either end up as food for larger species like fish or sink to the bottom at the end of the plankton's life.

Figure 3.1 gives a schematic view of the global ocean circulation pattern. The pattern is driven by density contrasts. Since the density is a function of temperature and salinity, the pattern is known as the *thermohaline circulation* (THC). The basic structure consists of a loop. In the Antarctic polar region, cold water is driven eastward in deep currents toward the Pacific Ocean. It upwells en route and is then transported back in near-surface currents through the Indian Ocean, around the Cape, and into the Atlantic Ocean. As it reaches the Norwegian–Greenland Sea and the Labrador Sea, it overturns to return as deep-water currents to the Antarctic region. This structure, which is known as the *conveyor belt*, was first suggested by Broecker [6] and is very important for the climate system. It provides a stabilizing effect on climate because of its long time scale (several decades), but can also cause abrupt climate change in the space of a few decades if it is disturbed in certain ways. The conveyor belt is, of course, a gross simplification. Various modifications have been proposed since it was first presented in 1987; Figure 3.2 gives one such refinement [86].

The strength of the oceanic flow is measured in sverdrups (Sv). The sverdrup is not an SI unit; 1 Sv is equivalent to one million cubic meters per second. The entire global

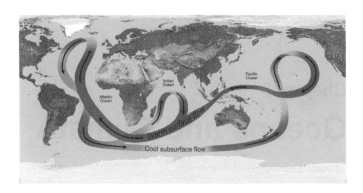

Figure 3.1. *The conveyor belt. Reprinted courtesy of NASA.*

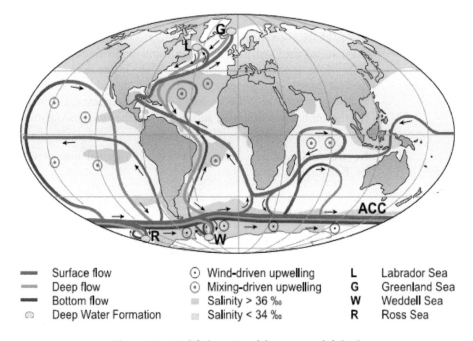

━━	Surface flow	⊙	Wind-driven upwelling	**L**	Labrador Sea
═══	Deep flow	⊚	Mixing-driven upwelling	**G**	Greenland Sea
━━	Bottom flow	▪	Salinity > 36 ‰	**W**	Weddell Sea
⌒	Deep Water Formation	▪	Salinity < 34 ‰	**R**	Ross Sea

Figure 3.2. *Modified version of the conveyor belt* [86].

input of fresh water from rivers to the ocean is about 1 Sv; the current strength of the pole-to-equator flow is about 20 Sv.

Figure 3.3 is a map of the global surface circulation pattern. It shows average conditions for winter months in the Northern Hemisphere; there are local differences in the summer, particularly in regions affected by monsoonal circulations. Subtropical *gyres* rotate anticyclonally (clockwise in the Northern Hemisphere, counterclockwise in the Southern Hemisphere) in each of the main ocean basins. The western margins of each of these gyres have strong poleward currents, such as the Gulf Stream in the North Atlantic. In the Northern Hemisphere there is some evidence for cyclonic subpolar gyres, where the westerly winds change to polar easterlies. In the Southern Hemisphere the water is able to flow around the entire globe. The Antarctic Circumpolar Current (ACC) is the longest current system in the world; it is driven by strong westerly winds, transporting

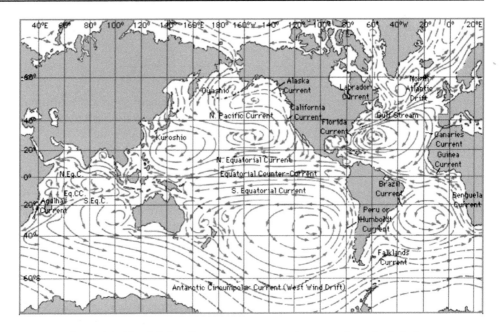

Figure 3.3. *The global surface current system; cool currents shown by dashed arrows, warm currents by solid arrows. Reprinted with permission from FAS.*

about 130 Sv on average through the Drake Passage. Subpolar gyres exist in the Weddell Sea and Ross Sea, poleward of the ACC.

The THC plays a major role in maintaining the Earth's overall energy balance, so a change of the circulation pattern will have a dramatic impact on the global climate. There are indeed indications in several paleoclimatic data sets that changes in the patterns or collapses of the THC have coincided with large variations in climate, especially in the North Atlantic region. It is therefore important to have some idea about the stability of the current configuration, as exemplified in the conveyor belt mechanism, and to know whether the ocean circulation system admits multiple equilibrium configurations and an exchange of stability might occur under certain forcing scenarios.

The THC is driven by density gradients. Since the density varies with temperature and salinity, we devote brief sections to these dependencies before we go into the mathematical modeling.

3.2 ▪ Temperature

The temperature profile of the oceans is rather simple. Radiative heat from the Sun is significant only in a relatively thin top layer of a few meters. Since mixing is strong due to wind and waves, the temperature in this *mixing layer* is essentially constant. An intermediate layer, the *thermocline* region, separates the mixing layer from the deep *abyssal zone*, which is entirely decoupled from the direct influence of the Sun. In the thermocline region, the temperature decreases more or less linearly; in the abyssal zone, the temperature is again approximately constant, a few degrees above freezing. The abyssal zone comprises about 98% of the total volume of the Earth's oceans.

The temperature in the thermocline changes principally by upwelling of cold water (advection) and diffusion due to small-scale eddies. A simple one-dimensional advection-diffusion equation for the temperature T as a function of depth z is

$$\frac{\partial T}{\partial t} = -w\frac{\partial T}{\partial z} + x\frac{\partial^2 T}{\partial z^2}, \tag{3.1}$$

where w is the advection rate (the upwelling velocity) and x the diffusion coefficient of the fluid. At steady state, the left-hand side of the equation is zero, T depends only on z, and the differential equation for T can be integrated. We find

$$T(z) = T_0 + T_1 e^{-z/z^*}, \tag{3.2}$$

where T_0 and T_1 are constants and $z^* = x/w$ is a characteristic depth for the temperature profile. Typical order-of-magnitude estimates are $x \sim 10^{-2}\,\mathrm{m^2 s^{-1}}$ and $z^* \sim 10^2\,\mathrm{m}$, so $w \sim 10^{-4}\,\mathrm{m\,s^{-1}}$. The upwelling velocity is very small indeed.

The depth of the thermocline depends strongly on latitude. Figure 3.4 gives a schematic view of a cross-section of the Atlantic Ocean from the North Pole (left) to the South Pole (right), with the equator in the middle. The mixing layer and the thermocline essentially vanish near both the poles, where the abyssal zone extends to the surface at latitudes near the edge of the polar ice caps. The ice that forms at these latitudes is nearly fresh, leaving an increased salt content in the remaining water and reducing its freezing point. The cold, salty brine is extremely dense and sinks to the bottom, where it fills the basins at the bottom of the polar seas and eventually the rest of the deep ocean. Centuries later, this water returns to the surface at lower latitudes.

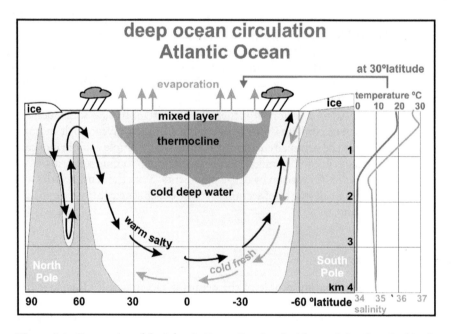

Figure 3.4. *Cross-section of the Atlantic Ocean. Reprinted with permission from Seafriends.*

3.3 ▪ Salinity

The most critical aspect of ocean composition is its salinity, the dissolved salts that have a large effect on water density. These salts are due to weathering of rocks in the early stages of the formation of oceans and continents. Sea water is about 3.5% salt by weight on average. The salt is composed primarily of sodium chloride ($\sim$ 85%), with smaller amounts of magnesium chloride, magnesium sulfate, and calcium carbonate. One of the remarkable aspects of ocean salinity is how constant the relative ratios of these components are, no matter where in the ocean one looks or what the value of the salinity is. This indicates that the oceans are well mixed on the time scale of salt input.

Salinity is measured in "practical salinity units" (psu), which is a ratio of conductivities and therefore a dimensionless unit. Salinity ranges from approximately 31 to 39 in the mixed layer (depending mainly on the balance between precipitation and evaporation) and remains close to 35 in the abyssal zone; see Figure 3.4.

The density of sea water depends on its temperature and salinity. The dependence is given by the equation of state, but unlike in the atmosphere, there is no simple law that can be derived from first principles. In climate studies, one usually applies empirical formulas that have been established from experimental observations. Figure 3.5 shows the density of water as a function of temperature and salinity. The global mean density of the oceans is about 1.035 kg m^{-3}.

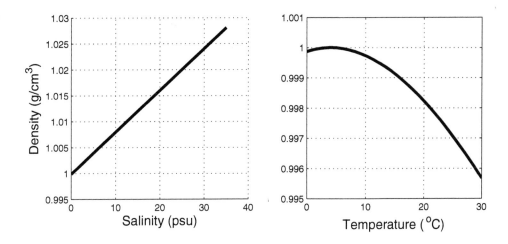

Figure 3.5. *Density as a function of salinity and temperature.*

3.4 ▪ Box Models

Modeling the THC is a formidable challenge, not only because of the lack of a universal equation of state which connects the density of the water to temperature and salinity, but also because of the complicated form of the domain, which is bounded by the edges of the various continents. But we can make some headway if we take a system-level approach. Considered as a system, the ocean is just a reservoir filled with salt water, where the circulation is driven by density differences.

Figure 3.6 shows a simple two-box model which can be used to study the overturning circulation in the North Atlantic. The North Atlantic is represented by two boxes; Box 1 represents the polar region (high latitudes) and Box 2 the equatorial region (low latitudes). The water in each box is well mixed, so the temperature and salinity are uniform in each box but not necessarily the same in the two boxes. Any difference results in different densities, and this difference drives a flow through the capillary pipe which connects the boxes at the bottom. A compensating flow at the surface ensures that the volume of

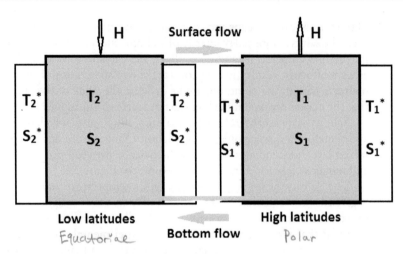

$T_2 > T_1$

Equatorial   Polar

Figure 3.6. *Two-box model of the North Atlantic with evaporation and precipitation.*

water in each box does not change. The combination of the flow through the capillary pipe at the bottom and the compensating flow at the surface represents the overturning circulation. Each box also exchanges heat and salinity with a surrounding basin, which represents the atmosphere and neighboring oceans. The temperature and salinity in the surrounding basins are maintained at constant values. The strength of the exchange flow is proportional to the difference between the temperature and salinity inside and outside the box. External wind forces and Coriolis effects are ignored.

A "virtual salt flux" at the surface of the boxes accounts for the effects of evaporation, precipitation, and runoff from continents. The flux is virtual, because no salt is flowing into or out of the boxes. But here is the explanation. The temperature in Box 2 (the equatorial region) is higher than the temperature in Box 1, so evaporation dominates in Box 2, and there is a net loss of salt-free moisture into the atmosphere and therefore a compensating virtual salt flux into Box 2. On the other hand, precipitation and runoff dominate at the lower temperature in Box 1, so there is a net gain of salt-free moisture from the atmosphere and therefore a compensating virtual salt flux out of Box 1. We assume that the salt flux into Box 2 matches the salt flux out of Box 1, so the overall salt flux into the two boxes is zero.

Having established our "ocean model," we now turn to the mathematics. Denote the temperature and salinity by T_1 and S_1 in Box 1, and T_2 and S_2 in Box 2; these are all functions of time (denoted by t, as usual). In the "normal" state, we have $T_2 > T_1$, a positive salt flux H into Box 2, and a compensating salt flux $-H$ out of Box 1, as shown in Figure 3.6.

The (constant) temperature and salinity in the basins surrounding the boxes are T_1^* and S_1^* (Box 1), and T_2^* and S_2^* (Box 2).

The capillary flow (the "overturning circulation") is driven by the pressure difference between the two boxes at their respective bottoms. This difference is in turn proportional to the difference between their densities. If there is a difference in salinity and both boxes have the same temperature, the water in the saltier box has a higher density, and as a result there is a bottom flow from the saltier box to the less salty one, with a compensating surface flow in the other direction. This is a "salinity" engine, which converts a difference of salinities into mechanical work. Similarly, a temperature difference between two boxes with the same salinity leads to a bottom flow from the colder to the warmer box, since cold water has a higher density than warm water.

Let ρ_1 and ρ_2 be the density in Box 1 and Box 2, respectively. The flow q in the capillary pipe is driven by the difference $\rho_1 - \rho_2$,

$$q = k \frac{\rho_1 - \rho_2}{\rho_0}. \tag{3.3}$$

Here, ρ_0 is a reference density to be defined below. The sign of q does not matter, since a bottom flow is always matched by a surface flow in the opposite direction; hence, a reversal of the flow direction results in the same effective exchange of heat and salinity. The convention is to have $q > 0$ when the flow through the capillary goes in the direction of the equator as a result of higher densities at high latitudes. The hydraulic constant k is a parameterization of various conditions for the flow, such as bottom friction and wind-driven turbulent mixing near the surface. Its value is chosen to obtain a reasonable match with the actual value of the overturning circulation, and a typical value is $k = 1.5 \cdot 10^{-6}\,\mathrm{s}^{-1}$. Note that q and k have the same dimension.

To connect the density to the temperature and salinity, we need an *equation of state*— that is, a function $\rho : (T, S) \mapsto \rho(T, S)$. Figure 3.5 shows graphs of this function for fixed S and for fixed T. Using Taylor's theorem, we may assume that ρ varies linearly with T and S near their average values. Since water expands as its temperature increases, its density decreases with increasing temperature; and since water gets heavier as its salinity increases, its density increases with increasing salinity. We therefore use the approximate equation of state

$$\rho = \rho_0(1 - \alpha(T - T_0) + \beta(S - S_0)). \tag{3.4}$$

Here, T_0 and S_0 are average values for the temperature and salinity; ρ_0 is the density if $T = T_0$ and $S = S_0$, which is the reference density used in Eq. (3.3); α and β are average values of the thermal contraction coefficient and saline expansion coefficient, respectively. Typical values are $\alpha = 1.5 \cdot 10^{-4}\,\mathrm{deg}^{-1}$ and $\beta = 8 \cdot 10^{-4}\,\mathrm{psu}^{-1}$; cf. Figure 3.5. Substituting the expression (3.4) into Eq. (3.3), we find that q is a linear function of the differences in temperature and salinity in the boxes,

$$q = k(\alpha(T_2 - T_1) - \beta(S_2 - S_1)). \tag{3.5}$$

Note that the two terms have opposite signs, which indicates that temperature and salinity have opposite effects on the overturning circulation. In a sense, they are competing, and at this point we have no way to tell which way the flow is going.

The governing equations for temperature and salinity are essentially the conservation laws for heat (thermal energy) and salinity. Accounting for all sources and sinks, we obtain a system of four ordinary differential equations,

$$\frac{dT_1}{dt} = c(T_1^* - T_1) + |q|(T_2 - T_1),$$

$$\frac{dT_2}{dt} = c(T_2^* - T_2) + |q|(T_1 - T_2),$$

$$\frac{dS_1}{dt} = -H + d(S_1^* - S_1) + |q|(S_2 - S_1),$$

$$\frac{dS_2}{dt} = H + d(S_2^* - S_2) + |q|(S_1 - S_2), \tag{3.6}$$

where q is given by Eq. (3.5). The constants c and d determine the characteristic time scale for the evolution of temperature and salinity, respectively. The symbol H stands for the virtual salt flux, $H > 0$.

Consider the averages of temperature and salinity, $T_0 = \frac{1}{2}(T_1 + T_2)$ and $S_0 = \frac{1}{2}(S_1 + S_2)$, and similarly for starred quantities. It follows from Eq. (3.6) that T_0 and S_0 satisfy the equations

$$\frac{dT_0}{dt} = c(T_0^* - T_0),$$
$$\frac{dS_0}{dt} = d(S_0^* - S_0). \tag{3.7}$$

The first equation shows that the average temperature of the water in the ocean tends to the average temperature of the surrounding basins as time goes to infinity; the second equation shows the same for the average salinity.

This result suggests that we take the average temperature and salinity of the surrounding basins as reference values and introduce the temperature and salinity *anomalies*,

$$\bar{T}_1 = T_1 - T_0^*, \quad \bar{T}_2 = T_2 - T_0^*; \quad \bar{S}_1 = S_1 - S_0^*, \quad \bar{S}_2 = S_2 - S_0^*. \tag{3.8}$$

Note that $T_2 - T_1 = \bar{T}_2 - \bar{T}_1$ and $S_2 - S_1 = \bar{S}_2 - \bar{S}_1$. Furthermore,

$$T_1^* - T_1 = -\tfrac{1}{2}(T_2^* - T_1^*) - \bar{T}_1, \quad T_2^* - T_2 = \tfrac{1}{2}(T_2^* - T_1^*) - \bar{T}_2;$$
$$S_1^* - S_1 = -\tfrac{1}{2}(S_2^* - S_1^*) - \bar{S}_1, \quad S_2^* - S_2 = \tfrac{1}{2}(S_2^* - S_1^*) - \bar{S}_2.$$

These expressions simplify if we introduce the quantities, T^* and S^*,

$$T^* = \tfrac{1}{2}(T_2^* - T_1^*), \quad S^* = \tfrac{1}{2}(S_2^* - S_1^*), \tag{3.9}$$

so

$$T_1^* - T_1 = -T^* - \bar{T}_1, \quad T_2^* - T_2 = T^* - \bar{T}_2;$$
$$S_1^* - S_1 = -S^* - \bar{S}_1, \quad S_2^* - S_2 = S^* - \bar{S}_2.$$

We now rewrite the system (3.6) in terms of the anomalies and drop the bars,

$$\frac{dT_1}{dt} = c(-T^* - T_1) + |q|(T_2 - T_1),$$
$$\frac{dT_2}{dt} = c(T^* - T_2) + |q|(T_1 - T_2),$$
$$\frac{dS_1}{dt} = -H + d(-S^* - S_1) + |q|(S_2 - S_1),$$
$$\frac{dS_2}{dt} = H + d(S^* - S_2) + |q|(S_1 - S_2). \tag{3.10}$$

Here, q is given by the same expression (3.5), where the T and S now stand for temperature and salinity anomalies. The system (3.10) is not much different from the original system (3.6), but the advantage is that we are now dealing with a single constant T^* instead of two constants, T_1^* and T_2^*, and similarly for S^*. In the absence of the virtual salt flux, Eq. (3.10) reduces to Stommel's box model [104], which we will discuss in Chapter 6.

3.5 ▪ One-Dimensional Model

We now ignore the temperature equations and focus on the salinity equations, taking $T_1(t) = -T^*$ and $T_2(t) = T^*$ for all t in the expression (3.5) for q. This approximation assumes that, on the time scale of interest for the THC, the temperature in each box equilibrates almost instantaneously with the temperature in the surrounding basin and that the difference between the temperatures in the two boxes is small. We also simplify the salinity equations by assuming that the exchange of salinity between a box and its surrounding basin is negligible, setting $d = 0$. With these approximations, the system (3.10) reduces to

$$\frac{dS_1}{dt} = -H + |q|(S_2 - S_1),$$
$$\frac{dS_2}{dt} = H + |q|(S_1 - S_2),$$

(3.11)

where $q = k(2\alpha T^* - \beta(S_2 - S_1))$. This two-box model is sometimes referred to as Stommel's model [109], but while Stommel's original model is also a two-box model [104], it differs in several respects from the present model, as we will see in Chapter 6.

Since $S_1 + S_2$ is constant and the temperature difference is fixed, $\Delta T = T_2 - T_1 = 2T^*$, the only variable remaining is the salinity difference, $\Delta S = S_2 - S_1$, which satisfies the equation

$$\frac{d\,\Delta S}{dt} = 2H - 2k\,|\alpha\,\Delta T - \beta\,\Delta S|\,\Delta S.$$

(3.12)

3.5.1 ▪ Dynamical System

K - length/sec
β - length/psu
α - length/deg

We render the problem dimensionless by introducing

ΔS psu
ΔT deg.

$$x = \frac{\beta\,\Delta S}{\alpha\,\Delta T}, \quad t' = 2\alpha k|\Delta T|t, \quad \lambda = \frac{\beta H}{\alpha^2 k\Delta T|\Delta T|}.$$

Notice units cancel!

The variable t' is just a rescaled version of t; without loss of generality we may drop the prime $'$. The parameter λ corresponds to a dimensionless surface salinity flux; since $\Delta T = T_2 - T_1$ is positive, we have $\lambda > 0$.

The dynamics of the two-box model are now described by a scalar ordinary differential equation for the function $x : t \mapsto x(t)$ in $\mathbb{R}$,

$$\dot{x} = \lambda - |1 - x|x.$$

(3.13)

The dot $\dot{}$ denotes differentiation with respect to t. Note that the right-hand side of the equation is Lipschitz continuous but not continuously differentiable.

3.5.2 ▪ Equilibrium States

Constant Solns

The equilibrium states are found by setting the right-hand side equal to zero,

$$|1 - x|x = \lambda.$$

(3.14)

Since $\lambda > 0$, any root x^* of this equation is necessarily positive. After a bit of algebra, we find that there is one root for all $\lambda > 0$, namely $x_1^* = \frac{1}{2}(1 + \sqrt{1+4\lambda})$. If $\lambda \in (0, \frac{1}{4})$, there are two more roots, namely $x_2^* = \frac{1}{2}(1 + \sqrt{1-4\lambda})$ and $x_3^* = \frac{1}{2}(1 - \sqrt{1-4\lambda})$.

To investigate the stability of the equilibrium solutions, we could use the same type of arguments as in Section 2.5, which are based on the sign of the function in the right-hand side of the differential equation (3.13). Here we employ a different argument based on linearization, in view of later generalizations to multidimensional systems of equations.

We linearize Eq. (3.13) near a critical point x^*, distinguishing between values of x greater than and less than 1 because of the absolute value sign. Taking $x = x^* + y$ and ignoring terms of order y^2 and higher, we obtain the linearized equations

$$\dot{y} = \pm(2x^* - 1)y, \quad x^* \lessgtr 1. \tag{3.15}$$

At $x^* = x_1^*$, the coefficient of y is $-(2x_1^* - 1) = -\sqrt{1+4\lambda} < 0$, so y decays exponentially as $t \to \infty$ and the equilibrium solution is stable; at $x^* = x_2^*$, the coefficient of y is $(2x_2^* - 1) = \sqrt{1-4\lambda} > 0$, so y grows exponentially as $t \to \infty$ and the equilibrium solution is unstable; and last, at $x^* = x_3^*$, the coefficient of y is $(2x_3^* - 1) = -\sqrt{1-4\lambda} < 0$, so y decays exponentially as $t \to \infty$ and the equilibrium solution is again stable.

Note that the nature of the two stable solutions is quite different. Since $x_1^* > 1$, salinity dominates the density difference between the two boxes. The equilibrium x_1^* is therefore referred to as the S-mode. In the S-mode, the overturning circulation is driven by salinity, $q < 0$, the surface flow is directed toward the equator, and the bottom flow is directed toward the pole. On the other hand, since $x_3^* < 1$, temperature is the dominant agent at x_3^*. The equilibrium solution x_3^* is referred to as the T-mode. In the T-mode, temperature drives the overturning circulation, $q > 0$, the surface flow is directed toward the pole, and the bottom flow is directed toward the equator.

3.5.3 ▪ Bifurcation

Figure 3.7 shows the graphs of x_1^* (upper branch), x_2^* (middle branch), and x_3^* (lower branch) as functions of the bifurcation parameter λ; solid lines (x_1^* and x_3^*) indicate stability, and the dashed line (x_2^*) indicates instability of the corresponding equilibrium.

It is interesting to compare the bifurcation diagrams of Figures 2.6 and 3.7. The diagrams are similar, but while the former is smooth, the latter is only piecewise smooth. This is due to the fact that the forcing in Eq. (3.13) is Lipschitz continuous but not continuously differentiable. While Figure 2.6 shows hysteresis with two distinct paths, one leading from right to left and the other leading from left to right, indicating that the system can transit into and out of each of the two stable states, the situation in Figure 3.7 is quite different. If the system is initially at equilibrium on the upper branch (S-mode), it stays on the upper branch as the surface salinity flux (λ) decreases or increases quasi-statically. On the other hand, if the system is initially at equilibrium on the lower branch (T-mode), it stays on the lower branch as the surface salinity flux increases quasi-statically, until the surface salinity flux reaches a tipping point at the critical value $\lambda = \frac{1}{4}$, when a transition occurs to an S-mode. Once it is in the S-mode, the system stays in the S-mode and never returns to the T-mode.

The fact that the system is unable to jump down from the upper to the lower stable branch when λ reaches zero is a consequence of the fact that λ is restricted to positive values. If negative values were allowed, we would find that the solution x_3^* of Eq. (3.14) could be continued for $\lambda < 0$, so the system could jump from the upper to the lower stable branch at $\lambda = 0$ and we would have a complete hysteresis loop. But the laws of physics would be violated.

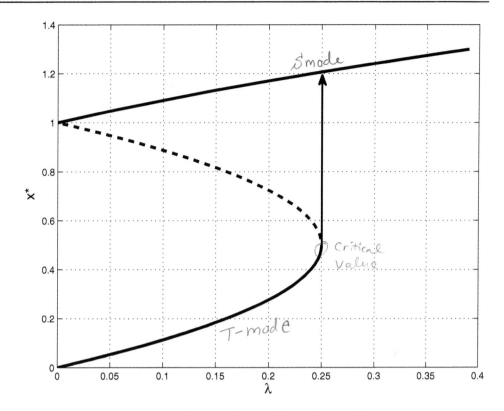

Figure 3.7. *Bifurcation diagram of the two-box model* (3.13).

3.6 ▪ Exercises

1. Determine the mean outflow of the 10 largest rivers in the world, in Sv units, from publicly available sources.

2. What happens to the steady-state temperature profile (3.2) if the upwelling velocity w increases? Identify an oceanic region with strong upwelling and make a prediction about its temperature profile. Check your prediction using publicly available sources.

3. Derive differential equations for $\Delta T = T_2 - T_1$ and $\Delta S = S_2 - S_1$ from the system of equations (3.10) for the temperature and salinity anomalies.

4. Show that if the salt flux H in the system (3.10) vanishes, then the quantities ΔT and ΔS defined in the previous exercise tend to zero as $t \to \infty$. Explain what this means in language that is accessible for a nonmathematician.

5. Consider the system of equations (3.10) for the temperature and salinity anomalies. Assume as in Section 3.5 that the temperature anomalies are equal to the values in the surrounding basins, $T_1(t) = -T^*$ and $T_2(t) = T^*$ for all t, but $d > 0$.

   (i) Derive the analogue of Eq. (3.12).

   (ii) Render the problem dimensionless as in Section 3.5.1 and show that the dimensionless salinity x satisfies the differential equation

   $$\dot{x} = \lambda - (\mu + |1 - x|)x.$$

   Find the formula for μ.

(iii) Show that this equation always has at least one stationary solution $x^* > 0$.

(iv) Show that if $\mu < 1$ and $\mu < \lambda < \frac{1}{4}(\mu+1)^2$, then the equation has three positive stationary solutions x_1^*, x_2^*, and x_3^* which satisfy the inequalities $x_1^* < x_2^* < 1 < x_3^*$.

(v) Determine the stability of each of the three positive stationary solutions x_1^*, x_2^*, and x_3^*.

Chapter 4

Dynamical Systems

In this chapter we give a brief introduction to the theory of autonomous differential equations and dynamical systems. We recall the fundamental theorems on the existence, uniqueness, and regularity and show how the solution of an autonomous differential equation gives rise to the concept of a dynamical system. We highlight equilibrium solutions and periodic solutions and define what it means for these solutions to be stable. We show how the local structure of the flow near an isolated critical point can be found by linearization and discuss the details for the planar case.

Keywords: Autonomous equation, vector field, dynamical system, trajectory, orbit, phase portrait, attractor, limit set, critical point, equilibrium solution, periodic solution, stability, linearization, node, saddle point, spiral, center, stable and unstable manifold.

4.1 ▪ Autonomous Differential Equations

In this chapter we give an introduction to the theory of *autonomous differential equations*—ordinary differential equations (ODEs), where the forcing term is independent of time. The general form of such an equation is

$$\dot{x} = f(x). \tag{4.1}$$

Here, x is an unknown vector whose components $x_1, \ldots, x_n$ are real-valued functions of t. The dot denotes differentiation with respect to t, and the components of $\dot{x}$ are the derivatives $\dot{x}_1, \ldots, \dot{x}_n$. The forcing term f is a given function, which is defined on an open set $D \subset \mathbb{R}^n$ and takes values in $\mathbb{R}^n$, so $f : D \to \mathbb{R}^n$ (in words, "f maps D into $\mathbb{R}^n$"). We note that D need not be the maximal domain of definition for f. It could be that the argument of f must be restricted to a range of physically realistic values or to a range where f is a valid description of a physical process.

A scalar autonomous equation of order n is a special case of the first-order equation (4.1), provided it can be written in the form

$$x^{(n)} = g\left(x, x^{(1)}, \ldots, x^{(n-1)}\right). \tag{4.2}$$

(The superscript $^{(k)}$, $k = 1, \ldots, n$, denotes the kth derivative with respect to t.) In that case, the vector $x = (x_1, \ldots, x_n)$ with $x_k = x^{(k-1)}$ for $k = 1, \ldots, n$ satisfies Eq. (4.1), where

$f(x) = (x_2, x_3, \ldots, g(x_1, \ldots, x_n))$. Depending on the problem of interest, we may use either the scalar form (4.2) or the vector form (4.1).

The more general nonautonomous differential equation

$$\dot{x}(t) = f(x(t), t) \tag{4.3}$$

can be reduced to the form (4.1) by introducing the new independent variable y by the definition $y(t) = (x(t), t)$ and the new right-hand side $\tilde{f}$ by the definition $\tilde{f}(y) = (f(x, y_{n+1}), 1)$. Then any solution x of Eq. (4.3) becomes a solution of the autonomous equation $\dot{y} = \tilde{f}(y)$ and vice versa.

Autonomous differential equations arise naturally in many applications. The energy balance models (EBMs) discussed in Chapter 2 and the box models of ocean circulation discussed in Chapter 3 are examples of autonomous differential equations arising in the context of climate. Two standard examples in $\mathbb{R}^2$ are the *pendulum equation* and the *Van der Pol equation*.

(i) The planar *pendulum equation*,

$$\ddot{x} + \sin x = 0. \tag{4.4}$$

The variable x is the angle of deviation from the vertical. The equation is equivalent to the system

$$\begin{aligned} \dot{x}_1 &= x_2, \\ \dot{x}_2 &= -\sin x_1. \end{aligned} \tag{4.5}$$

(ii) The *Van der Pol equation*,

$$\ddot{x} + x = \mu(1 - x^2)\dot{x}, \quad \mu > 0. \tag{4.6}$$

The equation was proposed by the Dutch physicist BALTHASAR VAN DER POL (1889–1959) to model relaxation-oscillation phenomena in electrical circuits. The Van der Pol equation is equivalent to the system

$$\begin{aligned} \dot{x}_1 &= x_2, \\ \dot{x}_2 &= -x_1 + \mu(1 - x_1^2)x_2, \end{aligned} \tag{4.7}$$

and also to the system

$$\begin{aligned} \dot{x}_1 &= x_2 - \mu(\tfrac{1}{3}x_1^3 - x_1), \\ \dot{x}_2 &= -x_1. \end{aligned} \tag{4.8}$$

The latter is the *Liénard form* of the Van der Pol equation. The example shows that there may be more than one way to reduce an nth order scalar differential equation to a system of n first-order equations.

Autonomous differential equations are idealizations, often the result of significant simplifications, and more relevant as conceptual models than true-to-life models of physical phenomena. For example, in the pendulum equation the assumption is that there is no friction, the pendulum is massless, and the pendulum bob is just a single point mass. Nevertheless, the pendulum equation, like the Van der Pol equation, is a useful model for studying fundamental phenomena in nonlinear dynamics.

The theory of autonomous differential equations is universal, in the sense that it applies to classes of natural phenomena. Therefore, the equations are normally taken to be dimensionless, and in dealing with particular applications it is good practice to work with dimensionless variables. To emphasize this aspect of universality, we use the generic symbol x to denote the unknown(s) and t to denote the independent variable (time). Our convention is to write $x : t \mapsto x(t)$ to indicate explicitly that x is a function of t. (In words, "x maps t into $x(t)$.") Anticipating the fact that we will encounter systems of coupled differential equations in later chapters, we focus the discussion on equations for vector-valued functions in $\mathbb{R}^n$ ($n = 1, 2, \ldots$).

The theory of autonomous differential equations and the related theory of dynamical systems are quite extensive. In this chapter we can only touch on those aspects that seem most relevant in the context of climate. For a more in-depth treatment we refer the reader to the textbooks [3, 13, 33, 34, 37] and the research monographs [7, 32, 116].

Definition 4.1. *A function φ is a* solution *of the differential equation $\dot{x} = f(x)$ on an interval $I \subset \mathbb{R}$ if $\varphi : I \to D \subset \mathbb{R}^n$ is differentiable and satisfies the equation $\dot{\varphi}(t) = f(\varphi(t))$ for all $t \in I$.*

Sometimes we are interested in a specific solution of Eq. (4.1) which satisfies, in addition to the differential equation, an *initial condition* $x(t_0) = x_0$, where $t_0 \in I$ and $x_0 \in D$ are given. This type of problem is referred to as an *initial value problem* (IVP). Our first theorem addresses the issue of existence of a solution. Proofs of this and the following theorems can be found in the literature cited above.

Theorem 4.1 (Existence). *If $f : D \to \mathbb{R}^n$ is continuous, then there exists, for any $t_0 \in \mathbb{R}$ and $x_0 \in D$, an interval $I(x_0) \subset \mathbb{R}$ which contains t_0 in its interior and a solution $\varphi : t \mapsto \varphi(t)$ of the differential equation $\dot{x} = f(x)$ on $I(x_0)$ which satisfies the initial condition $\varphi(t_0) = x_0$.*

The domain of φ can be quite complicated, as the time interval $I(x_0)$ depends on x_0. The largest possible interval $I(x_0)$ is called the *maximal interval of existence* of φ; it also varies with x_0. If $I(x_0) = (\alpha, \beta)$ is the maximal interval of existence for φ and α is finite, then $\lim_{t \downarrow \alpha} \varphi(t)$ does not exist in D; similarly, if β is finite, then $\lim_{t \uparrow \beta} \varphi(t)$ does not exist in D.

To emphasize the dependence on the initial condition, we sometimes write $\varphi(t; t_0, x_0)$ instead of $\varphi(t)$. In this case, $\varphi(t_0; t_0, x_0) = x_0$.

The following examples show that an IVP may have a solution for all times, a solution that exists only on a limited time interval, or multiple solutions.

(i) $\dot{x} = -x$, $x(t_0) = x_0$.
  The function $\varphi : t \mapsto \varphi(t) = x_0 e^{-(t-t_0)}$ is a solution; it is defined for all $t \in \mathbb{R}$, and there are no other solutions.

(ii) $\dot{x} = x^2$, $x(t_0) = x_0$.
  The function $\varphi : t \mapsto \varphi(t) = x_0/(1 - x_0(t - t_0))$ is a solution. If $x_0 > 0$, this solution is defined for all $t \in (-\infty, t_0 + 1/x_0)$, blowing up as $t \uparrow t_0 + 1/x_0$; if $x_0 < 0$, it is defined for all $t \in (t_0 - 1/|x_0|, \infty)$, blowing up as $t \downarrow t_0 - 1/|x_0|$; and if $x_0 = 0$, it is defined for all $t \in \mathbb{R}$. There are no other solutions.

(iii) $\dot{x} = \sqrt{x}$, $x(t_0) = x_0$, subject to the condition $x(t) \geq 0$.
  For any $x_0 \geq 0$, the function $\varphi : t \mapsto \varphi(t)$ defined by $\varphi(t) = 0$ for $t \in (-\infty, t_0 - 2\sqrt{x_0}]$ and $\varphi(t) = \frac{1}{4}(t - t_0 + 2\sqrt{x_0})^2$ for $t \in (t_0 - 2\sqrt{x_0}, \infty)$ is a solution for all $t \in \mathbb{R}$.

The example (iii) shows that something more than continuity of f is required to guarantee uniqueness of a solution.

The function $f : D \to \mathbb{R}^n$ is said to be *Lipschitz continuous* (or simply *Lipschitz*) if it satisfies a *Lipschitz condition*—that is, if there exists a constant k such that $\|f(x_1) - f(x_2)\| \le k\|x_1 - x_2\|$ for every $x_1, x_2 \in D$. Here, $\|\cdot\|$ can be any norm in $\mathbb{R}^n$. The constant k is called the *Lipschitz constant*.

Note that Lipschitz continuity implies continuity. If f is differentiable and its partial derivatives $\partial f / \partial x_k$ $(k = 1, \dots, n)$ are continuous, then f is Lipschitz continuous, so this gives a simple criterion to determine whether a function is Lipschitz continuous.

Theorem 4.2 (Uniqueness). *Let $\varphi : t \mapsto \varphi(t)$ and $\psi : t \mapsto \psi(t)$ be two solutions of $\dot{x} = f(x)$ defined on a common interval I. If $f : D \to \mathbb{R}^n$ is Lipschitz continuous with Lipschitz constant k, then*

$$\|\varphi(t) - \psi(t)\| \le e^{k|t - t_0|}\|\varphi(t_0) - \psi(t_0)\|, \quad t \in I. \tag{4.9}$$

In particular, if $\varphi(t_0) = \psi(t_0)$, then $\varphi(t) = \psi(t)$ for all $t \in I$.

For any physically realistic system, "existence and uniqueness" of a solution of an IVP simply means that the data are neither overspecified (so that a solution exists) nor underspecified (so that the solution is uniquely determined).

The following theorem expresses the fact that smooth functions f yield solutions of the IVP that depend smoothly on the initial data. The theorem requires some notation. For $k = 1, 2, \dots$, $C^k(D)$ denotes the set of all functions $f : D \to \mathbb{R}^n$ that are k times differentiable and whose kth derivatives are continuous in D. If no confusion is possible, we may say that f is a C^k function or simply f is C^k on D.

Theorem 4.3 (Regularity). *If $f : D \to \mathbb{R}^n$ is C^k, then there exists, for any $t_0 \in \mathbb{R}$, $x_0 \in D$, and $t \in I(x_0)$, a neighborhood U of x_0 such that the function $\phi_t : x \mapsto \varphi(t; t_0, x)$ is C^k on U.*

An important property of autonomous differential equations is given in the following lemma.

Lemma 4.1 (Translation property). *If $\varphi : t \mapsto \varphi(t)$ is a solution of $\dot{x} = f(x)$ and t_0 is a constant, then $\psi : t \mapsto \psi(t) = \varphi(t - t_0)$ is also a solution.*

Proof. Since f does not depend on t, a change of variables from t to $t - t_0$ does not change the differential equation. $\qquad\square$

Because of the translation property, the choice of the initial time t_0 in the specification of an IVP is immaterial. Therefore, there is no loss of generality if we take $t_0 = 0$. With this choice of t_0, we write $\varphi(t, x_0)$, instead of $\varphi(t; 0, x_0)$.

4.2 ▪ Geometrical Objects

In one and two dimensions, one can get an impression of the action of an autonomous differential equation by drawing the *vector field* of f—that is, the set of vectors from x to $x + f(x)$ at points x where $f(x)$ is defined. Figure 4.1 shows the vector field of the linear harmonic oscillator.

An autonomous differential equation gives rise to a dynamical system. The concept of a dynamical system is a very general one that extends well beyond autonomous differential

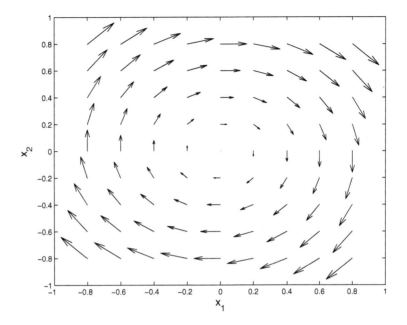

Figure 4.1. *Vector field of the linear harmonic oscillator;* $\dot{x}_1 = x_2, \dot{x}_2 = -x_1$.

equations. It takes the notion that the solution of an autonomous differential equation carries initial data forward or backward in time and broadens it from the action on specific initial data to the action on sets of initial data. By varying x_0 in the solution $\varphi : t \mapsto \varphi(t, x_0)$ of the IVP $\dot{x} = f(x), x(0) = x_0$ over the set of all admissible initial data, we obtain the *dynamical system* $\phi = \{\phi_t\}_t$, where $\phi_t : x \mapsto \phi_t(x) = \varphi(t, x)$. Thus, a dynamical system is a family of maps that describes the action of a vector field on sets of initial data.

A number of geometrical objects are associated with a dynamical system.

Definition 4.2. (i) *The* trajectory *of a point* $x \in D$ *is the set* $\{(t, \phi_t(x)) : t \in I(x)\} \subset \mathbb{R}^{n+1}$.
(ii) *The* orbit $\gamma(x)$ *of* x *is the projection of its trajectory onto* $\mathbb{R}^n$ *along the t-axis,* $\gamma(x) = \{\phi_t(x) : t \in I(x)\} \subset \mathbb{R}^n$. *If* $I(x) = (\alpha, \beta)$ *with* $\alpha < 0 < \beta$, *then the* positive orbit $\gamma^+(x)$ *and* negative orbit $\gamma^-(x)$ *of* x *are the subsets of* $\gamma(x)$ *obtained by restricting* t *to positive and negative values, respectively.*

Because time has been eliminated, an orbit does not reflect the progression or regression of time. To compensate for this loss of time parameterization, it is common practice to insert arrows on the orbit $\gamma(x)$ indicating the direction in which $\phi_t(x)$ changes as t increases. Figure 4.2 shows a solution, a trajectory, and an orbit for the linear harmonic oscillator, $\dot{x}_1 = x_2, \dot{x}_2 = -x_1$.

The collection of all the orbits together with their direction arrows is called the *phase portrait*. Figure 4.3 shows the phase portrait of the planar pendulum equation, $\dot{x}_1 = x_2, \dot{x}_2 = -\sin x_1$. Notice the variety of orbits, which include isolated points on the x_1-axis, closed orbits, and nonclosed curves on which x_2 does not change sign. Notice also the special orbits which separate the closed orbits from the nonclosed ones. Such an orbit is called a *separatrix*.

Since the planar pendulum has the property that values of x_1 that differ by a multiple of 2π correspond to identical positions of the pendulum, it is more natural to choose the

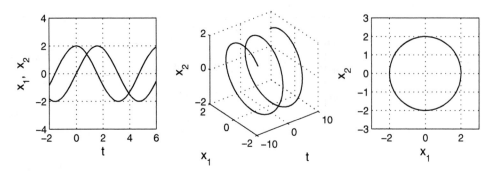

Figure 4.2. *Linear harmonic oscillator,* $\dot{x}_1 = x_2, \dot{x}_2 = -x_1$; *left: graphs of the solution,* $\{x_1(t), t \in I\}$, $\{x_2(t) : t \in I\}$; *middle: a trajectory in* (t, x_1, x_2)-*space; right: an orbit in the* (x_1, x_2)-*plane.*

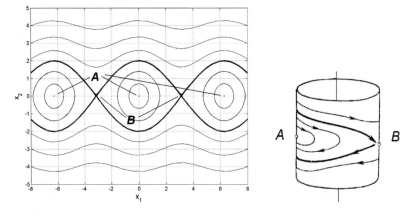

Figure 4.3. *Phase portrait of the pendulum equation,* $\dot{x}_1 = x_2$, $\dot{x}_2 = -\sin x_1$; *multiple points A, B on the left correspond to single points A, B on the right.*

cylinder $(x_1 \bmod(2\pi), x_2)$ rather than the (x_1, x_2)-plane as the phase space. The portrait on the right in Figure 4.3 is the result of wrapping the portrait of the figure on the left around the cylinder.

Definition 4.3. *A subset* $U \subset \mathbb{R}^n$ *is said to be* invariant *if* $\phi_t(U) = U$ *for all* $t \in \mathbb{R}$.

There are various kinds of invariant sets, but the most interesting ones are the so-called α- and ω-limit sets, named after the first and last letters of the Greek alphabet.

Definition 4.4. *The* α-*limit set* (ω-*limit set*) *of x for* ϕ *is the set of accumulation points of* $\phi_t(x)$ *as* $t \to -\infty$ ($t \to \infty$).

The limit sets are obviously closed; they are also invariant, as is proved in most textbooks on dynamical systems. The α- and ω-limit sets are visited infinitely often as $t \to \pm\infty$ and therefore look like what one might call an "attractor." Yet, for real attractors we need an extra property, namely that the set of initial states for which the α-limit set (ω-limit set) is the same must be "large enough," so it is associated with the dynamical system and not with just a special point. This makes it difficult to give a precise definition of an attractor, and there is no consensus in the literature on how to do this. Here is one possible definition.

Definition 4.5. *The ω-limit set $\omega(x)$ is a* positive attractor *if there exists an arbitrarily small neighborhood U of $\omega(x)$ such that $\omega(y) \subset \omega(x)$ for any $y \in U$ and $\phi_t(U) \subset U$ for all sufficiently large positive values of t. If the α-limit set $\alpha(x)$ has this property for all sufficiently large negative values of t, then $\alpha(x)$ is a* negative attractor.

Figure 4.4 shows the phase portrait of the Van der Pol equation, $\dot{x}_1 = x_2, \dot{x}_2 = -x_1 + \mu(1 - x_1^2)x_2$. The equation has a periodic orbit, which is a positive attractor, and any orbit which is not closed approaches the periodic orbit arbitrarily closely without ever touching it.

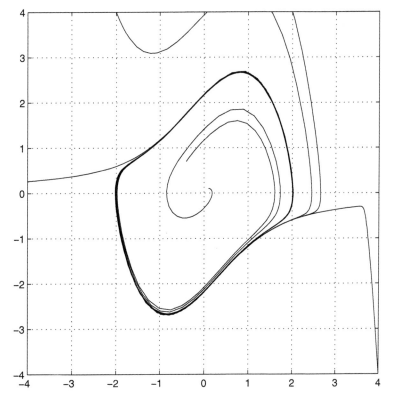

Figure 4.4. *Phase portrait of the Van der Pol equation, $\dot{x}_1 = x_2, \dot{x}_2 = -x_1 + (1 - x_1^2)x_2$.*

If a dynamical system has an attractor, then practically all orbits which start near it approach the attractor in the course of time. In the long run, the evolution of the system takes place entirely in this neighborhood, energy stored in short-term variations of the system is dissipated, and a description in terms of some "average" state makes sense. All this is rather intuitive and imprecise, but it offers an interesting perspective on the feasibility of climate modeling. Since climate is "averaged weather," the question is whether the equations that model the Earth's weather system admit an attractor. If so, then climate modeling could be viewed as weather modeling on or near this attractor. Of course, this argument is purely speculative, especially since it is not at all certain that the equations relevant for weather modeling (such as the Navier–Stokes equation, discussed in Chapter 14) admit an attractor. In fact, this is one of the hardest open problems in the theory of differential equations.

In this context we should also mention the term *strange attractor*, which was coined by Ruelle and Takens [93, 94] to indicate an attracting set with a complicated topological structure. A typical example is the Lorenz attractor, which we discuss in Chapter 7. There exist definitions of "strange attraction," but they are difficult to apply to specific cases. Numerical experiments suggest the existence of attracting sets with noninteger dimensions for several dynamical systems, but it is not necessarily the case that these are strange attractors. The notion of a strange attractor is closely related to the concept of *chaos*. For a thorough discussion, we refer the reader to [7, 32].

4.3 ▪ Critical Points

There are two types of distinguished orbits, namely critical points, discussed in this section, and periodic orbits, discussed in the next section.

Definition 4.6. *A point $x^* \in D$ is a* critical point *of ϕ if $\phi_t(x^*) = x^*$ for all t.*

The orbit of a critical point consists of a single point, namely the critical point itself. If x^* is a critical point, then $f(x^*) = 0$, so a critical point of the dynamical system corresponds to an equilibrium solution of the underlying ODE. Equilibrium solutions are defined for all times.

Equilibrium solution of the differential equation can be stable or unstable. A pendulum has two equilibrium states, one where the pendulum is hanging down and the other where it is standing up. If the pendulum is in the downward position and it experiences a small perturbation, it will oscillate and stay near the downward equilibrium position. But if it is in the upward position, an arbitrarily small perturbation will move the pendulum away from its equilibrium. The downward vertical position is "stable," and the upward vertical position "unstable." An unstable equilibrium is not easily realized; because there are always minor perturbations, any system will be driven away from an unstable equilibrium and we have to make a special effort to realize it.

Since stability plays such a fundamental role in the theory of dynamical systems, it is important that we have more than an intuitive understanding of it.

Assume that $x^* \in D$ is an isolated critical point. The concept of stability refers to the behavior of solutions that are "close" to x^*.

Definition 4.7. *A critical point x^* of ϕ is called* Lyapunov stable *or* stable *if, for every neighborhood V of x^*, there is a neighborhood $V_1 \subset V$ and a $T > 0$ such that $\phi_t(x) \in V$ for all $x \in V_1$ and $t \geq T$. It is called* unstable *if it is not Lyapunov stable.*

The trivial solution of the pendulum equation and the trivial solution of the equation for the harmonic oscillator are both Lyapunov stable. If the solutions not only stay near the critical point but approach it as time progresses, then we are dealing with a stronger type of stability, and there is a special term for it.

Definition 4.8. *A critical point x^* of ϕ is called* asymptotically stable *if x^* is Lyapunov stable and if, in addition, the neighborhood V_1 in Definition 4.7 can be chosen so that $\phi_t(x) \to x^*$ as $t \to \infty$.*

The two types of stability of a critical point—Lyapunov stability and asymptotic stability—are illustrated in Figure 4.5.

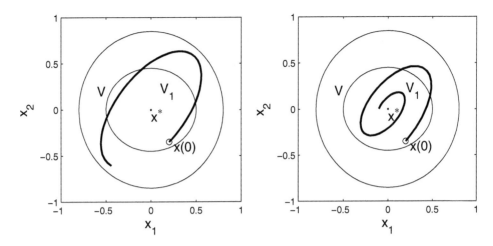

Figure 4.5. *Lyapunov stability (left) and asymptotic stability (right).*

One might conjecture that a critical point that satisfies the definition of an attractor is asymptotically stable. But here is an example to show that this is not necessarily the case,

$$\dot{r} = r(1-r), \quad \dot{\theta} = \sin^2(\tfrac{1}{2}\theta). \tag{4.10}$$

Here, r and θ are polar coordinates in $\mathbb{R}^2$. This system has two critical points which, in Cartesian coordinates, are $x_1^* = (0,0)$ and $x_2^* = (1,0)$. Every orbit which starts outside the origin eventually winds up at $(1,0)$, so $(0,0)$ is unstable and $(1,0)$ is a positive attractor. However, in every neighborhood of $(1,0)$ there are solutions which leave the neighborhood (although only temporarily), so $(1,0)$ is also unstable. Fortunately, this example is somewhat pathological and can be ruled out with some additional technical conditions.

4.4 • Periodic Orbits

Definition 4.9. *A solution $\varphi : t \mapsto \varphi(t)$ of $\dot{x} = f(x)$ is called* periodic *with period T or T-periodic if $\varphi(t+T) = \varphi(t)$ for all t. If, moreover, $\varphi(t+\tau) \neq \varphi(t)$ for any $\tau \in (0,T)$, then T is called the* minimal period.

Like equilibrium solutions, periodic solutions are defined for all times. A periodic solution corresponds to a bounded closed orbit. The converse is also true but more difficult to prove: every bounded closed orbit corresponds to a periodic solution. Such an orbit is also called a *cycle*.

Periodic solutions can be stable or unstable, just like equilibrium solutions, but in this case the concept is more subtle. Consider again the pendulum equation, $\ddot{x} + \sin x = 0$, or the equivalent system $\dot{x}_1 = x_2$, $\dot{x}_2 = -\sin x_1$ on the cylinder $(x_1 \bmod (2\pi), x_2)$. The system has two critical points, $(0,0)$ and $(\pi,0)$. The former, labeled A in Figure 4.3, is a stable equilibrium point; the latter, labeled B in Figure 4.3, is an unstable equilibrium point. Any solution starting in a neighborhood of B will leave the neighborhood in the course of time. Solutions starting in a neighborhood of A, say at $(x_1, x_2)(0) = (a,0)$ for some $a \in (0,\pi)$, are periodic and return periodically to this neighborhood. This might be an argument for stability. But the period of these solutions depends on where they start (unlike in the case of the harmonic oscillator, where the period is the same for all solutions). In fact, one can show that the period is given by the integral $T = 4 \int_0^a (2\cos x - 2\cos a)^{-1/2} dx$, which varies with a. As a consequence, neighboring solutions have different periods, and the position of two pendulums that start out close together will eventually diverge.[2] Should we still call the trivial solution stable?

The definition of Lyapunov stability given for critical points can be extended to periodic solutions, but the extension is rather restrictive and therefore not very useful. It requires that phase points, once they start out close, must stay close—a requirement that can be met only if all solutions have the same period. As we have seen, the requirement is not met in the case of the pendulum equation, so periodic solutions of the pendulum equation are not stable in the sense of Lyapunov. For periodic solutions, which are associated with closed orbits in the state space, it makes more sense to define stability in terms of their orbital properties.

Let φ be a T-periodic solution, and let γ be the closed orbit in D defined by φ. Take a *transverse manifold* Σ—that is, an $(n-1)$-dimensional surface in $\mathbb{R}^n$ that is punctured by γ and is nowhere tangent to it. If γ has multiple intersections with Σ, shrink Σ until there is only one intersection. Let p be the (unique) point where γ intersects Σ, and let $U \subset \Sigma$ be a neighborhood of p.

Consider an orbit $\gamma(q)$ which starts at $q \in U$ and follow it until it returns to U (see Figure 4.6). If it does not return to U, choose another starting point q that is closer to p. Because $\phi_t : x \mapsto \phi_t(x)$ is continuous, there certainly is a $q \in U$ sufficiently near p for which the trajectory will return to U. The map P that maps the point $q \in U$ into the next point of intersection with U is called the *Poincaré map* or *return map*. Note that the time it takes for the orbit $\gamma(q)$ to first return to Σ depends on q and need not be equal to the period $T = T(p)$ of γ; however, it approaches T as q approaches p.

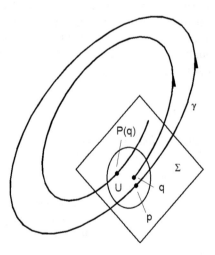

Figure 4.6. *Orbital stability.*

If the Poincaré map can be applied repeatedly, we obtain a sequence $\{P^n q : n = 1, 2, \ldots\}$ of points in U. The point p is a fixed point of P.

Definition 4.10. *A periodic solution φ of $\dot{x} = f(x)$ is called* orbitally stable *or* stable *if every successive return map of any point near p stays near p—that is, if, for every neighborhood V of p in U, there is a neighborhood $V_1 \subset V$ such that, for every $q \in V_1$, the iterates $P^n q$ are defined and lie in V for $n = 1, 2, \ldots$. It is called* unstable *if it is not orbitally stable.*

---

[2]This was a major problem in the design of pendulum clocks, which was eventually solved by the Dutch mathematician, astronomer, physicist, and horologist CHRISTIAAN HUYGENS (1629–1695) in the 1650s.

The periodic solutions of the pendulum equation are orbitally stable. Take for V any segment of the interval $(0, \pi)$ with a nonempty interior; the Poincaré map is the identity map.

Just as for equilibrium solutions, there is a stronger type of stability for periodic solutions. The definition refers again to the Poincaré map.

Definition 4.11. *A periodic solution φ is called* asymptotically stable *if it is orbitally stable and if, in addition, the neighborhood V_1 in Definition 4.10 can be chosen so that $P^n q \to p$ as $n \to \infty$.*

4.5 ▪ Dynamics near Critical Points

If the critical points of a dynamical system are all isolated, we can analyze the trajectories in a neighborhood of each critical point and learn more about the global structure. Under the appropriate conditions, the local structure can be found by applying the process of *linearization*.

4.5.1 ▪ Linearization

If x^* is an isolated critical point of ϕ, then $f(x^*) = 0$. Assume that f is sufficiently smooth (for example, C^2) near x^*. The first nonzero term in the Taylor expansion of f near x^* is $Df(x^*)(x - x^*)$, where $Df(x^*)$ is the $n \times n$ matrix of partial derivatives of f, $Df = (\partial f_i / \partial x_j)_{i,j=1}^n$, evaluated at x^*. To exclude degeneracies, we assume that $\det(Df(x^*)) \neq 0$.

We rewrite the differential equation $\dot{x} = f(x)$,

$$\dot{x} = Df(x^*)(x - x^*) + g(x), \tag{4.11}$$

where $g(x) = f(x) - Df(x^*)(x - x^*)$. Linearizing the equation $\dot{x} = f(x)$ at x^* means that we interpret g as a (nonlinear) perturbation, ignore g, and consider the linear system

$$\dot{x} = Df(x^*)(x - x^*). \tag{4.12}$$

Without loss of generality we can shift the point x^* to the origin by taking $\bar{x} = x - x^*$. Since the matrix $Df(x^*)$ is constant, $\bar{x}$ satisfies the differential equation $\dot{\bar{x}} = Df(x^*)\bar{x}$. Omitting the bar and using the notation $A = Df(x^*)$, we obtain the linear system

$$\dot{x} = Ax, \quad \det(A) \neq 0. \tag{4.13}$$

If A were a scalar, say a, where a is a real number, we would know how to solve the equation in terms of exponentials. The solution of the differential equation $\dot{x} = ax$ which satisfies the initial condition $x(0) = x_0$ is $x(t) = e^{ta}x_0$, so the dynamical system which carries the initial data forward or backward in time is $\phi_t = e^{ta}$. The situation with respect to Eq. (4.13) is more complicated but similar, once we have defined the exponential of an $n \times n$ matrix.

Let $L(\mathbb{R}^n)$ be the space of matrices of order n with the uniform norm, and let M be any element of $L(\mathbb{R}^n)$. The repeated product of M with itself is again a matrix of order n, so M^k is an element of $L(\mathbb{R}^n)$ for $k = 0, 1, \ldots$. Now consider the finite sum $\sum_{k=0}^N M^k / k!$. This is also an element of $L(\mathbb{R}^n)$ which, as $N \to \infty$, converges uniformly to the element $\sum_{k=0}^\infty M^k / k! \in L(\mathbb{R}^n)$. This matrix is the *exponential* of M, $e^M = \sum_{k=0}^\infty M^k / k!$.

Now that we know how to exponentiate a matrix, we can define the exponential e^{tA},

$$e^{tA} = \sum_{k=0}^{\infty} \frac{t^k A^k}{k!}. \tag{4.14}$$

This matrix is essentially an ordered set of n^2 real-valued functions of t, whose derivatives with respect to t are well defined. The derivative of e^{tA} is nothing but the matrix of the derivatives of each of these functions, and it can be shown that

$$\frac{d}{dt} e^{tA} = A e^{tA} = e^{tA} A. \tag{4.15}$$

The order of the matrices e^{tA} and A is immaterial. The following theorem is the fundamental theorem of linear differential equations with constant coefficients.

Theorem 4.4. *Let A be a constant matrix of order n. Then the solution of the IVP $\dot{x} = Ax$, $x(0) = x_0 \in \mathbb{R}^n$ is $e^{tA} x_0$, and there are no other solutions.*

In other words, the dynamical system generated by Eq. (4.13) is $\phi = \{\phi_t\}_t$, where $\phi_t(x) = e^{tA} x$. The matrix representation of ϕ_t involves the eigenvalues of A and the eigenvectors, depending on the multiplicity of the eigenvalues. If the distinct eigenvalues of A are $\lambda_1, \ldots, \lambda_s$ and λ_i has (algebraic) multiplicity m_i (so $m_1 + \cdots + m_s = n$), then the matrix representing ϕ_t involves terms of the form $t^k e^{\lambda_i t}$, where $k = 1, \ldots, m_i$. Hence, if the real parts of all the eigenvalues are negative, every solution of Eq. (4.13) tends to zero as $t \to \infty$ and the origin is a *positive attractor*. A positive attractor is associated with stable behavior as $t \to \infty$. On the other hand, if the real parts of all the eigenvalues are positive, every solution of Eq. (4.13) tends to zero as $t \to -\infty$, and the origin is a *negative attractor*. A negative attractor is associated with stable behavior as $t \to -\infty$.

If none of the eigenvalues of A is located on the imaginary axis, then the origin is called a *hyperbolic critical point* The eigenvalues can then be separated into two disjoint groups, one group being located in the left half and the other in the right half of the complex plane. The eigenvectors and generalized eigenvectors associated with the former span a subspace E^s of $\mathbb{R}^n$ and those associated with the latter a subspace E^u of $\mathbb{R}^n$, and since there are no other eigenvalues, the two subspaces together span the entire space, $\mathbb{R}^n = E^s \oplus E^u$. Hence, every vector in $\mathbb{R}^n$ can be decomposed uniquely into two components, one in E^s and the other in E^u. Linear algebra tells us that E^s and E^u are invariant under the action of $\dot{x} = Ax$. Since E^s is associated with stable directions as $t \to \infty$, E^s is called the *stable manifold* associated with the critical point x^*; E^u being associated with unstable directions as $t \to \infty$ is called the *unstable manifold*.

4.6 ▪ Planar Case

In the planar case ($n = 2$), A is a 2×2 matrix,

$$A = \begin{pmatrix} a & b \\ c & d \end{pmatrix}, \tag{4.16}$$

with real constants a, b, c, and d. Since we have assumed that the determinant of A does not vanish ($ad - bc \neq 0$), the origin is the only critical point of $\dot{x} = Ax$.

The eigenvalues of A are the roots of the characteristic equation,

$$\lambda^2 - T\lambda + D = 0,$$

where $T = \text{trace}(A) = a + d$ and $D = \det(A) = ad - bc$. The eigenvalues are

$$\lambda_1 = \tfrac{1}{2}(T + \sqrt{T^2 - 4D}), \quad \lambda_2 = \tfrac{1}{2}(T - \sqrt{T^2 - 4D}). \tag{4.17}$$

According to the fundamental theorem of linear differential equations with constant coefficients (Theorem 4.4), the system $\dot{x} = Ax$ generates a dynamical system ϕ_t which is given by the exponential e^{tA}. In the planar case it is possible to give a complete classification of ϕ_t by considering all possible forms of the coefficient matrix A. We will do so in the remainder of this chapter; the results will be illustrated on several conceptual climate models in later chapters.

4.6.1 ▪ Distinct Real Eigenvalues

From linear algebra we know that there exists a nonsingular constant matrix F—that is, a change of coordinates in the plane—such that the matrix $J = FAF^{-1}$ is diagonal,

$$J = \begin{pmatrix} \lambda_1 & 0 \\ 0 & \lambda_2 \end{pmatrix}.$$

Then

$$e^{tJ} = \begin{pmatrix} e^{\lambda_1 t} & 0 \\ 0 & e^{\lambda_2 t} \end{pmatrix}$$

and $e^{tA} = F^{-1} e^{tJ} F$.

(i) λ_1 and λ_2 have the same sign.

Suppose $\lambda_2 < \lambda_1 < 0$. The orbits of $\dot{y} = Jy$ are curves in the (y_1, y_2)-plane given by the equation $|y_1|^{|\lambda_2|} = C|y_2|^{|\lambda_1|}$ for some constant C; they resemble parabolas. The origin is a positive attractor.

A critical point with this type of behavior is called a *node*. All the trajectories except one approach the origin tangent to the manifold spanned by the eigenvector belonging to the eigenvalue with the smaller absolute value (the *slow manifold*). The one exception is the trajectory which approaches the origin along the manifold spanned by the eigenvector belonging to the eigenvalue with the larger absolute value (the *fast manifold*).

If $0 < \lambda_2 < \lambda_2$, the arguments are entirely similar. The origin is a negative attractor.

Figure 4.7 shows the orbits for $J = \begin{pmatrix} -4 & 0 \\ 0 & -8 \end{pmatrix}$ (left) and $A = \begin{pmatrix} -7 & 1 \\ 3 & -5 \end{pmatrix}$ (right).

(ii) λ_1 and λ_2 have opposite signs.

The origin is a hyperbolic critical point. The orbits of $\dot{y} = Jy$ are curves given by the equation $|y_1|^{|\lambda_2|}|y_2|^{|\lambda_1|} = C$ for some constant C; they resemble hyperbolas. The origin is unstable.

A critical point with this type of behavior is called a *saddle point*. The eigenvector corresponding to the negative eigenvalue spans the *stable manifold* E^s, and the eigenvector corresponding to the positive eigenvalue spans the *unstable manifold* E^u. Together, E^s and E^u span the entire state space, $\mathbb{R}^2 = E^s \oplus E^u$. A typical trajectory approaches the unstable manifold as $t \to \infty$ and the stable manifold as $t \to -\infty$.

Figure 4.8 shows the orbits for $J = \begin{pmatrix} -2 & 0 \\ 0 & 2 \end{pmatrix}$ (left) and $A = \begin{pmatrix} 1 & -1 \\ -3 & -1 \end{pmatrix}$ (right).

4.6.2 ▪ Complex Conjugate Eigenvalues

The eigenvalues of A are $\lambda_1 = \alpha + i\beta$ and $\lambda_2 = \alpha - i\beta$ with $\beta \neq 0$. From linear algebra we know that there exists a real nonsingular constant matrix F such that

$$J = FAF^{-1} = \begin{pmatrix} \alpha & \beta \\ -\beta & \alpha \end{pmatrix}.$$

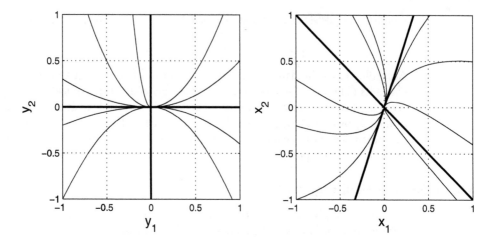

Figure 4.7. *Node.*

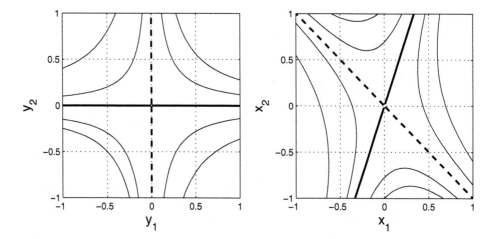

Figure 4.8. *Saddle point.*

Then

$$e^{tJ} = e^{\alpha t} \begin{pmatrix} \cos \beta t & \sin \beta t \\ -\sin \beta t & -\cos \beta t \end{pmatrix}$$

and $e^{tA} = F^{-1} e^{tJ} F$.

(i) $\alpha \neq 0$.

Solutions of $\dot{y} = Jy$ are oscillatory with a decreasing amplitude if $\alpha < 0$ or an increasing amplitude if $\alpha > 0$. The orbits are spirals. All solutions converge to the origin as $t \to \pm\infty$ if $\alpha \lessgtr 0$, so the origin is a positive attractor if $\alpha < 0$ and a negative attractor if $\alpha > 0$. A critical point with this type of behavior is called a *spiral point* or a *focus*.

Figure 4.9 shows the orbits for $J = \begin{pmatrix} -0.1 & -1 \\ 1 & -0.1 \end{pmatrix}$ (left) and $A = \begin{pmatrix} 0.61 & -1.08 \\ 1.39 & -0.81 \end{pmatrix}$ (right).

(ii) $\alpha = 0$.

The expression for e^{tJ} reduces to the parametric equation of the unit circle centered at the origin. The origin is called a *center*.

Figure 4.10 shows the orbits for $J = \begin{pmatrix} 0 & -2 \\ 2 & 0 \end{pmatrix}$ (left) and $A = \begin{pmatrix} -1 & 1 \\ -5 & 1 \end{pmatrix}$ (right).

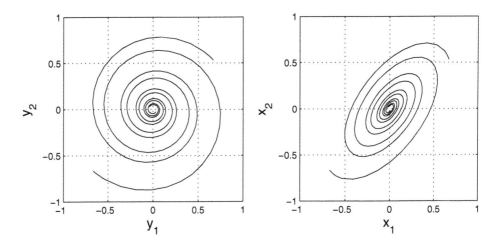

Figure 4.9. *Spiral point.*

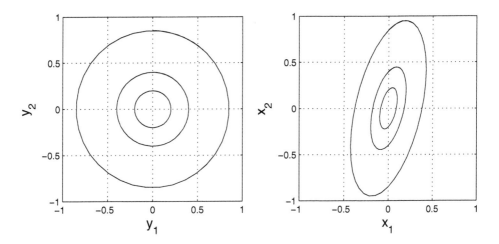

Figure 4.10. *Center.*

4.6.3 ▪ Eigenvalues with Algebraic Multiplicity Two

The eigenvalues of A coincide, $\lambda_1 = \lambda_2 = \lambda$. The eigenspace of λ is either one-dimensional or two-dimensional—that is, the geometric multiplicity of λ is either one or two.

(i) λ has geometric multiplicity 1.

The eigenspace associated with λ is one-dimensional. There exists a real nonsingular constant matrix F such that $J = FAF^{-1} = \left(\begin{smallmatrix} \lambda & 0 \\ 1 & \lambda \end{smallmatrix}\right)$. The critical point is a *degenerate node*. At the critical point, there is only one special direction, namely the direction of the eigenvector associated with λ. All trajectories have the same limiting direction at the critical point as $t \to \infty$ (stable case) or as $t \to -\infty$ (unstable case).

Figure 4.11 shows the orbits for $J = \left(\begin{smallmatrix} -1 & 0 \\ 1 & -1 \end{smallmatrix}\right)$ (left) and $A = \left(\begin{smallmatrix} -1.21 & -0.42 \\ 0.11 & -0.79 \end{smallmatrix}\right)$ (right).

(ii) λ has geometric multiplicity 2.

The eigenvectors associated with λ span $\mathbb{R}^2$, and every vector is an eigenvector with the same eigenvalue λ. The matrix A must be a multiple of the identity, $A = \left(\begin{smallmatrix} \lambda & 0 \\ 0 & \lambda \end{smallmatrix}\right)$, $\lambda \neq 0$.

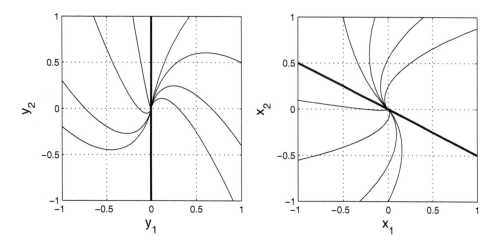

Figure 4.11. *Degenerate node.*

All trajectories are straight lines through the origin. The origin is a positive attractor if $\lambda < 0$ and a negative attractor if $\lambda > 0$. This type of critical point is called a *star node*. We do not include a figure for this case.

There is a simple way to classify the various cases if we go back to the expressions (4.17) for the eigenvalues of A,

$$\lambda_1 = \tfrac{1}{2}(T + \sqrt{T^2 - 4D}), \quad \lambda_2 = \tfrac{1}{2}(T - \sqrt{T^2 - 4D}),$$

where $T = \mathrm{trace}(A) = a + d$ and $D = \det(A) = ad - bc$. By assumption, $D \neq 0$.

If $D < 0$, the eigenvalues are real and have opposite signs; hence, the critical point is a saddle point.

If $D > 0$ and $T^2 > 4D$, the eigenvalues are real with the same sign, and the critical point is a node. If $D > 0$ and $T^2 < 4D$, the eigenvalues are complex conjugate, and the critical point is a spiral point or a center. The parabola $T^2 = 4D$ thus separates nodes from spirals. A critical point is asymptotically stable if $T < 0$ and unstable if $T > 0$. If $T = 0$, the critical point is a center, which is stable but not asymptotically stable.

Figure 4.12 summarizes these results. The large open regions in the (D, T)-plane are filled with the most common types of critical points: saddle points ($D < 0$), nodes ($D > 0$, $T^2 > 4D$), and spirals ($D > 0$, $T^2 < 4D$). Transitions occur when we cross the borders between the regions. These transitions will be explored in the exercises.

4.7 ▪ Nonlinear Systems

We return to the nonlinear system $\dot{x} = f(x)$ in $\mathbb{R}^n$. Suppose $f \in C^1$ near x^*, $f(x^*) = 0$. Without loss of generality, we may assume that x^* has been translated to the origin. Then the equation can be rewritten in the form

$$\dot{x} = f(x) = Ax + g(x), \tag{4.18}$$

where $A = Df(x^*)$ and $g(x) = f(x) - Df(x^*)x$. If A is nonsingular, the origin is an isolated critical point for the nonlinear system (4.18)—that is, there exists a neighborhood of the origin where this is the only critical point. Moreover, since $f \in C^1$, Eq. (4.18) has a unique solution for any initial value $x(0)$ near 0.

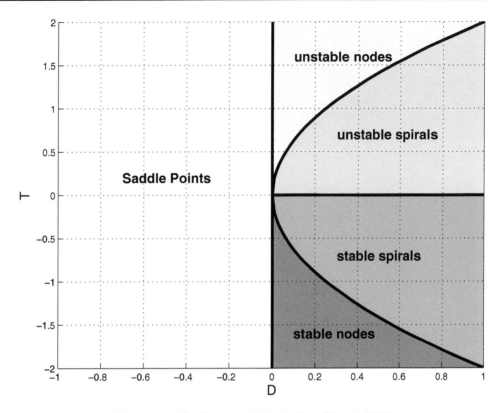

Figure 4.12. *Classification of critical points of $\dot{x} = Ax$* [105].

Note that $g(0) = 0$. Actually, we can be more specific. Since $f \in C^1$, $g(x)$ goes to zero faster than x as $x \to 0$,

$$g(x) = o(\|x\|) \text{ as } x \to 0.$$

Here, $\| \cdot \|$ denotes any norm, for example the Euclidean norm, in $\mathbb{R}^n$.

Intuitively speaking, since the perturbation g is "small" compared to any linear term, we expect that the dynamics of (4.18) near the origin are similar to the dynamics of the linearized version $\dot{x} = Ax$. This is in fact generally true; in particular, if the origin is a positive (negative) attractor for the linearized system, then it is also a positive (negative) attractor for the nonlinear system. On the other hand, if A has an eigenvalue with a positive real part, then the critical point $x = 0$ cannot be a positive attractor for the nonlinear system.

Moreover, in the hyperbolic case (where the eigenvalues of A all have nonzero real parts), there exist in the neighborhood of the critical point stable and unstable manifolds W^s and W^u with the same dimensions as the stable and unstable manifolds E^s and E^u of the linearized system $\dot{x} = Ax$, and E^s and E^u are tangent to W^s and W^u, respectively, at $x = 0$. In other words, the flows generated by the nonlinear dynamical system $\dot{x} = Ax + g(x)$ and its linearization $\dot{x} = Ax$ are locally similar. (In technical terms, they are *homeomorphic* in the neighborhood of the critical point $x = 0$. A homeomorphism is a continuous mapping which has a continuous inverse.) This result is illustrated in Figure 4.13, where $A = \left(\begin{smallmatrix} 1 & -1 \\ -3 & -1 \end{smallmatrix} \right)$ (the matrix that gave the saddle point in Figure 4.8) and the

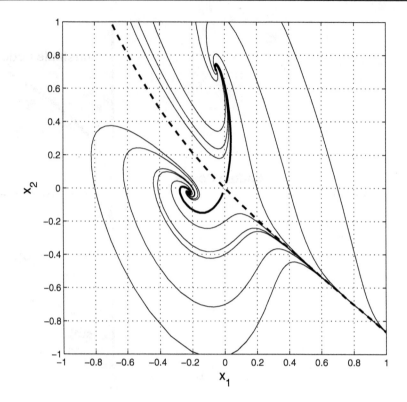

Figure 4.13. *Local stable and unstable manifolds.*

nonlinear perturbation is given by

$$g(x) = \begin{pmatrix} (x_1 + 2x_2)(3x_1 + x_2) \\ -4x_1(3x_1 + 4x_2) \end{pmatrix}.$$

The figure clearly shows that the homeomorphism is local and confined to a small neighborhood of the critical point. Away from the origin, the figure shows two unstable spiral points. Orbits emanating from the unstable spiral points become part of the stable manifolds of the saddle point at the origin.

Stable and unstable manifolds play an important role in the global structure of the flow of an autonomous system. When their dimensions match, they enable a trajectory to pass from one critical point to a neighboring critical point, or to return to the originating critical point. An orbit which connects two distinct critical points is called a *heteroclinic orbit*; an orbit which connects a critical point to itself is called a *homoclinic orbit*.

4.8 ▪ Exercises

1. Discuss the dynamics of the equation $\ddot{x} = x$. Sketch the solution, trajectories, and orbits in the appropriate spaces. Determine the attracting properties of any fixed points.

2. Consider the solution of the IVP $\dot{x} = |x|^\alpha$, $x(0) = 0$, for $0 < \alpha < 1$. Given $t_0 > 0$, write down a solution that is equal to 0 for $t < t_0$ and positive for $t > t_0$. (Hint: Use the translation property.)

3. Consider the pendulum equation, $\ddot{x} + \sin x = 0$. Find the critical points and discuss the linearization of the pendulum equation in the neighborhood of each critical point. Compare the phase portrait of the full nonlinear equation (Figure 4.3) with the phase portraits of the linearized equations.

4. Multiply both sides of the pendulum equation by $\dot{x}$ and show that the function $F(x, \dot{x}) = \frac{1}{2}\dot{x}^2 - \cos x$ is constant along trajectories. Generalize this observation to the scalar equation $\ddot{x} + f(x) = 0$. The function F thus obtained is called a *first integral*; it gives complete information about the phase portrait of the original equation.

5. For each of the equations

$$\ddot{x} + x + x^2 = 0, \quad \ddot{x} + x - x^2 = 0,$$
$$\ddot{x} + x + x^3 = 0, \quad \ddot{x} + x - x^3 = 0,$$

do the following:

   (i) Draw the level curves of the first integral.

   (ii) Find the critical points of the differential equation.

   (iii) Show that the critical points (in the calculus sense) of the first integral are precisely the critical points of the differential equation.

   (iv) Determine the stability of the critical points of the differential equation.

   (v) Classify the critical points of the first integral and compare the result to what you found under (iv) above.

6. Consider the equations of motion of a solid body rotating around a fixed point which coincides with its center of gravity,

$$a\dot{x} = (b - c)yz,$$
$$b\dot{y} = (c - a)xz,$$
$$c\dot{z} = (a - b)xy.$$

Here, x, y, and z are the projections of the angular velocity vector on the coordinate axes and a, b, and c are the moments of inertia, which are distinct, constant, and positive. Verify that $(x^*, 0, 0)$ is an equilibrium point for any $x^* > 0$.

   (i) Explain why we may assume $a < b < c$ without loss of generality.

   (ii) Verify that $V_1(x, y) = a(c - a)x^2 + b(c - b)y^2$ and $V_2(y, z) = b(b - a)y^2 + c(c - a)z^2$ are first integrals.

   (iii) Draw the level sets of V_1 and V_2 that pass through $(x^* + \delta x, \delta y, \delta z)$, where δx, δy, and δz are small relative to x^*, and describe (in words) their intersection. Explain (in words) why this shows that points that start near $(x^*, 0, 0)$ remain near $(x^*, 0, 0)$ for all times.

   (iv) Show that the function $V(x, y, z) = (V_1(x, y) - V_1(x^*, 0))^2 + V_2(y, z)$ is also a first integral. Hint: V is a differentiable function of two first integrals.

   (v) Show that V has a strict local minimum at $(x^*, 0, 0)$ and use this result to prove that $(x^*, 0, 0)$ is Lyapunov stable.

7. A system of differential equations of the form $\dot{p}_i = \partial H/\partial q_i$, $\dot{q}_i = -\partial H/\partial p_i$, $i = 1,\ldots,n$, is known as a *Hamiltonian system*; the function H of the $2n$ variables p_i, q_i is the *Hamiltonian* or *energy integral*. Show that H is constant along trajectories—that is, H is a first integral. Check that any system of the form $\ddot{x} + f(x) = 0$ is a Hamiltonian system.

8. Consider the system of Volterra–Lotka equations,

$$\dot{x} = ax - bxy,$$
$$\dot{y} = -cy + dxy,$$

where a, b, c, and d are positive constants.

   (i) Find all equilibrium points.

   (ii) Show that the transformation $(x,y) \mapsto (p,q)$, where $p = \ln x$, $q = \ln y$, leads to a Hamiltonian system, where $H = be^p - cp + be^q - aq$. Use this result to find a first integral (in terms of x and y) of the Volterra–Lotka equations.

   (iii) Consider the special Volterra–Lotka system $\dot{x} = x - xy$, $\dot{y} = -\delta y + xy$, where $\delta > 0$. Check that $(x^*, y^*) = (\delta, 1)$ is an equilibrium point, find the linearization of the system about this equilibrium point, and find the eigenvalues of the matrix A that occurs in the linearization.

9. The function $\varphi : \mathbb{R}^2 \to \mathbb{R}$ given by $\varphi(t,x) = \frac{3}{2}(1 - e^{-2t}) + xe^{-2t}$ defines a dynamical system $\phi_t : x \mapsto \phi_t(x) = \varphi(t,x)$. Find the linear first-order autonomous differential equation that is satisfied by all functions $t \mapsto \phi_t(x)$ for any $x \in \mathbb{R}$.

10. Consider the ODE $\dot{x} + \alpha x = \beta$, where α and β are real numbers. Given $x_0 \in \mathbb{R}$, describe the positive and negative orbits $\gamma^+(x_0)$ and $\gamma^-(x_0)$. (Consider the cases $\alpha > 0$, $\alpha = 0$, and $\alpha < 0$ separately.)

11. Suppose a dynamical system for the Earth's climate is shown to have an equilibrium point that is a positive attractor. Explain to somebody who does not know differential equations what this means.

12. Consider the second-order equation $\ddot{x} + \dot{x} - x + x^3 = 0$ or the equivalent system $\dot{x}_1 = x_2$, $\dot{x}_2 = x_1 - x_2 - x_1^3$.

   (i) Prove that the function $V(x) = -\frac{1}{2}x_1^2 + \frac{1}{4}x_1^4 + \frac{1}{2}x_2^2$ decreases along trajectories.

   (ii) Show that each orbit is bounded. Hint: Show that the level set $V^{-1}(c)$ is closed for c large.

   (iii) Show that the critical points $(-1,0)$ and $(1,0)$ are asymptotically stable while $(0,0)$ is a saddle point.

   (iv) Let $\mathscr{A} = W^u(0,0) \cup \{(-1,0)\} \cup \{(1,0)\}$. Prove that the α-limit set $\alpha(x_0)$ for any point $x_0 \in \mathscr{A}$ is an equilibrium point.

   (v) Show that the set $\mathscr{A}$ is the global attractor.

13. Consider the system $\dot{x}_1 = x_2$, $\dot{x}_2 = -x_1 + x_2(1 - x_1^2 - 2x_2^2)$. Show that the origin is the only critical point. Use the function $V(x_1, x_2) = \frac{1}{2}(x_1^2 + x_2^2)$ to show that any solution which starts in the annulus $\mathscr{D} = \{(x_1, x_2) : \frac{1}{2} < x_1^2 + x_2^2 < 1\}$ remains in $\mathscr{D}$ for all $t \geq 0$. Sketch the phase portrait near the boundary of $\mathscr{D}$ and discuss the long-time behavior of the solutions of the system.

14. A periodic solution of the autonomous differential equation $\dot{x} = f(x)$ corresponds to a closed orbit. Prove the converse: Every closed orbit corresponds to a periodic solution.

15. A set $S \subset \mathbb{R}^n$ is called an *invariant set* for the autonomous differential equation $\dot{x} = f(x)$ if the solution $\varphi(t, x_0)$ with $x_0 \in S$ is contained in S for all $t \in \mathbb{R}$. If this property holds only for $t \geq 0$ ($t \leq 0$), then S is called a *positive (negative) invariant set*.

  (i) Prove that every equilibrium point is an invariant set.

  (ii) Prove that every periodic solution defines an invariant set.

  (iii) Prove that every solution which exists for all times defines an invariant set.

16. (i) Prove that the α- and ω-limit sets are closed.

  (ii) Prove that the α-limit set (ω-limit set) is positive (negative) invariant. Hint: Use the translation property to show that $\varphi(t + t_n, x_0) = \varphi(t, \varphi(t_n, x_0))$. Keep t fixed and let $n \to \infty$.

  (iii) Prove that the α- and ω-limit sets contain only closed orbits.

17. (i) Consider the linear harmonic oscillator, $\ddot{x} + x = 0$. Show that $\omega(\gamma) = \alpha(\gamma) = \gamma$ for every orbit γ.

  (ii) Consider the pendulum equation $\ddot{x} + \sin x = 0$. For which orbits γ is it true that $\omega(\gamma) = \alpha(\gamma) = \gamma$?

18. Consider the nonautonomous differential equation $\dot{x} = -x + \cos t$.

  (i) Find the solution that satisfies the initial condition $x(0) = x_0$.

  (ii) Find the unique periodic solution of this equation.

  (iii) Show that all solutions approach this periodic solution as $t \to \infty$.

19. Consider the periodically forced damped harmonic oscillator,

$$\ddot{x} + 2\beta\dot{x} + x = \gamma \cos \omega t, \quad \beta \in (0, 1). \tag{4.19}$$

The period of the forcing term is $T = 2\pi/\omega$.

  (i) Show that the function x, defined by the expression

$$x(t) = a \cos \omega t + b \sin \omega t + e^{-\beta t}(c_1 \cos \lambda t + c_2 \sin \lambda t), \quad t \in \mathbb{R},$$

is a solution of Eq. (4.19) if and only if $\lambda = \sqrt{1 - \beta^2}$ and the constants a and b are given by

$$a = \frac{(1 - \omega^2)\gamma}{(1 - \omega^2)^2 + 4\beta^2\omega^2}, \quad b = \frac{2\beta\omega\gamma}{(1 - \omega^2)^2 + 4\beta^2\omega^2}.$$

  (ii) Prove that Eq. (4.19) has one and only one periodic solution. Show that all solutions approach this periodic solution as $t \to \infty$.

(iii) Determine the point $(x_0, 0)$, where the periodic solution intersects the positive x-axis in the $(x, \dot{x})$-plane. Consider a nonperiodic solution that starts from the point $(x_0 + \delta x, 0)$. Choose suitable values for β, γ, and ω and use MATLAB to explore the Poincaré map P by finding successive points of intersection of the orbit with the half-line $\{x > 0\}$. Verify numerically that these points converge to $(x_0, 0)$ and that the times between successive intersections converge to T.

20. It is possible to transform the general differential equation $\dot{x} = f(x)$ to an equivalent and "simpler" equation for a function y by a procedure known as *normalization*. This and the following exercise demonstrate the procedure for the scalar case. In the present exercise, the transformation maps the variable x into a new variable y that satisfies a linear equation. The procedure requires an integration, which may be difficult. In the following exercise, the transformation maps the forcing function f into a new forcing function which is close to linear. The procedure does not require integration, but its justification requires a certain degree of smoothness of the forcing function.

Assume that $f(0) = 0$. Then the Taylor series expansion of f in the neighborhood of the trivial solution $x^* = 0$ starts with the linear term, and if $f \in C^k$ for some $k \geq 2$, we have $f(x) = ax + x^2 g(x)$ for some $g \in C^{k-2}$. Assume that $a \neq 0$.

Prove that the transformation $\Psi : x \mapsto \Psi(x)$ given by

$$\Psi(x) = x \, \exp\left(-\int_0^x \frac{g(\xi)}{a + \xi g(\xi)} \, d\xi\right)$$

maps the function x into a function $y = \Psi(x)$ which satisfies the linear equation $\dot{y} = ay$.

21. Assume again that $f(0) = 0$ and that f admits a Taylor series expansion near $x^* = 0$, $f(x) = a_1 x + a_2 x^2 + a_3 x^3 + \cdots$, where $a_1 \neq 0$. Suppose x can be expanded formally in powers of y,

$$x = y + b_2 y^2 + b_3 y^3 + \cdots.$$

(i) Differentiate both sides of the formal expansion of x with respect to t, replace $\dot{x}$ by the power series expansion of $f(x)$ near $x^* = 0$, use the formal expansion of x to replace powers of x by powers of y, and solve the resulting equation for $\dot{y}$. Verify that y satisfies an equation of the form $\dot{y} = a_1 y + g(y)$, where $g(y)$ is the ratio of two infinite series.

(ii) Expand the ratio of the two infinite series formally in powers of y to obtain a series expansion of the form $g(y) = c_2 y^2 + c_3 y^3 + \cdots$ and express the first few coefficients c in terms of the coefficients a and b.

(iii) Show that, by a judicious choice of the coefficients b, the coefficients c can be eliminated one by one, starting with c_2. Verify that this procedure can be continued indefinitely to successively eliminate higher-order terms in the differential equation for y. If the procedure terminates, the resulting differential equation for y is linear, $\dot{y} = a_1 y$; if not, it is at least close to linear, $\dot{y} = a_1 y + g(y)$, where g is of higher order near $y = 0$, and therefore simpler to analyze in the neighborhood of the origin (which is a critical point for both the x and y equations).

The procedure is entirely formal, in the sense that we are not concerned with convergence of any series expansion and assume at all times that x and y are "sufficiently small."

Chapter 5

Bifurcation Theory

When an autonomous differential equation involves one or more parameters, it depends on the parameter values whether the equation admits equilibrium solutions, and if there are equilibrium solutions, their nature may vary with the values of the parameters. In fact, their nature may change quite dramatically if there are critical parameter values and a parameter crosses such a critical value. For example, equilibrium solutions may exchange stability, new equilibrium solutions may emerge, or existing solutions may vanish. Such phenomena fall under the rubric of bifurcation theory. In this chapter we give a number of examples of dynamical systems in one and two dimensions which, although relatively simple, are indicative of bifurcation phenomena in more general situations.

Keywords: Transcritical bifurcation, saddle-node bifurcation, fold bifurcation, hysteresis, pitchfork bifurcation, cusp bifurcation, Hopf bifurcation.

5.1 ▪ Bifurcation

Bifurcation theory is the study of equilibrium solutions of nonlinear (autonomous) differential equations that involve one or more parameters,

$$\dot{x} = f(\lambda, x). \tag{5.1}$$

As in the previous chapter, the function $x : t \mapsto x(t) \in \mathbb{R}^n$ is unknown. But now the vector field f depends not only on x but also on one or more real parameters $\lambda_1, \ldots, \lambda_m$, which are collectively represented by the symbol λ. Equilibrium solutions of Eq. (5.1) are found by solving the equation

$$f(\lambda, x) = 0, \tag{5.2}$$

so if x^* is an equilibrium solution, then x^* will, in general, depend on λ. Because f is a nonlinear function of x, this dependence can be quite complicated. For example, there may be more than one solution for a range of parameter values, and solutions may disappear, new solutions may emerge, solutions that were stable may become unstable or vice versa, and solutions may exchange stability as a parameter crosses a critical value. Collectively, these and similar phenomena are known as *bifurcation phenomena*, and the mathematical theory that deals with bifurcation phenomena is known as *bifurcation theory*.

The foundations of bifurcation theory were laid in the late 19th century by the French mathematician HENRI POINCARÉ (1854–1912), who studied a fluid between two rotating cylinders. Poincaré observed that the fluid split into two parts (bifurcated) when the relative velocity of the rotating cylinders reached a critical value and a new flow pattern emerged as the relative velocity increased beyond the critical value. Bifurcation phenomena occur in all areas of application where the underlying mathematical model is nonlinear. In fact, we have already seen several instances in previous chapters, when we discussed simple EBMs (Chapter 2) and a THC model (Chapter 3).

5.2 ▪ Examples

To show what bifurcation phenomena may look like and how they may arise, it is helpful to consider several examples. In the examples that follow, the vector field is a simple low-order, quadratic, or at most cubic polynomial function, with only one or at most two real parameters, so they can be analyzed in more or less detail. Though simple, the examples already indicate that equilibrium solutions of nonlinear differential equations can vary dramatically under the influence of just one or two parameters. As it turns out, the types of bifurcations associated with these low-order polynomial vector fields appear to be quite generic, in the sense that they are similar to the phenomena arising in situations where the nonlinearity of the vector field is clearly more complicated than quadratic or cubic. In fact, they appear to occur so often that their names have entered the language in which qualitative features of dynamical systems are described. If a dynamical system shows a sudden change of behavior, one often looks for one of the bifurcations described in this section to describe it.

5.2.1 ▪ Transcritical Bifurcation

The first example concerns a simple ODE with a quadratic nonlinearity for a scalar function $x : t \mapsto x(t)$,

$$\dot{x} = \lambda x - x^2. \tag{5.3}$$

As usual, the dot $\dot{}$ denotes differentiation with respect to t. The equation has two fixed points (equilibrium solutions), $x_1^* = 0$ (the *trivial solution*) and $x_2^* = \lambda$, which coincide if $\lambda = 0$. What can we say about the stability of x_1^* and x_2^*?

On the left in Figure 5.1 is shown the vector field (the right-hand side of the differential equation as a function of x) for $\lambda < 0$ (blue curve), $\lambda = 0$ (black curve), and $\lambda > 0$ (red curve).

First consider the case $\lambda < 0$. Then $x_2^* < x_1^*$. The vector field is negative for $x < x_2^*$, positive for $x_2^* < x < x_1^*$, and negative again for $x > x_1^*$. Hence, if $\varphi : t \mapsto \varphi(t)$ is a solution of Eq. (5.3), then φ is decreasing while $\varphi(t) < x_2^*$, increasing while $x_2^* < \varphi(t) < x_1^*$, and decreasing again while $\varphi(t) > x_1^*$. Consequently, solutions that start out near x_1^* converge to x_1^* as $t \to \infty$, while solutions that start out near x_2^* move away from x_2^* as $t \to \infty$. In other words, x_1^* is a stable equilibrium, while x_2^* is an unstable equilibrium.

Now consider the case $\lambda > 0$. Then $x_1^* < x_2^*$. Using the same arguments as before, we see that x_1^* is now an unstable equilibrium, while x_2^* has turned into a stable equilibrium. This exchange of stability occurred as we passed through $\lambda = 0$, when the two equilibrium solutions coincide, $x_1^* = x_2^* = 0$. Again using the same arguments as before, it is easy to see that this single solution is unstable.

These results are summarized in the diagram on the right in Figure 5.1. Here, the equilibrium solutions are plotted as functions of λ (λ is assigned to the horizontal axis, x^* to the vertical axis), so the diagram shows the graphs of the functions $x_1^* : \lambda \mapsto x_1^*(\lambda) = 0$

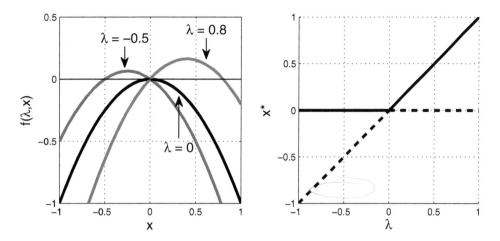

Figure 5.1. *Transcritical bifurcation, $\dot{x} = \lambda x - x^2$; vector field (left), bifurcation diagram (right).*

and $x_2^* : \lambda \mapsto x_2^*(\lambda) = \lambda$, following the convention that a solid line corresponds to stable equilibria and a dashed line to unstable equilibria.

This type of diagram is known as a *bifurcation diagram*. Here, it shows a *transcritical bifurcation*. A transcritical bifurcation is a local bifurcation in which an equilibrium solution exists for all values of the parameter and is never destroyed; however, such an equilibrium solution exchanges its stability with another equilibrium solution as the parameter is varied. In other words, both before and after the bifurcation, there is one unstable and one stable equilibrium solution. However, their stability is exchanged when they collide. The origin, where the exchange of stability occurs, is called a *bifurcation point*, and each portion of the bifurcation diagram where x^* can be represented as a function of λ is called a *solution branch*. For example, the segment $\{(\lambda, x^*) : x^* = \lambda, \lambda < 0\}$ is an unstable solution branch. Typically, solution branches meet and possibly gain or lose stability at bifurcation points.

It is possible to find bifurcation points in a systematic manner. A solution branch is a set of equilibrium points x^* which can be written as a smooth function of the bifurcation parameter λ—that is, a subset of $\{(\lambda, x) : f(\lambda, x) = 0\}$ which has the form $\{(\lambda, g(\lambda)) : \lambda \in (a, b)\}$ for some function g and some interval (a, b). By the implicit function theorem, this is possible near points (λ, x^*) where $(\partial f / \partial x)(\lambda, x^*) \neq 0$ and, of course, $f(\lambda, x^*) = 0$. Solution branches are expected to meet at points where $f(\lambda, x^*) = 0$ and $(\partial f / \partial x)(\lambda, x^*) = 0$. Such points are candidates for bifurcation points (but there is no guarantee that they actually are bifurcation points!). In the current example, the solution set of $f(\lambda, x) = 0$ is the set $\{(\lambda, x^*) : x^* = \lambda\} \cup \{(\lambda, x^*) : x^* = 0\}$. Also, $(\partial f / \partial x)(\lambda, x^*) = \lambda - 2x^*$, and this expression vanishes on the solution set precisely at the bifurcation point $(\lambda, x^*) = (0, 0)$.

5.2.2 ▪ Saddle-Node Bifurcation

The next example is again a scalar ODE with a quadratic nonlinearity,

$$\dot{x} = \lambda - x^2. \tag{5.4}$$

The vector field shifts up or down with λ, as shown on the left in Figure 5.2. If $\lambda < 0$ (blue curve), the equation has no equilibrium solution. If $\lambda > 0$ (red curve), it has two equilibrium solutions, $x_1^* = \lambda^{1/2}$ (stable) and $x_2^* = -\lambda^{1/2}$ (unstable). These two solutions

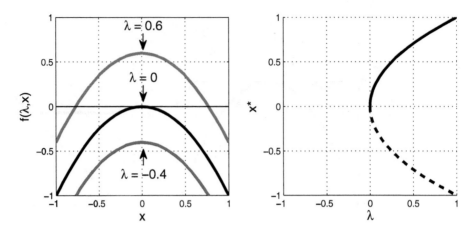

Figure 5.2. *Saddle-node bifurcation,* $\dot{x} = \lambda - x^2$; *vector field (left), bifurcation diagram (right).*

coincide if $\lambda = 0$, and the single trivial solution is unstable. The bifurcation diagram is shown on the right in Figure 5.2; as before, a solid line corresponds to stable equilibria and a dashed line to unstable equilibria. The origin is a bifurcation point; it is the only point where $f(\lambda, x^*) = \lambda - (x^*)^2 = 0$ and $(\partial f / \partial x)(\lambda, x^*) = -2x^* = 0$. This case is called a *saddle-node bifurcation* or a *fold bifurcation*. A saddle-node bifurcation is a local bifurcation in which two equilibrium solutions emerge out of nowhere. (For this reason, it is sometimes referred to as a *blue-sky bifurcation*.)

5.2.3 ▪ Hysteresis

Now consider a scalar ODE with a cubic nonlinearity,

$$\dot{x} = \lambda + x - x^3. \tag{5.5}$$

As in the previous example, the vector field shown on the left in Figure 5.3 shifts up or down with λ. The equation $f(\lambda, x) = 0$ has one solution for $\lambda < -\lambda^*$, three solutions for $-\lambda^* < \lambda < \lambda^*$, and one solution for $\lambda > \lambda^*$, where $\lambda^* = (\frac{4}{27})^{1/2} = 0.3849\ldots$. For $\lambda = \pm\lambda^*$, there are two solutions.

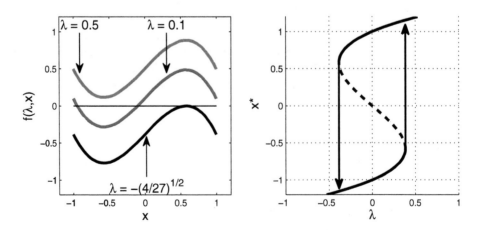

Figure 5.3. *Hysteresis,* $\dot{x} = \lambda + x - x^3$; *vector field (left), bifurcation diagram with hysteresis loop (right).*

Consider for a moment the special case $\lambda = 0$. In the language of dynamical systems, the equilibrium points are -1 (stable), 0 (unstable), and 1 (stable). The orbits are the open intervals $(-\infty, -1)$, $(-1, 0)$, $(0, 1)$, $(1, \infty)$ and the points $\{-1\}$, $\{0\}$, and $\{1\}$. The orbit structure is stable, in the sense that it persists qualitatively for all $\lambda \in (-\lambda^*, \lambda^*)$. For $\lambda = \pm\lambda^*$, the equation is at a bifurcation point, and the orbit structure changes. For $\lambda < -\lambda^*$ and $\lambda > \lambda^*$, the equation again has a stable orbit structure; in the case $\lambda > \lambda^*$, the orbit structure consists of the open intervals $(-\infty, x^*)$ and (x^*, ∞) and the single point $\{x^*\}$, where $\lambda = (x^*)^3 - x^*$.

It follows from the implicit function theorem that candidates for bifurcation points can be found by solving the system of equations

$$f(\lambda, x) = \lambda + x - x^3 = 0,$$
$$\frac{\partial f}{\partial x}(\lambda, x) = 1 - 3x^2 = 0.$$

It is easy to verify that there are two solutions, $\pm(\lambda^*, x^*)$, where $\lambda^* = (\frac{4}{27})^{1/2}$ and $x^* = (\frac{1}{3})^{1/2}$, in agreement with the previous analysis and Figure 5.3.

The dynamics of Eq. (5.5) are worth studying in more detail. Suppose we have the capability to continuously vary the value of the parameter λ, for example by changing some of the physics. (Think of emitting CO_2 into the atmosphere, or releasing large amounts of fresh water into the oceans by melting the polar ice cap.) If we start with a very large negative value of λ, then every solution will eventually wind up at some equilibrium state on the lower branch of the bifurcation diagram, regardless of its initial state. Now consider the system when it is near such an equilibrium state. As we change λ a little bit, the system will certainly stay near this equilibrium state; in fact, as we keep changing λ little by little, the system will "slide" along the stable branch of the bifurcation diagram. But once λ exceeds the value $\lambda^* = (\frac{4}{27})^{1/2}$, the system will "jump" (that is, transition very quickly) to an equilibrium state on the upper branch of the bifurcation diagram. As we change the value of λ further, the system will now "slide" along the stable upper branch. This transition from the stable lower branch to the stable upper branch cannot be reversed with small changes of λ, as indicated by the upward arrow in Figure 5.3. Similarly, if we start with a very large positive value of λ, the system will respond smoothly to small changes of λ until $\lambda \approx -\lambda^* = -(\frac{4}{27})^{1/2}$, where it can be expected to jump to an equilibrium state on the lower branch (see the downward arrow in Figure 5.3).

The example teaches us two lessons, namely that a system can jump from one equilibrium state to another as the value of a (physical) parameter changes and, furthermore, that the parameter value at which a jump occurs depends on the direction in which the parameter is varied. This phenomenon is called *hysteresis*, and the part in Figure 5.3 that resembles a parallelogram is called the *hysteresis loop*. We have encountered this situation in Chapter 2, when we studied the global EBM (2.8).

5.2.4 ▪ Pitchfork Bifurcations

Our next two examples involve a scalar ODE with a cubic nonlinearity but with opposite signs of the cubic term. The first equation is

$$\dot{x} = \mu x - x^3. \tag{5.6}$$

The parameter μ determines the slope of the vector field at the origin, which is shown for three different values of μ on the left in Figure 5.4. For all $\mu > 0$, the system has three equilibrium states, $x_1^* = -\mu^{1/2}$, $x_2^* = 0$, and $x_3^* = \mu^{1/2}$. Looking at the behavior of $f(\mu, x) = \mu x - x^3$, one sees that x_1^* and x_3^* are stable and x_2^* is unstable.

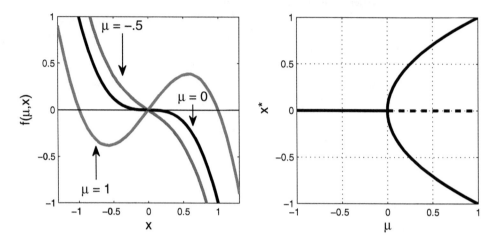

Figure 5.4. *Supercritical pitchfork bifurcation, $\dot{x} = \mu x - x^3$; vector field (left), bifurcation diagram (right).*

The orbit structure is stable for $\mu > 0$. At $\mu = 0$, the equilibrium states coalesce and a bifurcation occurs. For $\mu < 0$, the system has again a stable orbit structure, with one asymptotically stable equilibrium point $x_1^* = 0$ at the origin. The bifurcation diagram is shown on the right in Figure 5.4. The bifurcation is called a *supercritical pitchfork bifurcation*; "pitchfork" because of its appearance and "supercritical" because the additional equilibrium states which emerge at the bifurcation point exist for the values of the parameter ($\mu > 0$) at which the original equilibrium state $x^* = 0$ is unstable.

The second equation differs from the first equation in the sign of the cubic term,

$$\dot{x} = \mu x + x^3. \tag{5.7}$$

The equation has three equilibrium solutions for $\mu < 0$ and one for $\mu > 0$. The equilibrium solution $x^* = 0$ persists for all μ, and it is obviously stable for $\mu < 0$ and unstable for $\mu > 0$. None of the other equilibrium solutions are stable for any $\mu \neq 0$. This bifurcation is called a *subcritical pitchfork bifurcation*; see Figure 5.5.

5.2.5 ▪ Cusp Bifurcation

So far, all equations have involved only one parameter. Next, we consider a two-parameter equation with a cubic nonlinearity which generalizes Eq. (5.5),

$$\dot{x} = \lambda + \mu x - x^3. \tag{5.8}$$

Equilibrium solutions must lie on the critical manifold $\mathcal{M} = \{(\lambda, \mu, x) : \lambda + \mu x - x^3 = 0\}$ in $\mathbb{R}^3$. This manifold has a complicated structure, with sheets that are folded over one another, indicating that we may find more than one solution for a given set of parameters λ and μ; see Figure 5.6. To detect where solutions may appear or disappear, we use again the implicit function theorem. Uniqueness of an equilibrium solution as a function of the bifurcation parameters is lost when not only the vector field but also its gradient are

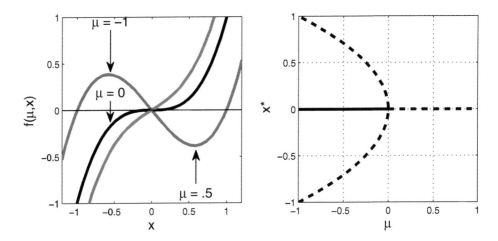

Figure 5.5. *Subcritical pitchfork bifurcation, $\dot{x} = \mu x + x^3$; vector field (left), bifurcation diagram (right).*

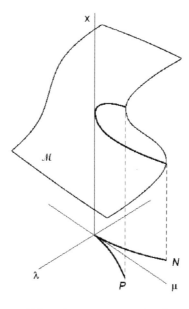

Figure 5.6. *Equilibrium manifold $\mathcal{M}$ for Eq. (5.8). Reprinted courtesy of John Guckenheimer.*

zero, so we need to solve the system of equations

$$\lambda + \mu x - x^3 = 0, \quad \mu - 3x^2 = 0.$$

Solving for λ and μ and eliminating x, we obtain a relation between λ and μ, namely $4\mu^3 = 27\lambda^2$. This is the equation of a *cusp* in the (λ, μ)-plane with a singularity at the origin and two branches P (positive) and N (negative) given by

$$P = \left\{ (\lambda, \mu) : \lambda = \frac{2}{3\sqrt{3}} \mu^{3/2}, \; \mu \ge 0 \right\}, \quad N = \left\{ (\lambda, \mu) : \lambda = -\frac{2}{3\sqrt{3}} \mu^{3/2}, \; \mu \ge 0 \right\}.$$

The cusp is the projection of the curves in $\mathbb{R}^3$ where $\mathcal{M}$ folds over, as shown in Figure 5.6. At a point (λ^*, μ^*) on either P or N, a jump to a different stable equilibrium on $\mathcal{M}$ occurs. The jump may be up or down, depending on where the equilibrium belongs initially, and since the jumps up and down happen at different points (λ^*, μ^*), we are seeing another scenario for hysteresis.

The local bifurcation diagram is shown in Figure 5.7. The point $(\lambda, \mu) = (0,0)$ is the origin of the two branches P and N, where saddle-node bifurcations occur. The two branches divide the parameter plane into two regions; inside the wedge there are always three equilibria, two stable and one unstable, and outside the wedge there is always a single stable equilibrium. If we approach the cusp point at the origin from inside the wedge, all three equilibria merge. The origin is a *cusp bifurcation*.

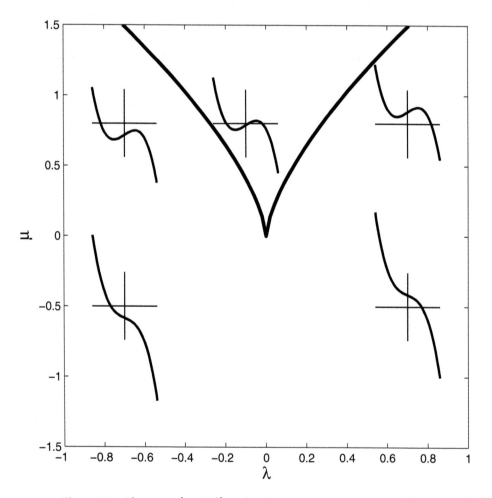

Figure 5.7. *The cusp $4\mu^3 = 27\lambda^2$ in the (λ, μ)-plane and some graphs of the vector field $x \mapsto \lambda + \mu x - x^3$. The center of the coordinate system for each small graph indicates the values of λ and μ.*

Generically, the cusp bifurcation is a bifurcation of equilibria in a two-parameter family of ODEs at which the critical equilibrium has one zero eigenvalue and the quadratic coefficient for the saddle-node bifurcation vanishes. At the cusp bifurcation point, two branches of saddle-node bifurcation curves meet tangentially, forming a semicubic parabola. For nearby parameter values, the system can have three equilibria, which collide and disappear pairwise via the saddle-node bifurcations. The cusp bifurcation implies the presence of a hysteresis phenomenon.

5.2.6 ▪ Two-Dimensional Example

To illustrate how techniques for finding solution branches and bifurcation points can be combined with the linear stability analysis of Chapter 4, we discuss a simple two-dimensional example, again involving no more than a quadratic nonlinearity,

$$
\begin{aligned}
\dot{x}_1 &= \lambda x_1 - x_1^2 + x_1 x_2, \\
\dot{x}_2 &= -2x_1 x_2 + x_1^2.
\end{aligned}
\tag{5.9}
$$

Denote the right-hand side of this system by $f(\lambda, x)$. We must first solve the system $f(\lambda, x) = 0$ for fixed λ to find equilibrium solutions. A bit of algebra shows that, for all $\lambda > 0$, there are two equilibrium solutions,

$$
x_{\pm}^* = \pm(x_1^*, x_2^*) = \pm(\sqrt{2\lambda}, \tfrac{1}{2}\sqrt{2\lambda}).
$$

These two solution branches meet at $(0,0)$ for $\lambda = 0$. Next, we compute the linearization of the right-hand side of Eq. (5.9),

$$
Df(\lambda, x) = \begin{pmatrix} -2x_1 + x_2 & x_1 \\ 2x_1 - 2x_2 & -2x_1 \end{pmatrix}.
$$

The determinant $\det(Df(\lambda, x)) = 2x_1^2$ vanishes only at the equilibrium solution $(0,0)$ for $\lambda = 0$. This is therefore our candidate for a bifurcation point.

To analyze the stability on each solution branch, we observe that, for $\lambda > 0$,

$$
\det(Df(\lambda, x_{\pm}^*)) = 4\lambda > 0, \quad \text{trace}(Df(\lambda, x_{\pm}^*)) = \mp\tfrac{7}{2}\sqrt{2\lambda} \lessgtr 0.
$$

The matrix $Df(\lambda, x_+^*)$ has two negative real eigenvalues, so the branch of x_+^*-solutions consists of stable nodes, while $Df(\lambda, x_-^*)$ has a negative and a positive real eigenvalue, so the branch of x_-^*-solutions consists of saddle points. The conclusion is that this is a saddle-node bifurcation. It is difficult to visualize the vector field for different values of λ simultaneously or to show a full bifurcation diagram in (λ, x_1, x_2)-space. In such a situation, one often shows the bifurcation by plotting a single scalar quantity (for example, a single coordinate of the family of equilibrium solutions or their norm), together with the bifurcation parameter in the same diagram. For this example, such plots in the (λ, x_1)- or (λ, x_2)-plane are scaled versions of the graph on the right in Figure 5.2.

5.2.7 ▪ Hopf Bifurcation

We conclude with another two-dimensional example,

$$
\begin{aligned}
\dot{x}_1 &= x_1(\lambda - x_1^2 - x_2^2) + x_2, \\
\dot{x}_2 &= x_2(\lambda - x_1^2 - x_2^2) - x_1.
\end{aligned}
\tag{5.10}
$$

Equilibrium solutions are found by setting both right-hand sides equal to zero,

$$
\begin{aligned}
x_1(\lambda - x_1^2 - x_2^2) + x_2 &= 0, \\
x_2(\lambda - x_1^2 - x_2^2) - x_1 &= 0.
\end{aligned}
\tag{5.11}
$$

This system is most easily solved if we multiply the first equation by x_1 and the second by x_2 and add the resulting two equations to obtain a single equation for the square of the amplitude, $r^2 = x_1^2 + x_2^2$, namely

$$r^2(\lambda - r^2) = 0.$$

The equation shows that $r_1^* = 0$ (the trivial solution) is a critical point for all values of λ and is the only critical point if $\lambda < 0$. If $\lambda > 0$, there is another critical point, $r_2^* = \lambda^{1/2}$, the meaning of which will become clear shortly. (Remember that the amplitude must be positive.) If we characterize an equilibrium solution by its amplitude and plot r^* vs. λ in a bifurcation diagram, then the trivial solution branch is represented by the λ-axis. Solutions defined by the second critical point are represented by a parabolic curve in the first quadrant.

Now consider the linearization of the system (5.10) near the origin. The coefficient matrix varies with λ,

$$A(\lambda) = \begin{pmatrix} \lambda & 1 \\ -1 & \lambda \end{pmatrix}.$$

Since $\det A(\lambda) = 1 + \lambda^2 \geq 1$ for all λ, $A(\lambda)$ is never singular, so one could suspect that there is no bifurcation point. Yet, it is readily verified that the origin is a stable spiral point for $\lambda < 0$ and an unstable spiral point for $\lambda > 0$, so the trivial solution branch appears to have lost stability at $\lambda = 0$ without meeting any other solution branch. What happened?

To solve this seeming paradox, we introduce polar coordinates, $r = (x_1^2 + x_2^2)^{1/2}$ and $\theta = \tan^{-1}(x_2/x_1)$, and rewrite Eq. (5.10) as a system for (r, θ),

$$\dot{r} = \lambda r - r^3, \quad \dot{\theta} = -1. \tag{5.12}$$

Notice that the radial component undergoes a supercritical pitchfork bifurcation at $\lambda = 0$ (cf. Figure 5.4). For $\lambda < 0$, there is only the stable equilibrium at the origin. But for $\lambda > 0$, the system has a special time-dependent solution,

$$r^* = \lambda^{1/2}, \quad \theta = -t + C, \tag{5.13}$$

where C is an arbitrary constant. In Cartesian coordinates, this solution is given by $x_1^* = \lambda^{1/2}\cos(t - C)$, $x_2^* = -\lambda^{1/2}\sin(t - C)$. The constant C contributes only a phase shift, so the critical point $r_2^* = \lambda^{1/2}$ that we found earlier represents a periodic solution,

$$x^*(t) = \lambda^{1/2}(\cos t, -\sin t). \tag{5.14}$$

In fact, since the radial component of this solution is stable, one can show with a bit of work that this periodic solution is orbitally stable. Figure 5.8 illustrates this phenomenon.

This type of bifurcation occurs when the coefficient matrix of the linearized system has a pair of complex conjugate eigenvalues which cross from the left-hand side of the complex plane into the right-hand side as λ passes through a critical value. In the case considered here, the eigenvalues of $A(\lambda)$ (for a fixed value of λ!) are $\lambda + i$ and $\lambda - i$ (i is the imaginary unit, $i^2 = -1$), so they cross the imaginary axis as λ passes through zero. In the language of dynamical systems, for $\lambda < 0$ there is a single stable orbit, namely the origin. This is also the ω-limit set of the system for all possible starting points. For $\lambda > 0$, the origin is still an orbit, but it is now unstable. The periodic solution given above is stable, and for all starting points except the origin the ω-limit set is a circle with radius $\lambda^{1/2}$.

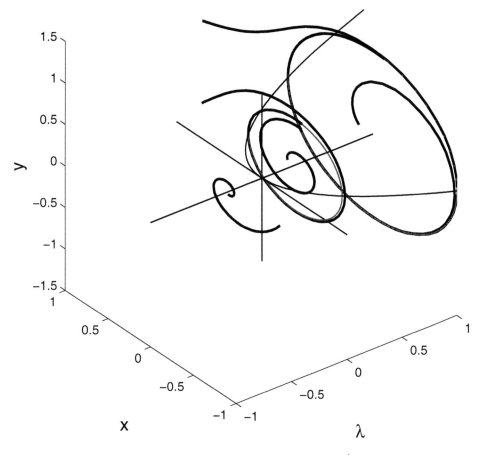

Figure 5.8. *Trajectories for the system* $\dot{r} = \lambda r - r^3$, $\dot{\theta} = -1$.

Bifurcations of this type were first investigated by Hopf in the 1940s [42], and this type of bifurcation is called a *Hopf bifurcation*. An extensive discussion can be found in [64].

5.3 ▪ From Examples to the General Case

It is somewhat remarkable that there are identifiable kinds of bifurcations that appear repeatedly in many problems. Pitchfork bifurcation and saddle-node bifurcation are indeed characteristic for one-dimensional equations with cubic nonlinearities. The question thus arises whether there is a way to classify all possible bifurcation diagrams. The answer is yes, but even a summary discussion would lead us well beyond the scope of the present text. Suffice it to say that there are many results if the dimensions are low: $m = 1$ (one parameter), $n = 1$ (one variable) or $n = 2$ (two variables), and a few results for $n = 3$ or more. As the dimensions increase, many new types of bifurcation appear, and a complete classification becomes increasingly involved. Nevertheless, the types of bifurcation that we have discussed here occur so often that their names have entered the language in which qualitative features of dynamical systems are described. They belong to a limited set of bifurcation scenarios that have been studied in great detail, and if a dynamical system shows a sudden change of behavior, one often looks for one of these bifurcations to describe it. Similarly, complex dynamic phenomena such as hysteresis or the sudden emergence of periodic solutions can also often be explained by bifurcation theory.

5.4 ▪ Bifurcation Points

We now return to the general differential equation (5.1) for the function $x : t \mapsto x(t) \in \mathbb{R}^n$. Without loss of generality, we may assume that $f(\lambda, 0) = 0$, so $x^* = 0$ is a steady-state solution (the *trivial solution*) of Eq. (5.1). We are interested in bifurcations of equilibrium solutions from the trivial solution.

The implicit function theorem tells us that Eq. (5.2) can be solved for x as a function of λ as long as the Jacobian $D_x f(\lambda, x)$ is nonsingular. The graph of each of these functions is a branch of equilibrium solutions of Eq. (5.1). At a point where $D_x f(\lambda, x)$ is singular, uniqueness is lost and several branches of equilibria may come together. Such a point is a candidate for a *bifurcation point*.

Since $f(\lambda, 0) = 0$, a Taylor series expansion of f in the neighborhood of the trivial solution $x^* = 0$ starts with the term $D_x f(\lambda, 0)x$. Separating out this linear term, we write Eq. (5.1) in the form

$$\dot{x} = A(\lambda)x + g(\lambda, x), \tag{5.15}$$

where $A(\lambda) = D_x f(\lambda, 0)$ and $g(\lambda, x) = f(\lambda, x) - D_x f(\lambda, 0)x$. In the neighborhood of the origin, g is a perturbation of the linear term. We assume that it satisfies the asymptotic estimate $g(\lambda, x) = o(\|x\|)$ as $x \to 0$, so $g(\lambda, x)$ goes to zero faster than x as $x \to 0$, uniformly in λ.

From the examples given in the preceding sections we observe that the nature of a bifurcation is determined essentially by the behavior of A or its eigenvalues as functions of λ.

5.5 ▪ Hopf Bifurcation Theorem

The examples in the earlier sections suggest that it should be possible to describe bifurcation scenarios locally—that is, in a neighborhood of a bifurcation point (λ^*, x^*)—based solely on information of the vector field and its derivatives at the bifurcation point. Here we give an example of such a result. It is a version of the Hopf bifurcation theorem for a planar dynamical system involving one real parameter. The presentation follows [64].

Theorem 5.1 (Hopf bifurcation in $\mathbb{R}^2$). *Let $f : (\lambda, x) \mapsto f(\lambda, x) \in \mathbb{R}^2$ be a C^4 vector field defined on $(-\delta, \delta) \times U$, where $U \subset \mathbb{R}^2$ is a neighborhood of the origin. Assume that* (i) *$f(\lambda, 0) = 0$ for all $\lambda \in (-\delta, \delta)$;* (ii) *$D_x f(\lambda, 0)$ has two nonreal eigenvalues, $\mu(\lambda)$ and $\overline{\mu}(\lambda)$ for all $\lambda \in (-\delta, \delta)$; and* (iii) *$\mu(0)$ is purely imaginary and $((d/d\lambda)\text{Re}(\mu))(0) \neq 0$. Then there exist an $\varepsilon > 0$, a neighborhood V of the origin, and a C^2-function $\Lambda : x_1 \mapsto \Lambda(x_1) \in \mathbb{R}$ defined on $(-\varepsilon, \varepsilon)$ and satisfying $\Lambda(0) = 0$ such that the following statement is true:*

If $\lambda = \Lambda(x_1)$ for some $x_1 \in (-\varepsilon, \varepsilon)$, then the point $(x_1, 0) \in \mathbb{R}^2$ is on a closed periodic orbit of Eq. (5.1) which is contained in V; the period π of this orbit satisfies $\lim_{x_1 \to 0} \pi(x_1) = 2\pi/\text{Re}(\mu)(0)$, and its amplitude converges to 0 as $x_1 \to 0$. Furthermore, all periodic orbits in V have these same properties.

The Hopf bifurcation theorem describes essentially the bifurcation diagram in the (λ, x_1)-plane. It requires two conditions, which have to hold at a single point in (λ, x)-space and concern the linearization at the equilibrium point: condition (ii) implies the existence of a center for the linearized problem, and condition (iii) is a nondegeneracy condition. The conclusion of the theorem is local in nature and holds in a small neighborhood in (λ, x)-space. It is possible to show that, for small x_1, the periodic orbit through $(x_1, 0)$ for $\lambda = \Lambda(x_1)$ "looks like" the periodic orbit of the linearized problem—it is "approximately" elliptical—but for larger orbits the shape is expected to change.

The Hopf bifurcation theorem does not say anything about the stability of the periodic orbit that branches off the equilibrium; however, conditions are known under which these orbits are stable. Nor does the theorem say anything about the amplitudes of these orbits. In the example given in Section 5.2.7, the amplitude behaves like $|\lambda|^{1/2}$, but it is easy to construct examples where the behavior is different (see the exercises). Typically, the amplitude of the periodic solution depends very sensitively on λ and behaves like some fractional power of $|\lambda|$.

5.6 ▪ Exercises

1. Draw a phase portrait and a bifurcation diagram in the manner of Figure 5.1 for the problem $\dot{x} = \lambda + x^2$.

2. Draw a phase portrait and a bifurcation diagram in the manner of Figure 5.1 for the problem $\dot{x} = \sin x - \lambda$ for $-4\pi \leq x \leq 4\pi$ and $-2 \leq \lambda \leq 2$.

3. Draw a phase portrait and a bifurcation diagram in the manner of Figure 5.1 for the problem $\dot{x} = \lambda x + x^3 - x^5$ for $-1 < \lambda < 1$ and $-2 < x < 2$ that show all the important features. Then describe how the equilibrium behavior of the system changes as λ varies from large negative to large positive values and back.

4. Draw a phase portrait and a bifurcation diagram in the manner of Figure 5.1 for the problem $\dot{x} = \sin x - \lambda x$ for $-4\pi \leq x \leq 4\pi$ and $\frac{1}{2} \leq \lambda \leq 2$. What happens in the interval $0 < \lambda < \frac{1}{2}$?

5. Figure 5.9 shows two bifurcation diagrams. As usual, solid lines indicate stable equilibria and dashed lines unstable equilibria.

   (i) Specify a differential equation $\dot{x} = f(\lambda, x)$ which gives rise to the bifurcation diagram on the left and draw its phase portrait.

   (ii) Specify a differential equation $\dot{x} = f(\lambda, x)$ which gives rise to the bifurcation diagram on the right and draw its phase portrait.

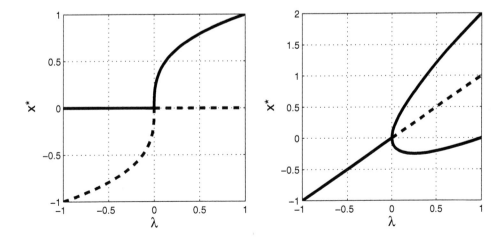

Figure 5.9. $\dot{x} = f(\lambda, x)$: *Bifurcation diagram.*

6. Consider the system of differential equations in $\mathbb{R}^2$ that is given by

$$\dot{r} = r(\lambda - r^4), \quad \dot{\theta} = -1 \tag{5.16}$$

in polar coordinates. Verify that the conditions of Theorem 5.1 are satisfied. Describe the function Λ whose existence is stated in that theorem. Find the amplitude of the periodic orbit as a function of λ.

7. Consider the system of differential equations in $\mathbb{R}^2$ that is given by

$$\dot{r} = r(\lambda - r^2), \quad \dot{\theta} = 1 + r^2 \tag{5.17}$$

in polar coordinates. Verify that the conditions of Theorem 5.1 are satisfied. Find the amplitude and the radius of the periodic orbit as a function of λ.

Chapter 6

Stommel's Box Model

In this chapter we take another look at the thermohaline circulation (THC) and discuss Stommel's two-box model using the methodology for dynamical systems developed in Chapters 4 and 5.

Keywords: Thermohaline circulation, Stommel's box model, dynamical system, bifurcation.

6.1 ▪ Stommel's Two-Box Model

In Chapter 3, we mentioned the difficulty of modeling the world's oceans and introduced the concept of a "box model." We considered, in particular, an ocean model consisting of two reservoirs connected by a capillary pipe at the bottom and an overflow mechanism at the surface, with a virtual salt flux at the surface to account for the effects of evaporation, precipitation, and runoff from continents. In Section 3.5 we introduced various simplifying assumptions to reduce the dynamics of the "overturning circulation" to a single ODE. We now return to the general two-box model (3.6) and introduce another set of simplifications which reduces the dynamics to a planar dynamical system. This is the famous two-box model due to HENRY STOMMEL (1920–1992) [104]. In this model, there is no salt flux at the surface; the forcing is due entirely to the exchange of heat and salinity with the atmosphere and neighboring oceans. We then apply the methodology of dynamical systems and bifurcations developed in Chapters 4 and 5 to Stommel's two-box model.

The variables are the temperature and salinity anomalies, which are measured relative to the equilibrium temperature and salinity of the surrounding basins; they are denoted by T_1 and S_1 in Box 1, and T_2 and S_2 in Box 2. The temperature and salinity anomalies are $-T^*$ and $-S^*$ in the basin surrounding Box 1, and T^* and S^* in the basin surrounding Box 2. The governing equations, Eq. (3.10), are

$$
\begin{aligned}
\frac{dT_1}{dt} &= c(-T^* - T_1) + |q|(T_2 - T_1), \\
\frac{dT_2}{dt} &= c(T^* - T_2) + |q|(T_1 - T_2), \\
\frac{dS_1}{dt} &= -H + d(-S^* - S_1) + |q|(S_2 - S_1), \\
\frac{dS_2}{dt} &= H + d(S^* - S_2) + |q|(S_1 - S_2),
\end{aligned}
\tag{6.1}
$$

where q is given by Eq. (3.5),

$$q = k(\alpha(T_2 - T_1) - \beta(S_2 - S_1)). \tag{6.2}$$

In his original article on box models [104], Stommel did not account for evaporation or precipitation, so there is no virtual salt flux, $H = 0$, but each box exchanges salinity with its surrounding basin, so $d > 0$. The system (6.1) reduces to

$$
\begin{aligned}
\frac{dT_1}{dt} &= c(-T^* - T_1) + |q|(T_2 - T_1), \\
\frac{dT_2}{dt} &= c(T^* - T_2) + |q|(T_1 - T_2), \\
\frac{dS_1}{dt} &= d(-S^* - S_1) + |q|(S_2 - S_1), \\
\frac{dS_2}{dt} &= d(S^* - S_2) + |q|(S_1 - S_2).
\end{aligned}
\tag{6.3}
$$

The mean temperature anomaly $\frac{1}{2}(T_1 + T_2)$ and the mean salinity anomaly $\frac{1}{2}(S_1 + S_2)$ both converge to zero, while the differences $\Delta T = T_2 - T_1$ and $\Delta S = S_2 - S_1$ satisfy the equations

$$
\begin{aligned}
\frac{d\,\Delta T}{dt} &= c(\Delta T^* - \Delta T) - 2|q|\,\Delta T, \\
\frac{d\,\Delta S}{dt} &= d(\Delta S^* - \Delta S) - 2|q|\,\Delta S,
\end{aligned}
\tag{6.4}
$$

where $\Delta T^* = 2T^*$ and $\Delta S^* = 2S^*$. The flow q through the capillary pipe is given by

$$q = k(\alpha \Delta T - \beta \Delta S). \tag{6.5}$$

Stommel's two-box model presents an opportunity to use the machinery for plane autonomous systems developed in Section 4.5 and is also interesting because of its historical importance. We closely follow the original presentation in [104].

6.2 · Dynamical System

We render the problem (6.4) dimensionless by introducing new variables,

$$\frac{dx}{dt'} = \frac{dx}{dt} \cdot \frac{dt}{dt'}$$

$$x = \frac{\Delta S}{\Delta S^*}, \quad y = \frac{\Delta T}{\Delta T^*}, \quad t' = ct.$$

The variable t' is just a rescaled version of t; without loss of generality we may drop the prime $'$. The system of equations (6.4) becomes

$$
\begin{aligned}
\dot{x} &= \delta(1 - x) - |f|x, \\
\dot{y} &= 1 - y - |f|y,
\end{aligned}
\tag{6.6}
$$

where $\delta = d/c$ and $f = -2q/c$. The relation $f = -2q/c$ can be written as

$$\lambda f(x, y) = Rx - y, \tag{6.7}$$

where

$$\lambda = \frac{c}{2\alpha k \, \Delta T^*}, \quad R = \frac{\beta \Delta S^*}{\alpha \Delta T^*}.$$

The system of equations (6.6) is a planar autonomous system with a Lipschitz continuous vector field and three parameters: δ, λ, and R. The constants c and d are inversely proportional to the relaxation times for heat and salinity exchange between the boxes and the surrounding basins. Since thermal energy generally exchanges on a faster time scale than salinity, the interest is focused on $\delta \in (0,1]$. The constants c and k both have the dimension of units per second and $\alpha \Delta T^*$ is dimensionless, so λ is dimensionless. The sign of λ is determined by the sign of ΔT^*, which can be positive or negative, depending on the configuration. The magnitude of λ is a measure of the strength of the THC. The dimensionless parameter R allows us to compare the effects of temperature and salinity differences between the external basins. We assume that ΔS^* and ΔT^* have the same sign, so R is positive; salinity differences dominate if $R > 1$, and temperature differences dominate if $R < 1$.

6.2.1 ▪ Equilibrium States

Equilibrium states correspond to critical points of the vector field, which occur if and only if both components of the vector field are zero. Thus, at a critical point (x^*, y^*) we have $\dot{x}, \dot{y} = 0$

$$x^* = \frac{\delta}{\delta + |f^*|}, \quad y^* = \frac{1}{1 + |f^*|}, \tag{6.8}$$

where $\lambda f^* = y^* - Rx^*$. Substituting x^* and y^* into the last relation, we see that f^* must satisfy the condition

$$\lambda f^* = \phi(f^*), \quad \phi(f^*) = \phi(f^*; R, \delta) = \frac{\delta R}{\delta + |f^*|} - \frac{1}{1 + |f^*|}. \tag{6.9}$$

This condition is most easily investigated graphically by plotting the graphs of $f \mapsto \lambda f$ and $f \mapsto \phi(f; R, \delta)$ and finding the points of intersection for different choices of the parameters. The procedure is illustrated in Figure 6.1 (adapted from [104]). In most cases there is only one point of intersection, which is always located in the right half-plane ($f^* > 0$), such as the points marked 'd', 'e', and 'g' in Figure 6.1. But if λ is sufficiently small, we may find three points of intersection for certain combinations of δ and R. Two points are necessarily in the left half-plane ($f^* < 0$), while the remaining third point is in the right half-plane ($f^* > 0$). A necessary condition for finding three points of intersection is that the graph of $f \mapsto \phi(f; \delta, R)$ dips below the horizontal axis, $\phi(f) = 0$ for some $f > 0$. If $\phi(f) = 0$, then $|f| = \delta(R-1)/(1-\delta R)$, so it must be the case that either $R > 1$ and $\delta R < 1$ or $R < 1$ and $\delta R > 1$. We assume henceforth that $R > 1$ and $\delta < \delta R < 1$.

Figure 6.1 shows three points of intersection for $\lambda = \frac{1}{5}$ (red line) and $R = 2$, $\delta = \frac{1}{6}$ (lower black curve). The points are marked 'a', 'b', and 'c'; the approximate values of f^* are $f_a^* = -1.068$, $f_b^* = -0.307$, and $f_c^* = 0.219$, respectively.

6.2.2 ▪ Stability

To investigate the stability of the equilibrium states, we linearize Eq. (6.6) near a critical point (x^*, y^*). After some manipulation, we obtain

$$\begin{pmatrix} \dot{\xi} \\ \dot{\eta} \end{pmatrix} = A \begin{pmatrix} \xi \\ \eta \end{pmatrix}, \tag{6.10}$$

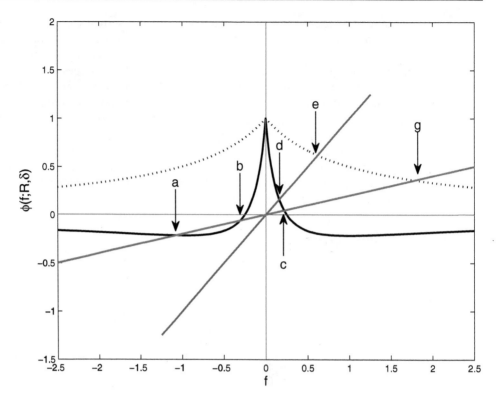

Figure 6.1. *Finding equilibrium states for Stommel's two-box model; $f \mapsto \lambda f$ is plotted for $\lambda = 1$ (blue) and $\lambda = \frac{1}{5}$ (red); $f \mapsto \phi(f;R,\delta)$ is plotted for $R = 2$ and $\delta = 1$ (dotted black) and $\delta = \frac{1}{6}$ (solid black).*

where the coefficient matrix A is given by

$$A = \begin{pmatrix} -(\delta + |f^*|) \mp \dfrac{Rx^*}{\lambda} \cdot & \pm \dfrac{x^*}{\lambda} \\ \mp \dfrac{Ry^*}{\lambda} & -(1 + |f^*|) \pm \dfrac{y^*}{\lambda} \end{pmatrix} \text{ if } f^* \gtrless 0. \qquad (6.11)$$

Here, $f^* = f(x^*, y^*) = (Rx^* - y^*)/\lambda$.

The trace and determinant of A determine the nature of the critical point (x^*, y^*). The trace of the matrix A in Eq. (6.11) is

$$T = \text{trace}(A) = -(1 + \delta + 3|f^*|),$$

which is negative for all values of f^*, and the determinant is

$$\begin{aligned} D = \det(A) &= (\delta + |f^*|)(1 + |f^*|) + |f^*|^2 \pm \frac{Rx^*}{\lambda} \mp \frac{\delta y^*}{\lambda} \\ &= (\delta + 2|f^*|)(1 + |f^*|) \pm (1 - \delta)\frac{y^*}{\lambda} \text{ if } f^* \gtrless 0. \end{aligned} \qquad (6.12)$$

If $f^* > 0$, then $D > 0$. Moreover,

$$T^2 - 4D = (1 - \delta - f^*)^2 - \frac{4(1-\delta)}{\lambda(1+f^*)} < 0.$$

Hence, the equilibrium solution corresponding to (x^*, y^*) is a *stable spiral point*. If $f^* < 0$,

$$D = (\delta - 2f^*)(1 - f^*) - (1 - \delta)\frac{y^*}{\lambda} = (\delta - 2f^*)(1 - f^*) - \frac{1 - \delta}{\lambda(1 - f^*)}.$$

This expression is decreasing for $f^* \in (-\infty, 0)$ and changes sign at some value between the two negative solutions of Eq. (6.9). Hence, if there are two negative solutions f_a^* and f_b^* with $f_a^* < f_b^* < 0$, then $D > 0$ at f_a^*, so 'a' is a stable equilibrium, and $D < 0$ at f_b^*, so 'b' is a *saddle point*. Since

$$T^2 - 4D = (1 - \delta + f^*)^2 + \frac{4(1 - \delta)}{\lambda(1 - f^*)} > 0$$

for all $f^* < 0$, 'a' cannot be a spiral point, so it must be a *stable node*.

These results are confirmed by the numerics. With $R = 2$, $\delta = \frac{1}{6}$, and $\lambda = \frac{1}{5}$ (Figure 6.1), we find that the matrix A has two negative eigenvalues (-3.596 and -0.764) for $f^* = f_a^* = -1.068$, one negative eigenvalue (-2.850) and one positive eigenvalue (0.760) for $f^* = f_b^* = -0.307$, and a pair of complex conjugate eigenvalues with negative real parts ($-0.911 \pm 1.822i$) for $f^* = f_c^* = 0.219$. Hence, 'a' is a *stable node*, 'b' is a *saddle point*, and 'c' is a *stable spiral point*. We have a temperature driven or T-mode at 'a' ($f^* < 0$), and a salinity driven or S-mode at 'c' ($f^* > 0$). Figure 6.2 shows the phase portrait for Stommel's two-box model in the (x, y)-plane, with the three equilibrium points and a few representative trajectories. The straight lines are equiflow lines (lines of equal f).

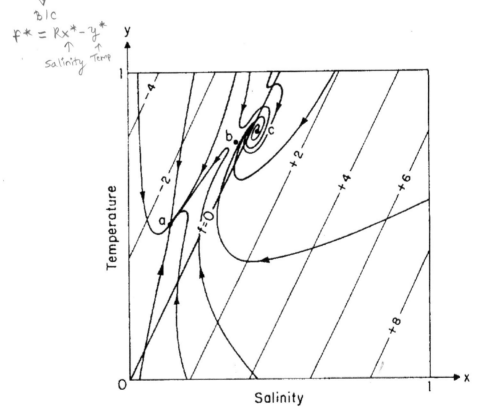

Figure 6.2. *Phase portrait for Stommel's two-box model* [104].

6.3 ▪ Bifurcation

To obtain a bifurcation diagram for the dynamical system (6.6), we need to select a scalar function of the equilibrium solutions and a bifurcation parameter. The discussion in the previous section suggests that the dimensionless flow rate f^* completely determines the character of the equilibrium solutions. Moreover, the sign of f^* tells us the direction of the flow through the bottom pipe (which represents the ocean circulation). It is therefore natural to use f^* to characterize the equilibrium solutions. Other choices are certainly possible; for example, we can also take either one of the dimensionless anomalies x^* or y^*.

Also, any one of the three dimensionless parameters $(\delta,\ \lambda,\ R)$ can be taken as the bifurcation parameter. Here, we select λ, keeping R and δ fixed. This means that we vary only the strength of the THC; the temperature and salinity anomalies of the basins surrounding the boxes and the ratio of salinity and temperature exchange rates between the boxes and their surrounding basins are to remain constant.

Figure 6.3 gives the bifurcation diagram for $R = \frac{3}{2}$, $\delta = \frac{4}{5}$; the dimensionless flow rate f^* at equilibrium is plotted on the left and the anomaly component y^* on the right. The lower solid portion of each curve is the stable T-mode, and the upper solid portion is the stable S-mode. We see a fold bifurcation at $\lambda = \frac{4}{5}$, $f^* = -\frac{1}{4}$, $y^* = \frac{4}{5}$. If the system is in the S-mode and λ is increased beyond the critical value $\frac{4}{5}$, the system will change to the stable T-mode and will remain in this mode even if λ is decreased again. As the left-hand side of the figure shows, the two solution branches do not meet for positive λ. It therefore becomes impossible or very difficult for the system to revert to the stable T-mode once it is in or near the S-mode. Again, there is a hysteresis phenomenon, a reversal of the flow, and an increase of the temperature anomaly. It turns out that the salinity anomaly will also increase. The bifurcation diagram on the right in Figure 6.3 is qualitatively similar to the bifurcation diagram in Figure 3.7. However, while both diagrams show the dimensionless salinity anomaly along the vertical axis, the bifurcation parameters λ along the horizontal axes have very different physical meanings.

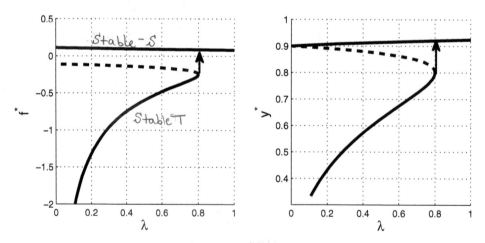

Figure 6.3. *Bifurcation diagram for Stommel's two-box model; $R = \frac{3}{2}$, $\delta = \frac{4}{5}$.*

6.4 ▪ Comments

There is an interesting bit of history behind the development of box models for the study of the THC. The two-box model (6.1) was first formulated in 1961 by Stommel [104] but went virtually unnoticed for 25 years. Meanwhile, in 1982, a three-box model was proposed independently by Rooth [90] that explained how a two-hemispheric THC symmetric about the equator might become unstable. This result inspired what is arguably the most influential study of the THC [9], but it faded out of the public eye because of the "rediscovery" of Stommel's paper [104]. It took more than 10 years before Rooth's model was applied to the steady-state pole-to-pole circulation [85, 100].

Clearly, two-box models are only a caricature of the THC. At best, they account for the pole-to-equator circulation in a single ocean basin (the North Atlantic). They certainly do not account for the fact that all the Earth's oceans are connected, nor for the fact that the oceans are coupled to the atmosphere and other components of the climate system. Nevertheless, the finding that such simple models predict the possibility of two distinct stable modes of circulation and that transitions from one mode to another can be induced by changing the forcing parameters has had a significant impact in oceanography and climate science.

To account for a larger ocean basin and interhemispheric flow, several authors have proposed three- and four-box models with various degrees of coupling among the boxes. We have already mentioned Rooth's three-box model; other examples can be found in [62, 84, 100].

Whether fully fledged climate models admit multiple equilibria of the THC and the possibility of transitions among the equilibria is an open issue. One can only speculate what the possible consequences would be for the Earth's climate if the system had multiple stable equilibria and was perturbed enough that the system would transit from one stable mode to another.

6.5 ▪ Exercises

1. Consider the function $\phi : f \mapsto \phi(f)$ defined in Eq. (6.9). Let ϕ_+ be the restriction of ϕ to nonnegative values of f,

$$\phi_+(f) = \frac{\delta R}{\delta + f} - \frac{1}{1+f}, \quad f \geq 0.$$

   Show in detail that

$$R > 1 \quad \Longrightarrow \phi_+(0) > 0,$$
$$R > \delta \quad \Longrightarrow \phi'_+(0) < 0,$$
$$R > 1 > \delta R \quad \Longrightarrow \phi_+(f) < 0 \text{ for large } f.$$

   Check that these properties are consistent with the phase portrait given in Figure 6.2.

2. Consider the function $\phi : f \mapsto \phi(f)$ defined in Eq. (6.9), where $\delta = \frac{1}{6}$ and $R = \frac{3}{2}$.

   (i) The equation $\lambda f = \phi(f)$ has exactly one solution $f = f^* < 0$ if $\lambda = \frac{4}{5}$. Find this solution.

   (ii) How many negative solutions does the equation $\lambda f = \phi(f)$ have if $\lambda \in (0, \frac{4}{5})$? Where are these solutions relative to the solution found in (i)?

   (iii) Draw a bifurcation diagram showing the negative solutions of $\lambda f = \phi(f)$ on the vertical axis and the parameter $\lambda > 0$ on the horizontal axis.

3. Suppose Eq. (6.9) has three solutions f_i^* $(i = 1, 2, 3)$, each leading to an equilibrium state (x_i^*, y_i^*), where $x_i^* = \delta /(\delta + |f_i^*|)$, $y_i^* = 1/(1 + |f_i^*|)$. Show that if $f_1^* < f_2^* < 0 < f_3^*$, then $x_1^* < x_2^* < x_3^*$ and $y_1^* < y_2^* < y_3^*$.

4. Show that Stommel's two-box model, Eq. (6.6), has only one equilibrium solution with $f^* > 0$ if $R > 1$, $\delta > 1$, $\lambda > 0$. Show that this solution is a stable node.

5. This exercise and the following ones concern a three-box model first proposed by Rooth [90] to analyze interhemispheric circulation. Figure 6.4 gives a schematic representation of the model, which consists of a northern high-latitude box (Box 1), a southern high-latitude box (Box 2), and a low-latitude box representing a tropical zone around the equator (Box 3). The tropical box is connected to both high-latitude boxes through surface flows, and the two high-latitude boxes are connected by a deep current passage (which bypasses the tropical box). The mass contained in the connecting passages is negligible. The model uses the same physical laws as the two-box model described in Section 3.5, except that the flow is driven by the pole-to-pole density difference. The high-latitude boxes are surrounded by basins with temperature T_P and salinity S_P, and the low-latitude box is surrounded by a basin with temperature T_E and salinity S_E. It is assumed that $T_E > T_P$. In addition, there are salinity fluxes H_N, H_S out of the high-latitude boxes and a balancing salinity flux $H_N + H_S$ into the low-latitude box. The magnitude of the flow is $|q|$. By convention, $q > 0$ means that the bottom flow is from north to south.

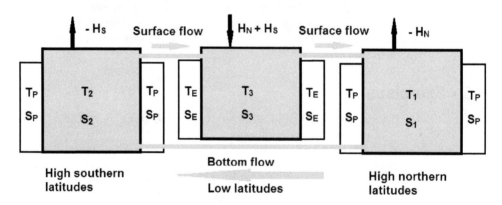

Figure 6.4. *A three-box model of the THC.*

(i) Use Figure 6.4 to derive the following differential equations for the temperature and salinity of the northern box, and derive the corresponding differential equations for the other two boxes:

$$\frac{dT_1}{dt} = c(T_P - T_1) + \begin{cases} q(T_3 - T_1) & \text{if } q > 0, \\ |q|(T_2 - T_1) & \text{if } q < 0, \end{cases}$$

$$\frac{dS_1}{dt} = -H_N + d(S_P - S_1) + \begin{cases} q(S_3 - S_1) & \text{if } q > 0, \\ |q|(S_2 - S_1) & \text{if } q < 0. \end{cases}$$

(ii) Recast the governing equations in terms of the temperature and salinity differences between the equatorial box and the high-latitude boxes,

$$T_N = T_3 - T_1, \quad T_S = T_3 - T_2, \quad S_N = S_3 - S_1, \quad S_S = S_3 - S_2.$$

6. If $q > 0$, one expects $T_P \le T_2 \le T_1 \le T_3 \le T_E$.

(i) Explain to somebody who does not know differential equations why this is expected to be the case.

(ii) Consider now a time interval $I = [t_0, t_1]$ such that $q(t) > 0$ for all $t \in I$ and

$$T_P < T_2(t_0) < T_1(t_0) < T_3(t_0) < T_E.$$

The goal of this exercise is to show that these (strict) inequalities must remain true for all $t \in I$. Suppose that some of the inequalities above do not hold at some $t \in I$. Then there is a smallest $t^* \in I$ where this is the case. Prove the following simple comparison argument: *If $f : [a, b] \to \mathbb{R}$ is differentiable such that $f(t) > 0$ for $t \in [a, b)$ and $f(b) = 0$, then $f'(b) \le 0$.* Show that, at t^*, one or more of the inequalities becomes an equality.

(iii) Assume that, at t^*, the leftmost inequality has become an equality, $T_2(t^*) = T_P$. Apply the comparison argument to $f = T_2 - T_P$ and use the differential equation for T_2 derived in Exercise 5 to show that, at t^*, $\dot{T}_2 = q(T_1 - T_2) \le 0$ and, hence, also $T_1(t^*) = T_2(t^*) = T_P$. Then apply the comparison argument to $f = T_1 - T_2$ and use the differential equation for T_1 to show that, at t^*, $\dot{T}_1 - \dot{T}_2 = q(T_3 - T_1) \le 0$ and, hence, also $T_3(t^*) = T_1(t^*) = T_P$. Apply the comparison argument once more to $f = T_3 - T_P$ to obtain the inequality $\dot{T}_3 = c(T_E - T_P) \le 0$ and arrive at a contradiction. (Recall that $c > 0$ and $T_E > T_P$.) Therefore, $T_2(t^*) > T_P$, and one of the other strict inequalities must be violated at t^*.

(iv) Use a similar set of arguments to show that also $T_2(t^*) < T_1(t^*)$.

(v) Prove that at t^*, all the other strict inequalities must also still hold. Therefore, t^* does not exist and the set of inequalities holds as long as $q > 0$.

(vi) How do these inequalities change if $q < 0$? How does one prove them?

(vii) Formulate and prove corresponding inequalities for the salinities in the boxes and basins.

7. (i) Use a computer algebra system to find the equilibrium values S_N^*, S_S^*, S_3^* of S_N, S_S, S_3 in terms of q, in the two cases $q < 0$ and $q > 0$.

   (ii) Show that as $d \to 0$, one obtains

$$S_N^* = \frac{H_N}{q}, \quad S_S^* = \frac{H_N + H_S}{q}, \quad S_3^* = \frac{2H_N + H_S + q(S_E + 2S_P)}{3q} \quad (q > 0),$$

$$S_N^* = \frac{H_N + H_S}{|q|}, \quad S_S^* = \frac{H_S}{|q|}, \quad S_3^* = \frac{H_N + 2H_S + |q|(S_E + 2S_P)}{3|q|} \quad (q < 0).$$

Note that this procedure gives an equilibrium value for S_3^* if there is no salt flux between the boxes and the surrounding basins. This value cannot be obtained by setting $d = 0$ first and then solving for the equilibrium values.

8. We now use Eq. (3.5) for the interhemispheric flow,

$$q = k(\alpha(T_1 - T_2) - \beta(S_1 - S_2)).$$

Consider the equilibrium values for the salinity that were derived in the previous exercise for the case $d = 0$. Also, assume that the temperature differences T_N and T_S are at equilibrium values T_N^* and T_S^*.

(i) Derive quadratic equations for the equilibrium fluxes q^*, separately for the cases $q^* > 0$ and $q^* < 0$.

(ii) Solve these quadratic equations, separately for the cases $q^* > 0$ and $q^* < 0$. Hint: Use the auxiliary variable $U = k\alpha(T_S^* - T_N^*)$.

(iii) Suppose the equilibrium flux q^* is positive and H_S is increased, for example by increased fresh water influx into the southern high-latitude box and/or more evaporation from the low-latitude box. What will happen to q^* when H_S is very large?

Chapter 7

Lorenz Equations

This chapter is devoted to the Lorenz equations, a set of three autonomous differential equations first proposed in the 1960s by the mathematician and meteorologist EDWARD N. LORENZ to model the circulation of the Earth's atmosphere. Although it was realized early on that the model is at best a caricature, it has had a significant influence on the subsequent development of the theory of dynamical systems. The Lorenz equations exhibit what is now called chaotic dynamics. The fact that their solutions lead to aesthetically pleasing pictures has further contributed to their popularity among scientists and the general public.

Keywords: Lorenz equations, Lyapunov function, strange attractor, chaotic dynamics.

7.1 ▪ Lorenz Model

We know from daily experience that atmospheric motion is driven primarily by gradients of the density and temperature (in addition to external forces like gravity and the Coriolis force). The physical laws that govern the density, velocity, and temperature of the atmosphere are the conservation laws of mass, momentum, and internal energy. These laws are formalized in the equations of hydrodynamics—the continuity equation, the Navier–Stokes equation, and the temperature equation—a set of partial differential equations (PDEs) expressing the variations in time (time derivatives) of these variables in terms of their variations in space (spatial gradients). Because PDEs do not fit into the framework of finite-dimensional dynamical systems discussed in Chapter 4, several substantial approximations are required to make them fit. Since not all approximations are created equal, their justification depends very much on the physical application of interest; this is where intuition and experience count.

In the 1960s, the mathematician and meteorologist EDWARD N. LORENZ (1917–2008) was interested in weather forecasting and questioned whether it is, in fact, feasible to make long-term weather predictions on the basis of a mathematical model. Inspired by earlier work of Saltzman [97], Lorenz [60] modeled the atmospheric circulation as a two-dimensional flow of an incompressible fluid in a three-dimensional layer which is heated from below. Starting from the hydrodynamic equations and following a systematic approximation procedure, Lorenz arrived at a set of three nonlinear autonomous differential equations,

$$\dot{x} = -\sigma x + \sigma y,$$
$$\dot{y} = \rho x - y - xz, \qquad\qquad (7.1)$$
$$\dot{z} = -\beta z + xy.$$

The procedure is outlined in Section 15.2, after the introduction of the hydrodynamic equations in Chapter 14. The equations are since known as the *Lorenz equations*. The system is sometimes referred to as the *Lorenz-63 model*, named after the year of publication, to distinguish it from two other atmospheric circulation models proposed by Lorenz, namely the Lorenz-84 model and the Lorenz-96 model.

The Lorenz equations involve three state variables, x, y, and z, representing, respectively, the spatial average of the hydrodynamic velocity, temperature, and temperature gradient. The dimensionless constants σ and ρ are related, respectively, to the Prandtl number and Rayleigh number of the fluid; their values are determined on the basis of physical considerations; β is a constant related to the aspect ratio of the domain under consideration. All three parameters are positive; σ and β are usually kept fixed with $\sigma > 1 + \beta$, and ρ is varied.

Lorenz showed that, for certain parameter regimes, the solution of the system of equations (7.1) never repeats its past history exactly; moreover, all approximate repetitions have finite duration. This phenomenon became known as *chaotic dynamics*. Lorenz's findings implied that there are inherent limitations to long-range weather forecasting. However, Lorenz also found that there are certain structures in the state space, now called *strange attractors*, that have noninteger dimensions and are involved in the long-term behavior of the dynamical system.

7.2 ▪ Preliminary Observations

The Lorenz equations define an autonomous dynamical system. The three variables x, y, and z jointly represent the state of the Earth's atmosphere, the state space is $\mathbb{R}^3$, and the evolution of the system is driven by a force field which is represented by the right-hand sides of Eq. (7.1). The following properties are readily verified.

- The Lorenz equations are symmetric with respect to reflection: If $\varphi : t \mapsto \varphi(t) = (x(t), y(t), z(t))$ is a solution, then $\psi : t \mapsto \psi(t) = (-x(t), -y(t), z(t))$ is also a solution.

- The z-axis is an invariant set: if $x = y = 0$ at some time t, then $x = y = 0$ at all times. Moreover, any solution with $x = y = 0$ tends to zero as $t \to \infty$.

An interesting property of the Lorenz equations is that all solutions are eventually trapped in a bounded region of the state space.

Definition 7.1. *A trapping set for a dynamical system in $\mathbb{R}^n$ is a closed connected set $\mathcal{D} \subset \mathbb{R}^n$ which, after a finite time T, is invariant with respect to the flow—that is, there exists a $T \geq 0$ such that $\phi_t(\mathcal{D}) \subset \mathcal{D}$ for all $t \geq T$.*

To find a trapping set, it is sufficient to show that the vector field is directed everywhere inward on the boundary of $\mathcal{D}$. In that case, we can define the associated attracting set as

$$\mathcal{A} = \bigcap_{t \geq 0} \phi_t(\mathcal{D}). \tag{7.2}$$

A trapping set is not unique, and it could consist of all of $\mathbb{R}^n$.

We will now show that, for the Lorenz system, there exists a bounded trapping set $\mathcal{D}$. This implies that all solutions eventually stay in a fixed neighborhood of the origin. To construct such a set, we use the *Lyapunov function* $V : \mathbb{R}^3 \to \mathbb{R}$,

$$V(P) = \tfrac{1}{2}\left(\rho x^2 + \sigma y^2 + \sigma(z - 2\rho)^2\right), \quad P = (x,y,z) \in \mathbb{R}^3. \tag{7.3}$$

As the point $P = (x,y,z)$ progresses along a trajectory of the Lorenz equations, $V(P)$ changes with time; its rate of change is found from the chain rule,

$$\begin{aligned}
\frac{d}{dt} V(\phi_t(P)) &= \frac{\partial V}{\partial x}\dot{x} + \frac{\partial V}{\partial y}\dot{y} + \frac{\partial V}{\partial z}\dot{z} \\
&= -\sigma\left(\rho x^2 + y^2 + \beta(z-\rho)^2 - \beta\rho^2\right).
\end{aligned} \tag{7.4}$$

Choose $\delta > 0$ and define the ellipsoid

$$\mathcal{E} = \{(x,y,z) \in \mathbb{R}^3 : \rho x^2 + y^2 + \beta(z-\rho)^2 - \beta\rho^2 < \delta\} \subset \mathbb{R}^3.$$

Clearly, the origin is inside $\mathcal{E}$. The computation in Eq. (7.4) shows that $(d/dt)V(\phi_t(P)) < -\sigma\delta$ if and only if $\phi_t(P) \notin \mathcal{E}$.

Let m be the maximum value of V in $\mathcal{E}$, and set

$$\mathcal{D} = \{P \in \mathbb{R}^3 : V(P) \leq m\}. \tag{7.5}$$

This is another ellipsoid, which contains $\mathcal{E}$. If $\phi_t(P) \notin \mathcal{D}$, then also $\phi_t(P) \notin \mathcal{E}$, so

$$(d/dt)V(\phi_t(P)) < -\sigma\delta.$$

It follows that the value of V along any orbit that starts at a point $P \notin \mathcal{D}$ decreases as long as $\phi_t(P) \notin \mathcal{D}$. After at most $(V(P) - m)/(\sigma\delta)$ time units, the orbit must enter $\mathcal{D}$ and remain inside $\mathcal{D}$ forever after.

A closed bounded trapping set like $\mathcal{D}$ must contain an attractor $\mathcal{A}$, which can be defined as in Eq. (7.2). Since $\mathcal{A}$ is the intersection of a decreasing family of compact nonempty connected sets, $\mathcal{A}$ itself is also compact, connected, and nonempty.

In the case of the Lorenz equations, we can say a bit more about $\mathcal{A}$. The divergence of the vector field is negative,

$$\frac{\partial}{\partial x}(-\sigma x + \sigma y) + \frac{\partial}{\partial y}(\rho x - y - xz) + \frac{\partial}{\partial z}(-\beta z + xy) = -(\sigma + 1 + \beta) < 0,$$

so ϕ_t is a *contracting flow* and the attractor $\mathcal{A}$ has zero volume. The actual structure of $\mathcal{A}$ remains to be investigated.

7.3 ▪ Equilibrium Solutions

If $0 < \rho < 1$, the Lorenz equations have only one equilibrium solution, which corresponds to the critical point $(0,0,0)$. This is the *trivial solution*—a pure conductive state, with zero velocity and a linear temperature gradient. In a neighborhood of the origin, the linearized system is

$$\begin{aligned}
\dot{x} &= -\sigma x + \sigma y, \\
\dot{y} &= \rho x - y, \\
\dot{z} &= -\beta z.
\end{aligned} \tag{7.6}$$

Here, the z-component is decoupled from the x- and y-components. Since $\beta > 0$, every solution decays to $z = 0$ as $t \to \infty$. The x- and y-components satisfy the system

$$\begin{pmatrix} \dot{x} \\ \dot{y} \end{pmatrix} = \begin{pmatrix} -\sigma & \sigma \\ \rho & -1 \end{pmatrix} \begin{pmatrix} x \\ y \end{pmatrix} = A \begin{pmatrix} x \\ y \end{pmatrix}. \tag{7.7}$$

The eigenvalues of A are real and negative,

$$\lambda_- < \lambda_+ < 0, \quad \lambda_\pm = -\tfrac{1}{2}(1+\sigma) \pm \tfrac{1}{2}\sqrt{(1+\sigma)^2 - 4(1-\rho)\sigma}.$$

Consequently, the origin is a positive attractor and is the only attractor.

As ρ passes through the value 1, the leading eigenvalue λ_+ passes through the origin, and a bifurcation occurs. For $\rho > 1$, we have $\lambda_- < 0 < \lambda_+$, so the origin has become a saddle. The presence of a positive eigenvalue implies that the trivial solution is no longer stable as $t \to \infty$. The unstable manifold is one-dimensional; it is spanned by the eigenvector associated with the positive eigenvalue λ_+.

For $\rho > 1$, the Lorenz equations admit two more equilibrium solutions, corresponding to the critical points $C_\pm = (\pm\sqrt{\beta(\rho-1)}, \pm\sqrt{\beta(\rho-1)}, \rho-1)$. The bifurcation that occurred as ρ passed through the value 1 is an example of a *pitchfork bifurcation*.

To investigate the stability of the two new equilibrium solutions, we linearize the equations in the neighborhood of $C_\pm$. The linearized system near C_+ is

$$\begin{pmatrix} \dot{\xi} \\ \dot{\eta} \\ \dot{\zeta} \end{pmatrix} = \begin{pmatrix} -\sigma & \sigma & 0 \\ 1 & -1 & -\sqrt{\beta(\rho-1)} \\ \sqrt{\beta(\rho-1)} & \sqrt{\beta(\rho-1)} & -\beta \end{pmatrix} \begin{pmatrix} \xi \\ \eta \\ \zeta \end{pmatrix} = A \begin{pmatrix} \xi \\ \eta \\ \zeta \end{pmatrix}. \tag{7.8}$$

The eigenvalues of A are the zeros of the characteristic polynomial,

$$p(\lambda) = \det(\lambda I - A) = \lambda^3 + (1+\beta+\sigma)\lambda^2 + \beta(\rho+\sigma)\lambda + 2\beta(\rho-1)\sigma. \tag{7.9}$$

The equation $p(\lambda) = 0$ has either three real roots or one real root and a pair of complex conjugate roots. The *Routh–Hurwitz criterion* [26] gives necessary and sufficient conditions for all these roots to be in the left half of the complex plane. For a cubic polynomial of the form $p(\lambda) = \lambda^3 + a_2\lambda^2 + a_1\lambda + a_0$ with real coefficients a_2, a_1, and a_0, the conditions are (i) $a_i > 0$ for $i = 0, 1, 2$, and (ii) $a_2 a_1 > a_0$. The first condition is clearly satisfied for the polynomial p defined in Eq. (7.9). The second condition is satisfied if

$$\rho < \rho_H = \frac{\sigma + \beta + 3}{\sigma - \beta - 1} = 1 + \frac{2(\beta+2)}{\sigma - \beta - 1}.$$

Since $\sigma > 1 + \beta$, the critical value ρ_H satisfies the inequality $\rho_H > 1$; hence, the second condition is satisfied for all $\rho \in [1, \rho_H)$. The eigenvalues of A are all in the left half of the complex plane, and the linearized system is stable at C_+.

As ρ passes through the critical value ρ_H, a *Hopf bifurcation* occurs. The real part of the complex conjugate eigenvalues of A changes sign, the eigenvalues cross the imaginary axis from the left half of the complex plane into the right half, and the linearized system (7.8) goes from stable to unstable. A simple calculation gives the crossing points, $\pm i\sqrt{2\sigma(\sigma+1)/(\sigma-\beta-1)}$. In this case, the unstable manifold is two-dimensional; it is spanned by the eigenvectors associated with the complex conjugate eigenvalues of A.

For ρ near ρ_H, there are periodic solutions centered roughly at C_+ whose amplitudes become small as $\rho \to \rho_H$. It is possible to show that this is a subcritical Hopf bifurcation—that is, these periodic solutions are also unstable.

Similar arguments apply to the equilibrium solution corresponding to the remaining critical point C_-. The behavior of the linearized system near the critical points mimics the behavior of the original nonlinear system, at least in sufficiently small neighborhoods of the critical points. We expect, therefore, that stability is lost near $C_\pm$ as ρ passes through its critical value ρ_H and, furthermore, that the dynamics of the Lorenz system change into some type of oscillatory behavior. This will be explored in more detail in the exercises, using numerical experiments.

7.4 ▪ Numerical Experiments

Since it is not easy to analyze the flow induced by the Lorenz equations by elementary means, much information has come from numerical experiments. Lorenz's original work was done for $\sigma = 10$ and $\beta = \frac{8}{3}$, so $\rho_H = \frac{470}{19} \approx 24.74$. Figure 7.1 shows the solution when the system (7.1) is integrated with $\rho = 28$, starting at the point $(3, 15, 1)$ at time $t = 0$. These figures can be generated with the MATLAB code given in Section C.1.

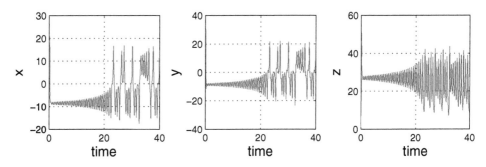

Figure 7.1. *Numerical solution of the Lorenz equations for $\sigma = 10$, $\beta = \frac{8}{3}$, $\rho = 28$. From left to right: $x(t)$, $y(t)$, $z(t)$ for $0 \le t \le 40$.*

Notice the growth of oscillations to an apparent threshold value, after which x and y show a sign change and oscillations grow once more. If we put the three coordinates together, we obtain a picture of the orbit in the state space; see Figure 7.2. The orbit is not closed; it consists of loops on the right- and left-hand sides of the origin. The loops form what looks like a branched surface, with the nontrivial saddles $C_\pm$ in the two holes of the surface. It turns out that the number of successive loops on the left and the right depends in a very sensitive way on the starting point: a small perturbation in the initial values of x, y, or z produces a markedly different series of loops, although the apparent surfaces covered by the orbits look quite similar. Numerical experiments suggest the existence of an attractor with a dimension slightly greater than 2, which is consistent with the earlier observation that the attractor has zero volume. The attractor may therefore contain very irregular orbits; it is what is now known as a *strange attractor*.

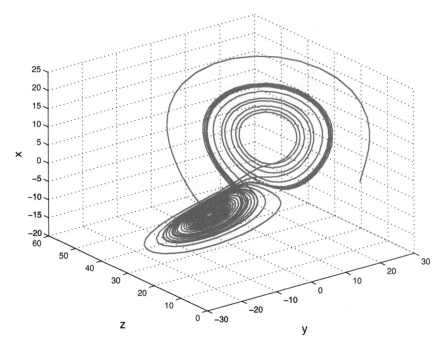

Figure 7.2. *Orbit of the Lorenz equations emanating from* $(3, 15, 1)$ *for* $\sigma = 10$, $\beta = \frac{8}{3}$, $\rho = 28$.

As a final remark, we caution the reader that the use of the Lorenz equations (7.1) as a model for the dynamics of the Earth's atmosphere is highly debatable. In fact, if one refines the model by including higher-order averages of the hydrodynamic velocity, temperature, and temperature gradient, one obtains quite different results. Nevertheless, we have presented the Lorenz system here in some detail, because it has contributed to a better understanding of dynamical systems, and the mathematics are interesting as well as beautiful.

7.5 • Exercises

1. Let $\mathscr{D}$ be a trapping set as defined in Eq. (7.2) for a general dynamical system. Prove the following:

   (i) For any $t \in \mathbb{R}$, $\phi_t(\mathscr{D})$ is also a trapping set.

   (ii) For any $s, t \in \mathbb{R}$,

   $$s < t \quad \Longrightarrow \quad \phi_t(\mathscr{D}) \subset \phi_s(\mathscr{D}).$$

   If $\mathscr{D}_1$ and $\mathscr{D}_2$ are trapping sets, is it true that $\mathscr{D}_1 \cap \mathscr{D}_2$ and $\mathscr{D}_1 \cup \mathscr{D}_2$ are trapping sets?

2. (i) Let $A : t \mapsto A(t)$ be a differentiable function defined on $(-1, 1)$ whose values are real matrices of order n, $A : (-1, 1) \to L(\mathbb{R}^n)$, satisfying the initial condition $A(0) = I$, I the identity matrix of order n. Prove that

   $$\left(\frac{d}{dt} \det A \right)(0) = \operatorname{trace} \left(\frac{dA}{dt} \right)(0).$$

(ii) Let $\phi_t(y)$ be the solution of the IVP $\dot{x} = f(x)$, $x(0) = y$, where f is C^2. Assume that this solution exists for all initial data $y \in \mathbb{R}^n$ and for all positive times. Let $D_y \phi_t(y)$ be the Jacobian. Prove that, for all $y \in \mathbb{R}^n$,

$$\left(\frac{d}{dt} \det \left(D_y \phi_t(y) \right) \right)(0) = \operatorname{div}_y f(y).$$

3. Let $U \subset \mathbb{R}^3$ be a set over which we can integrate—for example, a set that is defined by a finite set of inequalities involving differentiable functions—and let ϕ_t be as in Exercise 2 applied to the Lorenz equations. Use the results from Exercise 2 to prove that, for all t,

$$\frac{d}{dt} \operatorname{vol}(\phi_t(U)) = -(\sigma + 1 + \beta) \operatorname{vol}(\phi_t(U)).$$

Use this result to explain why the attractor $\mathcal{A}$ for the Lorenz equations cannot have a subset of positive volume.

4. Consider the Lorenz system with $\sigma = 10$ and $\beta = \frac{8}{3}$.

   (i) Verify numerically that the equilibrium point at the origin is stable for $0 < \rho < 1$ and unstable for $1 < \rho < \rho_H$ by computing several solutions with random initial data near this equilibrium and observing their behavior.

   (ii) Verify numerically that the equilibrium points C_+ and C_- are stable for $1 < \rho < \rho_H$.

   (iii) Verify numerically that, for $\rho > 1$ and ρ close to 1, solutions with initial data near the origin always converge to either C_+ or C_-. This illustrates that the unstable manifold of the system at the origin connects to these equilibrium points.

   (iv) There is a $\rho_0 > 1$ such that, for $\rho > \rho_0$, the behavior in (iii) no longer occurs. Try to determine ρ_0 to one decimal digit and describe the behavior of solutions starting near the origin if $\rho > \rho_0$.

5. Consider the Lorenz system with $\sigma = 10$ and $\beta = \frac{8}{3}$.

   (i) Verify numerically that the equilibrium point C_+ is unstable for $\rho > \rho_H$ by computing several solutions with random initial data near this equilibrium and observing their behavior.

   (ii) Verify numerically that, for $\rho < \rho_H$ and ρ close to ρ_H, solutions that start near C_+ converge to C_+ and solutions that start further away from it do not converge to C_+ or C_-. Describe the behavior of these nonconvergent solutions. This illustrates the fact that C_+ is locally stable but not globally stable.

6. Consider the Lorenz system with $\sigma = 10$ and $\beta = \frac{8}{3}$.

   (i) For very large values of ρ, say $\rho \approx 100$, it is known that there are stable periodic solutions of the Lorenz equations for certain initial values. Try to find such a solution by experimenting with different initial values.

   (ii) For a periodic solution as found in (i), decrease ρ gradually. What happens to the solution? What happens to its period?

7. Consider the system of differential equations

$$\dot{x} = -y - z,$$
$$\dot{y} = x + ay, \qquad\qquad (7.10)$$
$$\dot{z} = b + z(x - c),$$

where a, b, and c are positive parameters. This system of equations is due to Rössler [91, 92] and is known as the *Rössler system*. It shares many features with the Lorenz system of equations, the main ones being that it displays chaotic behavior and has a *strange attractor*; see Figure 7.3. Interesting parameter values are $a = 0.2$, $b = 0.2$, $c = 5.7$ and $a = 0.1$, $b = 0.1$, $c = 14$.

(i) Set $z = 0$ and analyze the dynamics of the resulting linear system $\dot{x} = -y$, $\dot{y} = x + ay$ in the (x, y)-plane.

(ii) Take $a = 0.2$, $b = 0.2$, $c = 5.7$ and show that the Rössler system (7.10) has two critical points, one in the center of the attractor loop and the other somewhat removed from the attractor.

(iii) Linearize the Rössler system in the neighborhood of each critical point and find the stable and unstable manifolds.

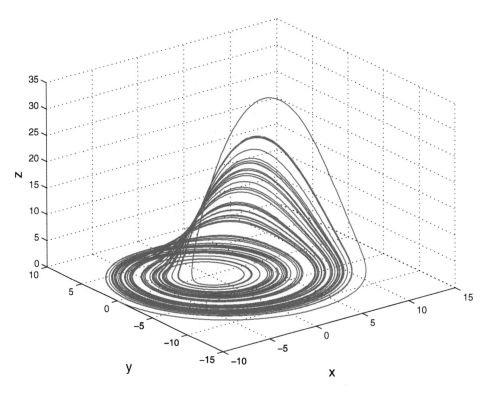

Figure 7.3. *The Rössler attractor for $a = 0.2$, $b = 0.2$, $c = 5.7$.*

Chapter 8

Climate and Statistics

In this introductory chapter we present several examples, without going into the technical details, of the uncommon challenges for statistics posed by climate science. We also introduce some of the terminology commonly used in climate statistics, where concepts like proxy data, reanalysis, and model skill play an important role.

Keywords: Data, heterogeneity, proxy data, reanalysis, model skill.

8.1 ▪ Challenges for Statistics

One of today's central public-policy questions—"Is the Earth's climate changing?"—is essentially a statistical question: Is the recent record of global weather data consistent with the inherent variability of the climate system, or is there evidence of long-term change, given the weather patterns of the past? As noted before, climate is sometimes defined as "average weather," so the statistical analysis of weather data—temperature, atmospheric pressure, wind patterns, precipitation, atmospheric particle count, and so on—plays a central role in climate science [127]. Statistical tools are used to prepare and clean large data sets, detect and extract patterns from data sets, validate models by comparing their predictions to new observations, reconstruct past climate states and forecast future climate scenarios, assess uncertainty of patterns and forecasts, and perform other tasks. In all this, computations play a central role, much more so than in traditional uses of statistics. The purpose of this chapter is to introduce some of the statistical challenges that arise in climate science and to familiarize the reader with the terminology. The discussion is kept at a nontechnical level; details will be taken up in subsequent chapters.

The special challenges presented by climate science for statistics are both methodological and practical. Here are the most obvious methodological challenges.

- In statistics, the "gold standard" for gathering data is to have randomized comparative experiments with sufficient replication. The experimenter compares several "treatments" by randomly assigning experimental conditions to a large set of subjects, while keeping close track of any factor that might influence the outcome of the experiment. Such a procedure is clearly impossible in climate science. We have only one Earth, and comparative field experiments are fundamentally out of the question. The only data we have are those from observations, and the only way to run an experiment is to do a numerical simulation using a mathematical model of the Earth's climate system.

- In inferential statistics, a standard technique is the testing of a *hypothesis*—that is, to find an answer to a "yes or no" question that is compatible with the observed data. However, in climate science it is difficult or even impossible to formulate clean statistical hypotheses. A hypothesis in climate science is usually the result of an observed pattern in the data. It is well known that the use of data to formulate a statistical hypothesis increases the likelihood of bias (nonrandom errors).

- In inferential statistics, one often builds a probability model to conduct "thought experiments" through mathematical analysis and then evaluates the results against observed data. But the climate system is extremely complicated, and even for simplified computational models there is usually no probabilistic model available that could be analyzed.

The practical challenges have to do with the implementation and application of statistical methods. Here are some specific examples.

8.1.1 ▪ Large Data Sets

Climate data sets come from observations and computations and are generally huge—much larger than those used in other statistical applications.

As a case in point, consider some of the data sets generated by the Global Precipitation Climatology Project (GPCP), which is part of the World Climate Research Programme [27]. Among its products are a compilation of precipitation estimates on a 2.5-degree grid over the entire globe at 5-day (pentad) intervals going back to January 1979 with more than $360 \times 180 \times 5000 \approx 3 \cdot 10^8$ data points, and a compilation of daily precipitation estimates on a 1-degree grid since October 1996 with more than $144 \times 72 \times 2100 \approx 2 \cdot 10^7$ data points. Keep in mind that these are highly aggregated data. The daily data for the 1-degree grid are available on the Web [73] in monthly data sets, each consisting of an unformatted binary file of about 7.7 MB. There are currently more than 180 such files.

8.1.2 ▪ Heterogeneous Data

Climate data tend to be heterogeneous, as they come from various sources and measurement stations that are operated in different ways and have different biases.

Here is a typical case having to do with the *urban heat island* effect. The effect is associated with relatively higher temperatures in a city compared with the surrounding rural areas. It is usually highly localized and depends strongly on such local factors as windiness, cloudiness, and proximity to the sea. Figure 8.1 shows two temperature records, one from an urban area (Sherbrooke) and the other from a nearby rural area (Shawinigan) in Québec. Both records cover the same 90-year period; the horizontal axis is time, and the vertical axis is the average daily mean temperature, shown as the deviation from the longtime average (the *temperature anomaly*) in degrees C. The graphs show, first of all, that the rural (Shawinigan) temperature data before 1916 are outliers, possibly because of observational errors, which should probably be ignored. The remaining rural temperature data seem to increase slowly, if at all, while the urban data seem to increase somewhat faster. But more interesting is the sudden drop in the temperature record for Sherbrooke around 1963. As it turns out, the location of the weather station was moved from downtown Sherbrooke to the airport, so a heterogeneity was introduced into the data. The data from Shawinigan were collected at a single location. In Chapter 9 we will discuss regression analysis and show how the effect of the move on the temperature record for Sherbrooke can be estimated and accounted for.

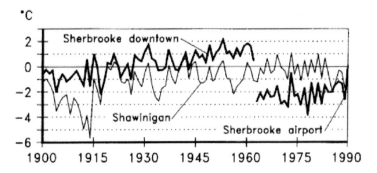

Figure 8.1. *Temperature records for two towns in Québec* [117].

8.1.3 ▪ Processed Data

Since it is impossible to travel into the past, the use of *proxy data* (inferred data from indirect observations) is common. Consequently, the data for statistical analysis are often themselves the result of prior statistical modeling (reanalysis).

Figure 8.2 shows annual global land-surface air temperature anomalies relative to the period 1961–1990. The black curve (CRUTEM3) is based on a recent analysis of data going back to 1850 by Brohan et al. [8]; the other curves are based on older data that are less complete. All the curves show decadal variations, and the long-term variations are in general agreement. But there are differences. Most of these differences arise from the diversity of spatial averaging techniques; for example, the recent global trends are largest in the curves marked CRUTEM3 (Hadley Centre, England) and NCDC (National Climate Data Center), which give more weight to the Northern Hemisphere, where recent trends have been greatest. Other differences are due to the different treatments of gaps in the data and the different number of observational stations used. The effects of different data preparation techniques are harder to incorporate into statistical analysis than data heterogeneities, which have well-documented causes.

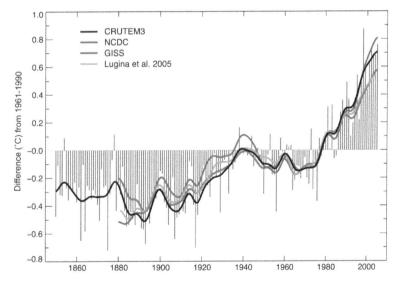

Figure 8.2. *Annual anomalies of global land-surface temperature, 1850–2005 relative to the 1961–1990 mean. Reprinted with permission from IPCC.*

8.1.4 · Incomplete Data

Temperature is one of the more obvious and easily measured climate variables, so it takes a central position in any debate over climate and climate change. But many other variables of interest also change when the climate system is forced, and their measurement may not be as straightforward. One of these variables is precipitation—a major climate variable, certainly at the regional level. If precipitation patterns shift, food production is affected in a major way, with potentially significant socioeconomic consequences. The continuity of precipitation records is not ideal, with many stations having fractured records or records covering only limited periods of time. Indeed, the "historical" records of precipitation are derived from land-based measurements and observations and do not provide any information on precipitation over the oceans—a not insignificant shortcoming, since the oceans cover the majority of the Earth's surface. Satellites are now providing enough data to observe precipitation patterns across the entire globe, but the record length is still relatively short, and any conclusion about possible trends is still subject to significant uncertainty. Filling in gaps in records of precipitation and other variables of interest is a continuing challenge for statistical climate modeling.

8.1.5 · Hard-to-measure Variables

For certain climate variables of interest it is difficult to develop both a definition to describe it and an index to measure it. A good example is the concept of drought. Many quantitative measures of drought have been developed, depending on the discipline affected, the region being considered, and the particular application. Common to all types of drought is the fact that they originate from a deficiency of precipitation resulting from an unusual weather pattern. They can be short-term or long-term, depending on the persistence of the underlying atmospheric circulation pattern and the duration of the precipitation deficit. To capture the essential elements of a drought in a quantifiable index is certainly nontrivial.

The Palmer Drought Severity Index (PDSI), developed by meteorologist WAYNE PALMER in the 1960s and known operationally as the Palmer Index, is based on a supply-and-demand model of soil moisture and attempts to measure the duration and intensity of the long-term drought-inducing circulation patterns. It uses a 0 as normal and quantifies drought severity by a negative number; for example, −2 is moderate drought, −3 is severe drought, and −4 is extreme drought. The Palmer Index can also reflect excess precipitation; +2 is moderate precipitation, etc. The Palmer Index can be applied to any site for which sufficient precipitation and temperature data are available, because it is standardized to local climate. Figure 8.3 shows the spatial pattern of the Palmer Index for the period 1900 to 2002. The lower panel shows the sign and the strength of the index over time. If the sign is positive, areas in red and orange in the top panel are drier than average, and areas in blue and green are wetter than average; if the sign is negative, the situation is reversed.

8.2 · Proxy Data

The climate system, like any complex dynamical system, displays chaotic fluctuations. The challenge for climate researchers is then to distinguish genuinely chaotic fluctuations from those that can be attributed to cycles or specific forcings. To this end, we would like to draw some inferences from the long-term record of the climate itself. But the problem is that actual data on climate fluctuations in the distant past are not readily available. There were no truly global measurements before the satellite era, and observations tend to be

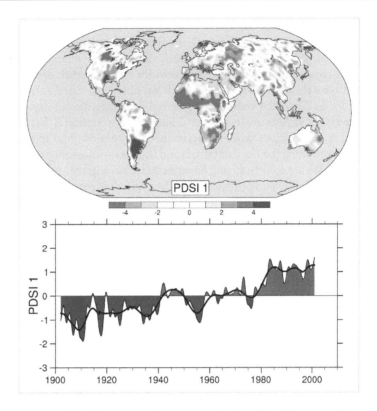

Figure 8.3. *The Palmer Index for the period* 1900–2002. *The smooth black curve shows decadal variations. Reprinted with permission from IPCC.*

less reliable as we go back in time. The first modern-style thermometer was developed around 1654, and before about 1800 there were virtually no measurements with carefully calibrated instruments. Consequently, we have to rely on *proxies* for the reconstruction of past climates. Proxy data fall roughly in one of three categories.

- *Geological proxies*, including isotope ratios in ice cores, lake and ocean sediments, and corals; pollen in sediment layers; temperatures in boreholes; and tree rings. To produce the most precise results, systematic cross-verification between proxy indicators is necessary for accuracy in readings and record-keeping. Geological proxies cover the Earth's history over a period of approximately one billion years.

- *Historical proxies*, like archival data, weather diaries, and literary reports on important events. Rough weather maps for Europe have been reconstructed from historical proxies from as far back as the winter of 1076–'77 (which was bitterly cold).

- *Phenological proxies*. These are observations of first occurrences of biological events in their annual cycle. Examples include the date of emergence of leaves and flowers, the first flight of butterflies, and the first appearance of migratory birds. In some cases, reconstructed temperature records go back more than 500 years.

We illustrate the use of phenological proxies with two examples, one involving the cherry-blossom flowering date as a proxy for late-winter temperatures and the other the date of the beginning of the grape harvest as a proxy for spring–summer temperatures.

The flowering date of the cherry tree (Prunus avium) has been recorded in Switzerland since about 1720, together with the average temperature (in degrees C) for the period from February through April [96]. Both data sets have a large degree of heterogeneity, since the temperature and the cherry-blossom flowering date depend strongly on altitude and also on the cherry variety, but both have been corrected as much as possible. The scatter plot in Figure 8.4 shows the corrected flowering date in days after January 1 against the average temperature in degrees C. Since the assumption is that the date at which the cherry blossom begins to flower is determined by the late-winter temperature, the temperature is plotted along the horizontal axis and the cherry-blossom flowering date along the vertical axis. We see a fairly strong linear negative association, with a correlation coefficient $r = -0.62$. For certain portions of the data set (when recordings are considered more reliable), the correlation is even stronger ($r = -0.82$). This indicates that good quality phenological data can be fairly reliable tools for reconstructing past temperature records.

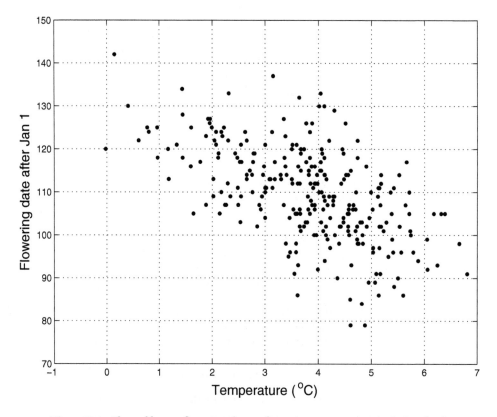

Figure 8.4. *Cherry-blossom flowering date vs. late-winter temperature in Switzerland.*

The second example is the date of the beginning of the grape harvest, which has been recorded in Burgundy (France) since 1370 [12] and on the Swiss Plateau since the mid-1500s. Again, the data are highly heterogeneous and have to be corrected to account for different recording locations, changes in cultivation methods, changes in grape varieties, and so on. Then a regression model is constructed from both harvest times and accurate temperature measurements for the period from April through August, where available. The reconstructed temperature anomalies are shown in Figure 8.5 (black), together with a smoothed version (red) and error margins of the reconstruction (blue). The figure shows

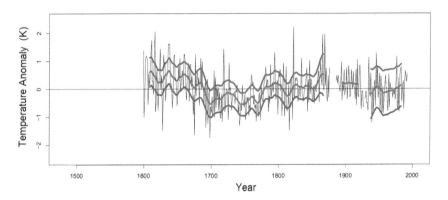

Figure 8.5. *April-to-August temperature anomalies obtained from grape harvest dates for the Swiss Plateau* [67].

a drop in temperature in the late 17th century and an increase in the middle of the 19th century. This cooling over about two centuries is known as the "Little Ice Age" and is well documented. Also visible are an oscillation with a period of about 25 years during the 17th century and an oscillation with about twice this period during the 18th century. It is unclear whether these are artifacts of the reconstruction or "real" phenomena. The reconstructed values for the period 1787–2000 correlate well with Paris instrumental data, with a correlation coefficient of approximately 0.75.

8.3 ▪ Reanalysis

Often, historical meteorological data find their way into climate studies through the process of *retrospective analysis* or *reanalysis*. Reanalysis is the evaluation of historical observations or data using a consistent statistical scheme to recreate complete, continuous, and physically consistent long-term climate data. The statistical methodology for such projects is known as *data assimilation*. Extra care is needed in the evaluation of historical data, because past observations were often done with different instruments or different methodologies, which may have introduced biases that may be difficult to assess. Missing data due to stations that went online or offline typically result in additional problems. Assembling data sources for reanalysis and assessing their quality often requires international collaboration.

Here is a quote from the NOAA website [72] on their 20th century reanalysis project: "Using a state-of-the-art data assimilation system and surface pressure observations, the *Twentieth Century Reanalysis Project* is generating a six-hourly, four-dimensional global atmospheric dataset spanning the entire 20th century, to place current atmospheric circulation patterns into a historical perspective." This project is capable of recreating a climatological description of events such as the dust bowl of the 1930s and the infamous Knickerbocker snowstorm of 1922, named after the collapse of a movie theater in Washington, DC, due to a wet snow load that killed 98 people, including former Congressman ANDREW JACKSON BARCHFELD from Pennsylvania.

The output of the 20th century reanalysis project is a description of the most likely global atmospheric state for the entire 20th century, with a time resolution of six hours and a spatial resolution of two degrees both in longitude and latitude, together with uncertainty estimates. The main data entering the model are global synoptic air pressure measurements, monthly sea temperature measurements, and measurements on the extent of sea ice. An example of the output is shown in Figure 8.6. The top left panel shows the sea-level pressure (SLP, in hectopascal, hPa) contour lines, together with the uncertainty (color coded); the 1,000 hPa and 1,010 hPa lines are thickened. A strong low pressure system off the mid-Atlantic coast is clearly visible. The top right panel shows the height at which the atmospheric pressure would be 500 hPa (Z500), again with color-coded uncertainty estimates; the 5,600 m line is thickened. The plot shows that there was a region of midaltitude low pressure right above the mid-Atlantic coast. The bottom left panel shows the recreated accumulated precipitation over the previous six hours, indicating that more than 10 mm of precipitation (more than 5 inches of new snow) fell during this time. The plot also shows much heavier precipitation further east over the Atlantic. The bottom right panel shows the estimated temperature at an altitude of 2 m, indicating that temperatures in the Washington area were just below freezing. All estimates were obtained with a method known as *ensemble Kalman filter* (EnKf).

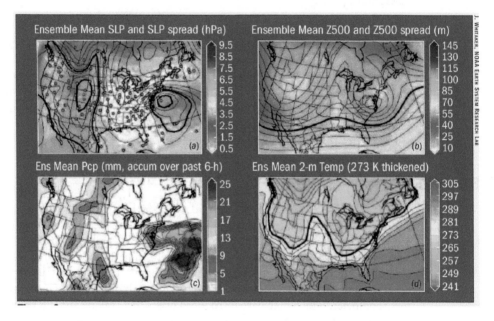

Figure 8.6. *Reanalysis of conditions on January 28, 1922, at 7 p.m., two hours before the collapse of the Knickerbocker movie theater* [15].

While the primary use of reanalysis is to reconstruct past climate states at a level of detail that was unavailable at the time, reanalysis can also be used to improve the assessment of current climate states and may thus provide a better background for current weather and climate prediction.

8.4 ▪ Model Skill

To assess the predictive or explanatory quality of a climate model, a simple quantity known as *skill* is often used. The definition of a model's skill requires three sets of data: a set $X = \{x_1, \ldots, x_n\}$ of actual observational data; a set $Y = \{y_1, \ldots, y_n\}$ of predictions of the same quantities over the same period of time, obtained with the climate model; and a set $Z = \{z_1, \ldots, z_n\}$ of long-term past averages of the same quantities over a period preceding the period of the actual observations. A "lazy" model would simply take the long-term averages and replicate them forward in time; its skill would be 0. On the other hand, an "oracle" would predict all observations exactly; its skill would be 1. A useful model would have a skill between 0 and 1, but it is also possible for a model to have negative skill—that is, the model provides predictions that are overall worse than the long-term averages. These arguments justify the definition of the skill S of a model,

$$S = 1 - \frac{\|X - Y\|^2}{\|X - Z\|^2}, \tag{8.1}$$

where $\| \cdot \|$ is some norm in $\mathbb{R}^n$. A common choice is the Euclidean norm, $\|X\|_2 = (\sum_{i=1}^{n} x_i^2)^{1/2}$ for any set $X = \{x_1, \ldots, x_n\} \in \mathbb{R}^n$. Other choices are also possible.

Today's climate models are typically "skillful" ($0 < S < 1$) if anthropogenic factors are included but have little or no skill ($S \approx 0$) if these factors are left out. This fact is not only an indication of the predictive quality of today's models, but can also be used as an argument to show that today's climate change is anthropogenic. On the other hand, standard climate models have very little skill when applied to a period of a decade or so; typically, they are incapable of predicting interannual global oscillations such as El Niño.

8.5 ▪ Exercises

1. While it is impossible to conduct experiments in climate science, sometimes nature conducts experiments for us by changing certain climate parameters while leaving other parameters constant. Examples include changes in solar activity and volcanic eruptions that change the aerosol content and the chemistry of the atmosphere. List three other such natural experiments, including at least one that influences oceans.

2. In statistics, *reliability* refers to the internal consistency of a method, while the *validity* of a method expresses the extent to which its results have anything to do with reality. Rate the reliability and the validity of the following measurements for the stated purposes (high, low, questionable, ...) and explain your answer.

   (i) Daily minimum temperatures in Washington, DC, as a measure for climate change in the mid-Atlantic region.

   (ii) Daily minimum temperatures in St. Michaels, MD, as a measure for climate change in the mid-Atlantic region.

   (iii) Daily minimum temperatures in Washington, DC, as a measure for climate change in North America.

   (iv) Distance traveled by a drifting instrument buoy in the Gulf Stream as a measure for oceanic flow.

   (v) Solar output measured by a satellite as a measure for insolation.

   (vi) Insurance claims per hurricane season as a measure for extreme weather.

   (vii) Number of hurricanes in Category 3 or higher per hurricane season as a measure for extreme weather.

3. When a new method is used to measure the same quantity, there can be systematic differences between the old and the new methods, as well as changes in accuracy and variability. To compare the old and the new methods, a graphical technique known as a *Bland–Altman plot* is used.

   (i) Look up how a Bland–Altman plot is constructed and interpreted.

   (ii) Find three scientific areas outside climate science where this technique is used.

(iii) To construct a Bland–Altman plot, both measurement methods must be used on a set of identical samples. This is not always possible in climate research, for example when a research vessel returns to an area of observation with a completely new set of instruments. Suggest a way to compare an old and a new measurement method when direct comparison on the same samples is impossible.

4. Is the Palmer Index a reliable measure of regional climate? Is it a valid measure?

5. Equation (8.1) can be used to define the skill of a weather prediction model. Consider a set Y of weather predictions for a one-week period, say, and compare it to the set of actual observations X and the set of predictions from a heuristic Z. Specifically, consider the persistence heuristic $Z = Z_p$, where all predictions for the next seven days are the same as the corresponding observations from today, and a climate model heuristic $Z = Z_C$, where all predictions come from a climate model.

- If Y comes from a weather forecast model, its predictive skill relative to Z_C typically is larger than its predictive skill relative to Z_p. Explain why this is so.

- The skill of Z_p relative to Z_C is typically positive. Explain why this is so.

- What is the skill of Z_C relative to Z_p?

Chapter 9

Regression Analysis

Regression analysis is a statistical technique for estimating relationships between observed variables. The technique can be used to reduce observational errors, to identify variables that influence a process in an essential way, to predict observations, or to reduce the heterogeneity of data from different sources. This chapter gives an outline of some basic regression techniques and of some diagnostic tools that are used in practice.

Keywords: Regression, objective function, least-squares, fit, coefficient of determination, linear regression, simple linear regression, regression diagnostics.

9.1 ▪ Statistical Modeling

Regression analysis is a statistical approach to modeling the functional relationship between a set $\{x_1, \ldots, x_p\}$ of *predictor variables* (also referred to as *covariates* or *regressors*) and a single *response variable y*. The relationship typically involves a set $\{\alpha_1, \ldots, \alpha_r\}$ of one or more *regression coefficients*, whose values are unknown and must be estimated from observations. The form of the functional relationship is assumed to be known; it is postulated, for example, on the basis of physical arguments or inferred from observational data and may be linear, quadratic, or more complicated. The estimation of the regression coefficients is usually done by comparing the observed and predicted values of the response at various combinations of the predictors and minimizing the difference in some metric.

Formally, the problem can be stated as follows. Let $\mathbf{x} = (x_1, \ldots, x_m)^T$ be a (column) vector of predictor variables and let $\alpha = (\alpha_1, \ldots, \alpha_p)^T$ be a (column) vector of regression coefficients. Given a predictor function $f : (\mathbf{x}, \alpha) \mapsto f(\mathbf{x}; \alpha)$ which depends parametrically on α and a set of data points $\{(\mathbf{x}_i, y_i) : i = 1, \ldots, n\}$, find α such that the set of *residuals* $\{r_i = y_i - f(\mathbf{x}_i; \alpha) : i = 1, \ldots, n\}$ is as close to zero as possible (in a metric to be specified).

Since there must be enough data available compared to the number of parameters to be estimated, one usually requires that $n \geq p$; otherwise, the problem does not have a unique solution. The relation between $\mathbf{x}$ and y is not an exact identity, so the usual notation is

$$y \sim f(\mathbf{x}; \alpha). \tag{9.1}$$

The *least-squares method* is a technique to find values of the parameters α_j that give the "best fit" in the sense that they minimize the least-squares *objective function*,

$$Q_2(\alpha) = \sum_{i=1}^{n} r_i^2 = \sum_{i=1}^{n} |y_i - f(\mathbf{x}_i; \alpha)|^2. \tag{9.2}$$

The size of the residuals gives an indication of how well the given model matches the observations for the particular choice of parameter values.

Sometimes it is desirable to reduce the influence of outliers. This can be accomplished by minimizing the quantity

$$Q_1(\alpha) = \sum_{i=1}^{n} |r_i| = \sum_{i=1}^{n} |y_i - f(\mathbf{x}_i; \alpha)|. \tag{9.3}$$

The price one pays is a somewhat more complicated numerical treatment. Another possibility is to introduce an additional parameter λ and minimize the quantity

$$Q(\alpha; \lambda) = \sum_{i=1}^{n} r_i^2 + \lambda \sum_{j=1}^{p} |\alpha_j|. \tag{9.4}$$

The value of the positive parameter λ must be chosen separately. The minimization tends to reduce the overall size of the residuals. The second term acts as a penalty; it prevents the parameters α_j from becoming too large. This approach can also be used to choose a model where many of the α_j are close to or equal to zero—that is, a model that can explain the observed data with fewer parameters. This allows one to come up with meaningful solutions even in the case where $n < p$.

Once values have been assigned to α, the right-hand side of Eq. (9.1) can be used to predict the value of the response variable at any $\mathbf{x}$. This value is called the *fit* at $\mathbf{x}$ and denoted by $\hat{y}$, so $\hat{y} = f(\mathbf{x}; \alpha)$, where α has been obtained by minimizing Q_2, Q_1, or $Q(\cdot; \lambda)$ for some choice of λ.

In this form, the problem is one of parameter fitting and is common in many areas of science. It becomes a statistical problem once the variability of observations and of parameters is incorporated with a suitable probabilistic model. Following are some illustrative examples from climate science.

9.1.1 ▪ Moisture and Biological Activity

In sediments taken from below a lake bed, the abundance of biological silicon y is measured together with a certain isotope ratio x. The isotope ratio is an indirect measure of moisture in the past (when the layer from which the sample was taken was at the bottom of the lake), while the silicon abundance is a measure for the biological activity in the lake at that time. A linear relation is posited between x and y, $y \sim \alpha_1 + \alpha_2 x$.

We can think of x as a predictor and y as the response, but formally, in the framework of regression analysis, it is more correct to think of two predictors, x_1 and x_2, where $x_1 = 1$ for all observations and $x_2 = x$, and a response y. The unknown parameters are α_1 and α_2, and it is important to note that they appear linearly in the relation $y \sim \alpha_1 x_1 + \alpha_2 x_2$. If this relation is close to an equality for given data, then y can be predicted also for values of x which have not been observed. In addition, similar regression models at other locations permit the comparison of the coefficients α_1 and α_2 between these locations, which gives insight into the sensitivity of biological life to changes in moisture at different locations. Note that we are not saying that a change of the isotope ratio x "causes" a change in biological activity; rather, x is a proxy for moisture, which certainly plays a role in biological activity.

9.1.2 ▪ Urban Heat Island

This example continues the discussion of statistical recalibration of temperature data begun in Section 8.1.2. We refer to Figure 8.1, where the temperature anomaly since 1900 is shown for two nearby locations, Shawinigan (rural) and Sherbrooke (urban), in Québec, with a gap in the record for Sherbrooke due to a move of the measuring station in 1963.

Let the index i denote the year since 1900 ($i = 1, \ldots, 90$) and denote the temperature anomaly in year i for Shawinigan by $T_{r,i}$ and for Sherbrooke by $T_{u,i}$. Since the record for Shawinigan is more or less homogeneous, we posit that $T_{r,i}$ is modeled by an expression of the form $T_{r,i} \sim \alpha_1 + \alpha_2 i$. To accommodate the jump in the record for Sherbrooke, we take $T_{u,i} \sim \beta_1 + \beta_2 i + \beta_3 c_i$, where $c_i = 0$ if $i = 1, \ldots, 62$ and $c_i = 1$ if $i = 63, \ldots, 90$.

A standard least-squares method results in the estimate $\hat{\beta}_3 \approx -3°C$, indicating that the observed temperature anomalies dropped by about 3°C due to the move. We then adjust the Sherbrooke temperature data by taking $\tilde{T}_{u,i} = T_{u,i} - \beta_3 c_i$ and apply a regression analysis to the linear models $T_{r,i} \sim \alpha_1 + \alpha_2 i$ and $\tilde{T}_{u,i} \sim \beta_1 + \beta_2 i$ to determine long-term trends in the temperature at both locations, to find out whether there is any trend at all for Shawinigan (that is, whether $\alpha_2 = 0$ or not), to find out whether there is evidence that the trends are different (that is, whether $\alpha_2 = \beta_2$ or not), and so on. The results of adjusting the observed Sherbrooke temperature data are given in Figure 9.1.

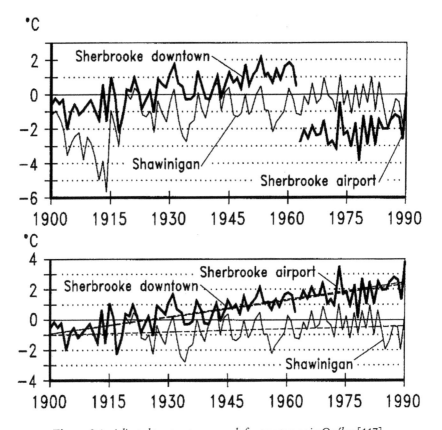

Figure 9.1. *Adjusted temperature records for two towns in Québec* [117].

9.1.3 ▪ Atmospheric CO_2 Concentration

In Chapter 10 we will discuss data from an ongoing series of observations measuring CO_2 concentrations in the atmosphere. The observational effort was initiated in 1958 by CHARLES DAVID KEELING (1928–2005) and is being carried out on Mauna Loa Volcano in Hawaii. The data show seasonal variations superimposed on what appears to be an increasing long-term trend. The relevant question for climate science is whether the long-term trend is statistically significant.

Let y be the average CO_2 concentration in a given year at a given location, and let x denote time measured in years, with $x = 0$ at some reference time, for example in 1950. We expect a relation of the form $y \sim \alpha_1 + \alpha_2 x$, or perhaps $y \sim \alpha_1 + \alpha_2 x + \alpha_3 x^2$. In the first case, there are two predictors, x_1 and x_2, where $x_1 = 1$ (constant for all observations) and $x_2 = x$. In the second case, there are three predictors, x_1, x_2, and x_3, where $x_1 = 1$, $x_2 = x$, and $x_3 = x^2$, so here one predictor (x_1) is constant, while another (x_3) is a nonlinear function of the second one (x_2). Again, the unknown parameters α_i appear linearly in both cases. The parameter α_2 corresponds to the average increase in CO_2 per year, and a standard statistical method (t-test) can be used to decide whether there is evidence that α_2 is positive, indicating an increasing long-term trend. If the data show that the parameter α_3 is positive, then there is evidence that this increase is accelerating. Note again that it is meaningless to say that an increase of x "causes" an increase of y.

9.1.4 ▪ Volcanic Eruptions and Global Temperatures

During a volcanic eruption, huge amounts of aerosols and trace gases are released. Eventually, the aerosol particles are washed out of the atmosphere and the trace gases disappear through natural chemical reactions. But as long as they stay in the atmosphere, the aerosols and trace gases influence cloud formation and radiative properties of the atmosphere and can thus have a significant effect on the global temperatures for several years. An important question for climate science is how to integrate the data into a mathematical model describing the behavior of temperature over time.

Let x be the time (measured in months) that has passed since such an eruption, and let y be the global temperature anomaly at that time. We are looking for a relation of the form $y \sim \alpha_1 - \alpha_2 e^{-\alpha_3 x}$, where $\alpha_2 > 0$ and $\alpha_3 > 0$. Note that we do not claim to have a physical model predicting the behavior of the temperature anomaly with time; rather, the relation is purely descriptive. The parameter α_1 describes any lasting effect (if there is one) and α_2 the initial effect of the eruption on global temperatures; α_3 tells us how fast the effects of the eruption disappear. All these quantities are obviously important for understanding the Earth's climate, but since they appear nonlinearly in the regression model, this problem is harder to solve than the problems in the previous examples. And of course, it makes no sense to say that changes in x cause changes in y.

9.2 ▪ Linear Regression

In *linear regression*, the mean (or, less commonly, the median or some other quantile) of the conditional distribution of y given X is assumed to be a linear function of the regression parameters $\alpha_1, \ldots, \alpha_p$. This is the case in the first three examples given above. As the second example shows, nonlinear relations between the response variable and one or more of the predictors are allowed and entirely compatible with the requirement of linearity for the parameters. As we have seen in the first two examples above, the predictor variables can be arbitrarily transformed, and in fact multiple copies of the same underlying predictor variable can be added, each one transformed differently, so without loss of generality we may assume a relation of the form

$$y_i \sim \alpha_1 x_{i,1} + \cdots + \alpha_p x_{i,p} = \mathbf{x}_i^T \boldsymbol{\alpha}, \quad i = 1, \ldots, n. \qquad (9.5)$$

The superscript ^{T} denotes the transpose, so $\mathbf{x}_i^T \boldsymbol{\alpha}$ is the inner product of the vector $\mathbf{x}_i$ and the vector $\boldsymbol{\alpha}$.

Linear regression is used extensively in practical applications, for the simple reason that linear models are easier to fit than nonlinear models. Also, the statistical properties of the resulting estimators are easier to determine, and in certain situations it is possible to give a detailed theoretical analysis and show that this approach is optimal in a suitable sense.

To account for all other factors that influence the response y_i other than the regressors $\mathbf{x}_i$, it is common to include an *error variable* in the predictor function. The error variable captures relationships that the linear regression model cannot describe (predictors and other effects not included in the model), as well as genuine stochastic effects (random observational errors for the y_i). Thus, a linear regression model takes the form

$$y_i = \alpha_1 x_{i,1} + \cdots + \alpha_p x_{i,p} + \varepsilon_i = \mathbf{x}_i^T \boldsymbol{\alpha} + \varepsilon_i, \quad i = 1, \ldots, n, \qquad (9.6)$$

or, in vector form,

$$\mathbf{y} = \mathbf{X} \boldsymbol{\alpha} + \boldsymbol{\varepsilon}, \qquad (9.7)$$

where

$$\mathbf{X} = \begin{pmatrix} \mathbf{x}_1^T \\ \mathbf{x}_2^T \\ \cdots \\ \mathbf{x}_n^T \end{pmatrix} = \begin{pmatrix} x_{1,1} & \cdots & x_{1,p} \\ x_{2,1} & \cdots & x_{2,p} \\ \vdots & \ddots & \vdots \\ x_{n,1} & \cdots & x_{n,p} \end{pmatrix}, \quad \mathbf{y} = \begin{pmatrix} y_1 \\ y_2 \\ \vdots \\ y_n \end{pmatrix}, \quad \boldsymbol{\alpha} = \begin{pmatrix} \alpha_1 \\ \alpha_2 \\ \vdots \\ \alpha_p \end{pmatrix}, \quad \boldsymbol{\varepsilon} = \begin{pmatrix} \varepsilon_1 \\ \varepsilon_2 \\ \vdots \\ \varepsilon_n \end{pmatrix}.$$

Numerous procedures have been developed for parameter estimation and inference. They vary in computational complexity, robustness with respect to heavy-tailed distributions, and theoretical assumptions needed to validate desirable statistical properties. The simplest and most common estimator is the *ordinary least-squares* (OLS) method, which minimizes the sum of squared residuals defined in Eq. (9.2). The usual first-order necessary conditions for a minimum can be written in the form

$$\mathbf{X}^T \mathbf{y} = \mathbf{X}^T \mathbf{X} \hat{\boldsymbol{\alpha}}. \qquad (9.8)$$

These equations are known as the *normal equations*. The OLS method leads to a closed-form expression for the estimated value $\hat{\boldsymbol{\alpha}}$ of the vector $\boldsymbol{\alpha}$,

$$\hat{\boldsymbol{\alpha}} = (\mathbf{X}^T \mathbf{X})^{-1} \mathbf{X}^T \mathbf{y}. \qquad (9.9)$$

The formula shows that the vector $\hat{\boldsymbol{\alpha}}$ can be obtained with linear algebra operations. In numerical implementations, one usually does not compute the inverse of $\mathbf{X}^T \mathbf{X}$ but rather uses suitable matrix factorizations such as QR-factorization to improve the numerical robustness of the method.

The estimation formula (9.9) has various desirable theoretical properties. Under suitable assumptions for the errors ε_i, the estimates $\hat{\boldsymbol{\alpha}}$ are unbiased (there is no systematic error) and consistent (they converge in a suitable sense to the true values as $n \to \infty$). They are also known to be optimal in certain situations, in the sense that they make the best use of the available information. It should be noted that these theoretical properties of the errors can sometimes be achieved in experiments, but in the case of observational data it is often unclear whether they hold, for various reasons. For example, the magnitudes of the errors could depend on the responses or the predictors (for example, in the case of observations that vary over several orders of magnitude), and the errors could be correlated. Often, it is necessary to cleanse the data and transform them before a linear regression approach can be applied.

Once the estimate $\hat{\alpha}$ has been obtained from Eq. (9.9), it is possible to compute predictions $\hat{\mathbf{y}} = \mathbf{X}\hat{\alpha}$ and residuals $\mathbf{r} = \mathbf{y} - \hat{\mathbf{y}}$. Let $\bar{\mathbf{y}}$ be the vector with identical components $(1/n)\sum_{i=1}^{n} y_i$. The quantity

Average of the data pts.

$$R^2 = \frac{\|\hat{\mathbf{y}} - \bar{\mathbf{y}}\|^2}{\|\mathbf{y} - \bar{\mathbf{y}}\|^2} \tag{9.10}$$

is called the *coefficient of determination*. As before, $\|\cdot\|$ is the Euclidean norm in $\mathbb{R}^n$. If R^2 is close to 1, almost all of the variation (the overall deviation from the mean) in the response $\mathbf{y}$ is explained by the regression model. One can show that $R^2 = 1 - \|\mathbf{r}\|^2/\|\mathbf{y} - \bar{\mathbf{y}}\|^2$; hence, R^2 is large if the residuals are small relative to the variations in the observations.

small $\qquad \|\hat{\mathbf{y}} - \bar{\mathbf{y}}\|^2$

9.3 • Simple Linear Regression

Simple linear regression is the special case of linear regression with a single predictor x. Simple linear regression with a least-squares estimator fits a straight line through a set of n points $\{(x_i, y_i) : i = 1, \ldots, n\}$ in such a way that the sum of squared residuals of the model—that is, the sum of the vertical distances between the points of the data set and the fitted line—is as small as possible. If the straight line is represented by the equation

$$y \sim \alpha_1 + \alpha_2 x, \tag{9.11}$$

the objective function (9.2) becomes

Minimize sum of squared residuals. $\longrightarrow$

$$Q_2(\alpha_1, \alpha_2) = \sum_{i=1}^{n} r_i^2 = \sum_{i=1}^{n} (y_i - \alpha_1 - \alpha_2 x_i)^2.$$

If there are reasons to believe that y is directly proportional to x, then the intercept term α_1 should be omitted. Minimization of the objective function yields the relations

$$\hat{\alpha}_1 = \bar{y} - r_{xy}\frac{s_y}{s_x}\bar{x}, \quad \hat{\alpha}_2 = r_{xy}\frac{s_y}{s_x}, \tag{9.12}$$

where $\bar{x}$ and $\bar{y}$ are the sample means of x and y, s_x and s_y their standard deviations, and r_{xy} the correlation coefficient.

The result (9.12) implies that

$\hat{y} - \bar{y} = \hat{\alpha}_1 + \hat{\alpha}_2 x - (\hat{\alpha}_1 + r_{xy}\frac{s_y}{s_x}\bar{x})$

$= r_{xy}\frac{s_y}{s_x}(x - \bar{x})$

$$\bar{y} = \hat{\alpha}_1 + \hat{\alpha}_2\bar{x}, \tag{9.13}$$

so the straight line obtained by simple linear regression passes through the *center of mass* $(\bar{x}, \bar{y})$ of the set of points $\{(x_i, y_i) : i = 1, \ldots, n\}$.

If we combine Eq. (9.13) with the relation $\hat{y} = \hat{\alpha}_1 + \hat{\alpha}_2 x$ and use the expressions for $\hat{\alpha}_1$ and $\hat{\alpha}_2$ given in Eq. (9.12), we find the following relation between the rescaled ("standardized") versions of the variables x and y:

$r_{xy} = \dfrac{y - \bar{y}}{x - \bar{x}} \cdot \dfrac{s_x}{s_y}$

$$\frac{y - \bar{y}}{s_y} = r_{xy}\frac{x - \bar{x}}{s_x}. \tag{9.14}$$

This relation shows that the coefficient of determination R^2 is simply the square of the correlation coefficient, $R^2 = r_{xy}^2$.

The construction of *confidence intervals* for the estimates $\hat{\alpha}_1$ and $\hat{\alpha}_2$ is relatively straightforward and is also implemented in all computer packages. We stress again that these confidence intervals are valid only if certain assumptions hold for the error terms, which may be quite restrictive and difficult to check for observational data.

9.4 ▪ Regression Diagnostics

A careful examination of the results of fitting a linear regression model is important, for observational data even more than for experimental data, since little information is available on the error terms and theoretical properties of the regression computation may therefore be in doubt. Techniques for doing this are generally known as methods for *regression diagnostics*. They typically involve a closer look at the residuals and usually require graphical methods, which are commonly implemented in computer packages.

We illustrate these techniques with the example of the cherry-blossom flowering dates described in Section 8.2. The purpose of the earlier discussion was to demonstrate that the cherry-blossom flowering date is a good proxy for late-winter temperature. Here, we want to establish a functional relationship between the two quantities.

Let x be the adjusted recorded flowering date, say in days after January 1, and let y be the average temperature for April through January. Note that we are deliberately reversing cause and effect; in all likelihood, the flowering date depends on the temperature and not the other way around. The reason is that the flowering date is the predictor and the temperature is the response: we would like to make inferences about past average temperatures based on recorded flowering dates. *Proxy data.*

The result of fitting a least-square model to the data is shown in Figure 9.2, where the flowering date x is plotted along the horizontal axis and the average $\hat{y}$ temperature y along

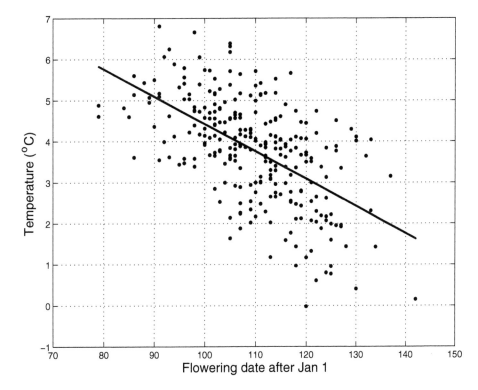

Figure 9.2. *Late-winter temperatures vs. corrected cherry-blossom flowering dates, with a fitted straight line.*

the vertical axis. The equation for the straight line turns out to be $\hat{y} = 11.09 - 0.0666\,x$, with a coefficient of determination $R^2 = 0.331$. The value of R^2 is not particularly high, but it certainly shows that there is an association between temperature and flowering dates. The correlation coefficient in this case is $r_{xy} = -0.575$.

We can now look at the residuals $r_i = y_i - \hat{\alpha}_1 - \hat{\alpha}_2 x_i$ in several different ways; see Figure 9.3.

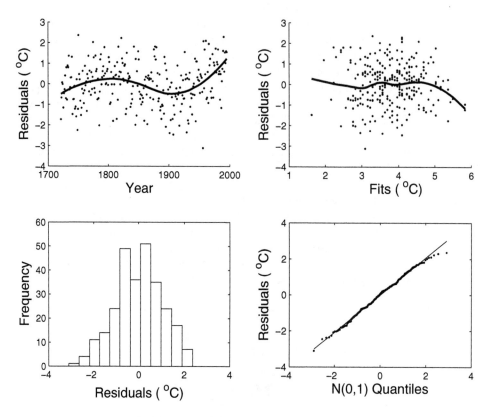

Figure 9.3. *Residual diagnostics for cherry-blossom flowering example: residuals against time (top left), residuals against fits (top right), histogram of residuals (bottom left), and normal probability plot of residuals (bottom right).*

The top left panel of Figure 9.3 gives a plot of the residuals against time, which may be useful for detecting possible changes in the measuring method or even a direct influence of time. The plot shows that the residuals have approximately the same magnitude throughout. The smooth curve through the data, which was obtained with a statistical smoothing method, shows an oscillatory pattern; the average residuals tend to be negative for years before 1800 and between 1850 and 1950 and positive after 1950. It is unclear whether this behavior is just a fluke, or what might be the cause of this pattern. Possible causes are changes in measurement methods for the temperature, changes in cherry tree varieties, or climate changes not related to temperature, for example changes in precipitation patterns.

The top right panel of Figure 9.3 gives a plot of the residuals against fits (predicted temperatures). Recall that simple linear regression postulates a linear relation between the predictor and response. This plot is useful for determining whether there is an additional nonlinear dependence of the response on the predictor. The smooth curve through the data indicates that there is no additional dependence for predicted temperatures in the intermediate range (between about 2 and 5 degrees C); however, residuals tend to be negative for higher temperatures (that is, for early flowering dates), so the straight-line prediction tends to be too high there. This is also visible in the scatter plot in Figure 9.2.

We can also look at purely statistical properties of the residuals. Graphical tools include a histogram and a normal quantile plot; both tell us whether the residuals are close to a normal distribution. The histogram in the bottom left panel of Figure 9.3 is approximately bell-shaped, and the normal quantile plot in the bottom right panel of Figure 9.3 is approximately a straight line, so the residuals appear to have a normal distribution. Note that the dependence of the residuals on time and on the predictor are ignored in these plots, while their relative magnitudes are shown. These two plots therefore contain genuinely new information.

All modern statistical packages can compute the coefficients α_j in linear regression problems using the least-squares approach. It is also easy to produce the residuals r_j and generate the plots shown in Figures 9.2 and 9.3. Modern packages can also be expected to compute regression coefficients with approaches different from the least-squares methodology and can even give results on certain nonlinear regression problems. In Section C.2 we give a few simple MATLAB scripts that allow a user to compute regression coefficients and residuals and do regression diagnostics.

9.5 ▪ Exercises

1. Suppose $\{y_i : i = 1, \ldots, N\}$ is a sequence of N annual averages of some meteorological quantity which has been observed at one location for $i = 1, \ldots, k$ and at another location for $i = k+1, \ldots, N$. We suspect that the move changed the average reading; however, we do not suspect a change of the reading over time (zero slope).

   (i) A simple model for the data is $y_i \sim \alpha_1 + \alpha_2 F_i$, where $F_i = 0$ for $i = 1, \ldots, k$ and $F_i = 1$ for $i = k+1, \ldots, N$. Then α_2 corresponds to the change in the reading that is due to the move. Set up the normal equations for the corresponding least-squares problem and give an explicit solution (that is, find explicit formulas for α_1 and α_2 in terms of the data).

   (ii) Another model for the data is $y_i \sim \alpha_1 G_i + \alpha_2 F_i$, where $G_i = 1$ for $i = 1, \ldots, k$ and $G_i = 0$ for $i = k+1, \ldots, N$ and the F_i are as in (i). Explain what the change in the reading due to the move is in this case. Then set up the normal equations for the corresponding least-squares problem and give an explicit solution.

2. This problem uses monthly averages of CO_2 data taken at the South Pole from November 1975 until December 2011. The data can be found on the Web [76]. A graph is given in Figure 9.4.

   (i) Download the data and turn it into a vector y of responses.

   (ii) Our predictor will be time. Fit a straight line to the data. What is the average *predicted increase per year*?

   (iii) Use the data to compute annual averages. Then fit a straight line to the annual averages. What is the average predicted increase per year from this method? Compare your answer to the one found in (ii).

   (iv) Plot the residuals, obtained by fitting a straight line to the data, against time. Does it look like the residuals are random? Are there any patterns in the residuals? If so, do you have an explanation for these patterns?

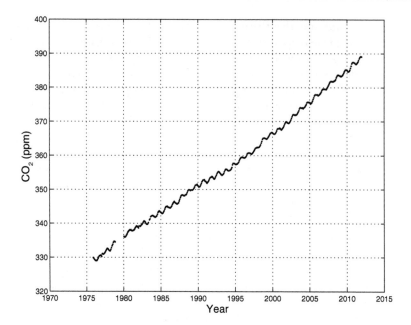

Figure 9.4. *South Pole* CO_2 *data.*

(v) Fit a quadratic function to the data and plot the residuals against time. Does it look like the residuals are random? Are there any patterns in the residuals? If so, compare them to the patterns found in (iv).

(vi) Compute the coefficients of determination from fitting a straight line and from fitting a quadratic function of time. Comment on the difference.

3. Tropical coral reefs can be used to obtain proxy data for ocean temperature, because their growth results in annual density bands that are clearly visible and because they incorporate certain isotopes whose abundances are known to be closely related to the ocean temperature. In [14], data have been obtained from a coral that grew (and still does grow) in Malindi Marine Park, Kenya (3° S, 40° E). The colony lies at a depth of approximately 6 m (low tide) and is about 4 m in height. Annual growth rates are roughly 12 mm/yr. The data set contains $\delta^{18}O$, a measure of the ratio of the stable oxygen isotopes ^{18}O and ^{16}O for the years 1801–1994, as well as mean sea surface temperature (SST) values in centigrade for 29 of the 44 years from 1951 to 1994, taken at 3° S, 41° E.

(i) Download the data set from the Web. Make a scatter plot of SST against coral $\delta^{18}O$ for those years for which both data are available. Compute the correlation coefficient of SST and coral $\delta^{18}O$ for these years. Do you think that the observed $\delta^{18}O$ data from this coral can be used as proxy data for the SST in this part of the Indian Ocean?

(ii) Compute a linear regression function, where coral $\delta^{18}O$ is used to predict SST.

(iii) Plot the residuals of the regression against years. Summarize your observations (magnitude and patterns of residuals, if applicable).

(iv) Predict SST for the years in which it is not available from the corresponding coral $\delta^{18}O$ data. Plot the predicted SST against the year and describe the main features of the plot.

(v) Plot the observed SST values for the years in which they are available in the same plot. How accurate are your predictions? Are the predictions from the linear regression model qualitatively correct?

Chapter 10

Mauna Loa CO$_2$ Data

In this chapter we show how statistical methods are used to cleanse raw observational data. The data come from monitoring the carbon dioxide (CO$_2$) concentration in the atmosphere on the Mauna Loa Volcano on Hawaii Island. The results are summarized in the so-called *Keeling curve*, which provides the most convincing evidence to date of anthropogenic effects on the Earth's climate.

Keywords: Carbon dioxide (CO$_2$), Keeling curve, cleansing raw data, trends, periodic components, residuals.

10.1 ▪ Keeling's Observational Study

In 1958, CHARLES DAVID KEELING (1928–2005) from the Scripps Institution of Oceanography began recording carbon dioxide (CO$_2$) concentrations in the atmosphere at an observatory located at about 3,400 m altitude on the Mauna Loa Volcano on Hawaii Island [53]. The location was chosen because it is not influenced by changing CO$_2$ levels due to the local vegetation and because prevailing wind patterns on this tropical island tend to bring well-mixed air to the site. While the recordings are made near a volcano (which tends to produce CO$_2$), wind patterns tend to blow the volcanic CO$_2$ away from the recording site. Air samples are taken several times a day, and concentrations have been observed using the same measuring method for over 60 years. In addition, samples are stored in flasks and periodically reanalyzed for calibration purposes. The observational study is now run by RALPH KEELING, Charles's son. The result is a data set with very few interruptions and very few inhomogeneities. It has been called the "most important data set in modern climate research." The raw data undergo rigorous statistical analysis before they are used, and the purpose of the present chapter is to show some of these analytical techniques. The results are summarized in the graph of Figure 10.1, which gives the monthly averages of CO$_2$ concentrations in parts per million by volume (ppmv, or ppm for short). The graph is referred to as the *Keeling curve*. A visual inspection shows a steadily increasing trend, together with a seasonal pattern that varies from year to year. A closer look also reveals that there are some months with missing data, notably in 1958 (when recordings were just beginning) and 1964.[3]

---

[3]Monitoring is science's Cinderella, unloved and poorly paid. Despite the import of the results, Keeling's work was often threatened, as is attested to by the missing data in 1964 when underfunding briefly halted measurement [71].

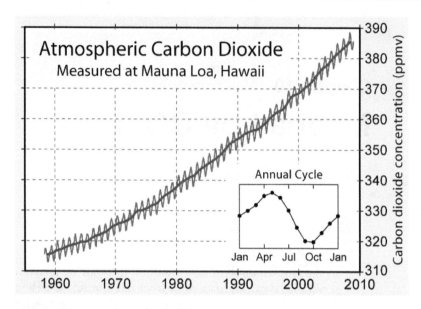

Figure 10.1. *Monthly average* CO₂ *concentrations (red), annual averages (blue), and mean annual cycles (black) on Mauna Loa since 1958. Image courtesy of Global Warming Art.*

10.2 ▪ Assembling the Data

The goal of the researchers at the Mauna Loa Observatory was and continues to be to collect air samples every hour in a manner that is as consistent over time as possible. Several measurements are taken and averaged to obtain an hourly data point. However, it is not always possible to get an average, due to instrument malfunction or equipment maintenance. Even if an average is obtained, it may be rejected for several reasons: too much variability among the measurements due to turbulent atmospheric conditions, too much change from the previous hour's average, strong local updrafts that bias the measurement, and so on. The hourly averages for the first week of January 2011 are plotted in Figure 10.2. Only data marked with solid circles were retained; all other data were rejected, and their symbols indicate the various rejection codes. For example, data marked with a + symbol were rejected because they were obtained during times of strong uphill winds; these winds bring CO₂-impoverished air from areas of dense vegetation. Data marked with a × symbol were rejected because they differed by more than 0.25 ppm from the previous measurement. In 2011, usable data were obtained only for about 62% of all possible times. The most common reasons for rejection were uphill winds and large variations from previous measurements.

As of November 2012 there are data for 656 months. The first month for which data are available is March 1958. No data are available for seven months between March 1958 and October 2012; for these months we will use interpolated values. The solid curve in Figure 10.1 has been obtained by averaging the monthly CO₂ data over several years.

10.3 ▪ Analyzing the Data

We use regression analysis to obtain a parametric fit of the data, decomposing the time series into a long-term trend given by a "simple" (linear or quadratic) function, a seasonal pattern, and a residual.

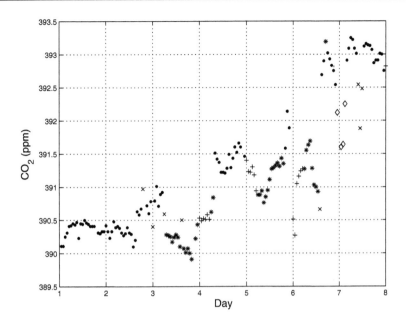

Figure 10.2. *Raw* CO_2 *data on Mauna Loa in early* 2011, *coded by rejection status.*

Let C_i be the average CO_2 concentration in month i ($i = 1, 2, \ldots$, counting from March 1958). We look for a description of the form

$$C_i = F(t_i) + P_i + R_i, \qquad (10.1)$$

where $F : t \mapsto F(t)$ accounts for the long-term trend; t_i is time at the middle of the ith month, measured in fractions of years after January 1, 1958; P_i is periodic in i with period 12, accounting for the seasonal pattern; and R_i is the remaining residual that accounts for all other influences. The decomposition is meaningful only if the range of F is much larger than the amplitude of the P_i and this amplitude in turn is substantially larger than the R_i.

10.3.1 ▪ Trend

A simple linear model is $F_1(t) \sim \alpha_1 + \alpha_2 t$, and a fit to the data gives $\alpha_1 = 308$ and $\alpha_2 = 1.468$ (omitting hats). In other words, the CO_2 content of the atmosphere near Mauna Loa increases by about 1.47 ppm per month on average. The coefficient of determination gives an indication of the quality of the fit; its value is $R^2 = 0.978$, which is very high. However, the errors are definitely not random; they still contain periodic components and, more importantly, the residuals $C_i - F_1(t_i)$ still show a long-term U-shaped pattern, as shown in the top left panel of Figure 10.3.

A quadratic model, $F_2(t) \sim \beta_1 + \beta_2 t + \beta_3 t^2$, results in the values $\beta_1 = 314$, $\beta_2 = 0.81$, $\beta_3 = 0.012$, and an even higher coefficient of determination, $R^2 = 0.991$. The interpretation is that the annual increase of the average CO_2 concentration has increased over the last 60 years as well. The U-shaped pattern is now gone, but the residuals, shown in the top right panel in Figure 10.3, still show the seasonal variation, and there is still a long-term oscillatory pattern with a period of perhaps 30 years. This pattern may be an artifact of using a quadratic regression over the entire length of the record.

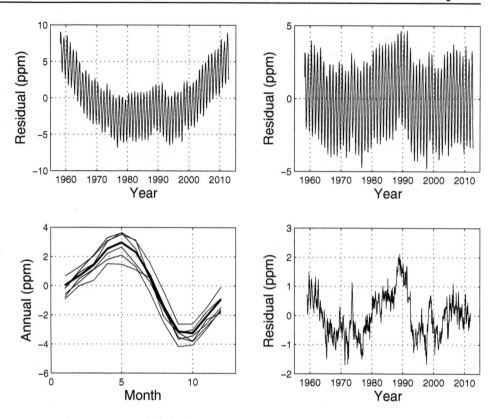

Figure 10.3. *Residuals for the Mauna Loa CO$_2$ data: residuals from linear regression (top left) and quadratic regression (top right); mean monthly residuals (bold) and a few sample monthly residuals (bottom left); remaining residuals (bottom right).*

10.3.2 ▪ Periodic Component

Next, we look for seasonal patterns in the data. These patterns appear in the residuals $C_i - F_2(t_i)$ and therefore should be extracted from them. Only full calendar years will be used, giving 636 months or 53 years from January 1959 until December 2011. Averaging these residuals for each month over the 53 years gives the oscillatory pattern shown in bold in the bottom left panel in Figure 10.3. The same panel also shows the observed monthly residuals for the years 1965, 1974, 1983, 1992, and 2011. The pattern is essentially the same as the one shown in the inset of Figure 10.1, with a peak in May and a minimum in September/October. That is, the observed CO$_2$ levels tend to exceed the annual average by about 3 ppm in May and are about 3 ppm below the annual average in September/October. This cycle is thought to be due to the cyclic update and release of CO$_2$ by vegetation in Asia; this CO$_2$ is carried by the prevailing winds to Hawaii, where it arrives at the observation site with a delay of several months.

10.3.3 ▪ Residuals

We now look at the residuals that remain after we subtract the long-term trend and the seasonal data,

$$\tilde{R}_i = C_i - F_2(t_i) - P_i .$$

These remaining residuals, shown in the bottom right panel of Figure 10.3, cannot be explained by the quadratic fit and a superimposed seasonal cycle. They appear as a jagged curve—that is, they appear to be essentially random. The changes from month to month are not particularly large, and their magnitude is less than the amplitude of the periodic component. A long-term oscillation with a period of about 30 years is still clearly visible, and there may be other patterns in these residuals.

Fourier analysis, which is the subject of the next chapter, gives us the tools to investigate the possible existence of periodic oscillations in the remaining residuals. Anticipating the discussion of Section 11.5, we mention that the Fourier transform decomposes a signal like the one represented in the bottom right panel of Figure 10.3 into its constituent components. Each component has a characteristic frequency, and its amplitude is a measure of the strength or *power* of the component. The resulting graph of power vs. frequency is the *power spectrum* of the signal.

The top plot in Figure 10.4 shows the power spectrum of the remaining residuals of the Mauna Loa data. The frequency is plotted on the horizontal axis as a multiple k of the base frequency $2\pi/656$ ($k = 1,\ldots,106$). (There are $N = 656$ points in the data set.) The strength of each component is plotted on the vertical axis. The highest frequency in this plot corresponds to a period of about half a year. There is a large peak at $k = 2$, corresponding to a period of about 25 years. This is the long-term oscillatory pattern observed in the residual plot of Figure 10.3. Its relevance is not obvious, and it may be an artifact. There are smaller peaks at $k = 6$, corresponding to a period of $656/6$ months or about 9 years and at $k = 15$, corresponding to a period of $656/15$ months or about 3.6 years. Their meaning is unclear, but there is an intriguing similarity to the typical

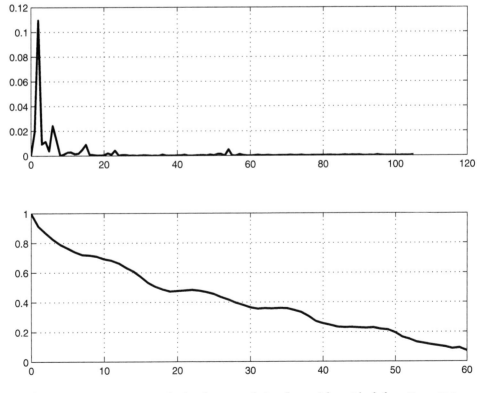

Figure 10.4. *Power spectrum (top) and autocorrelation (bottom) for residuals from Figure 10.3.*

period of 3–7 years of the ENSO. Finally, there is another small component at $k = 55$ which corresponds to a period of one year.

The bottom plot in Figure 10.4 shows the autocorrelation of the remaining residuals for lags up to 60 months or 5 years. The autocorrelation is quite large, even for sizable lags, and the autocorrelations for lags that are whole years are somewhat larger, resulting in a wavy pattern.

10.4 ▪ Exercises

1. Some people are skeptical about the Keeling curve, since its data were recorded on a volcano. They think that volcanic CO_2 emissions may have distorted the data. Comment on this criticism in light of the regression analysis for the South Pole CO_2 data in Exercise 2 of Chapter 9. Hint: The only active volcano in Antarctica is Mount Erebus, which is about 1,300 km from the South Pole.

2. Most CO_2 observation efforts record both *in situ* measurements (concentrations are measured at the time the samples are taken) and *flask* measurements (samples are stored and concentrations are measured at a later time). Suggest a possible reason for this procedure and then research the rationale, using available publications on the Internet.

3. This problem uses monthly averages of CO_2 data taken at the La Jolla pier in California since January 1969. The data can be found on the Web [101]. Use the monthly flask CO_2 data set and the monthly measured CO_2 data within that set.

   (i) Plot the data against time and fit a quadratic function to the data. Be sure to treat the missing data properly.

   (ii) Compare the coefficients of the quadratic fit to those for the Mauna Loa data. Suggest possible reasons for any discrepancies that you observe.

   (iii) Find the average monthly correction to the overall trend which you found in part (i). *Note that there are many years without complete records. Find a way to handle this problem!* Compare this monthly correction to the corresponding correction for the Mauna Loa data (amplitude, location of the maxima and minima).

   (iv) Compute and plot the remaining residuals—that is, "data minus quadratic fit minus average monthly correction." Compare your plot with Figure 10.3 (bottom right). Are the magnitudes comparable? Is the overall appearance similar?

Chapter 11

Fourier Transforms

This chapter is devoted to the Fourier transform and its applications. Starting with the basic notion of trigonometric interpolation, we introduce the discrete Fourier transform (DFT) and its numerical implementation in the fast Fourier transform (FFT). The FFT is the instrument of choice to compute power spectra and correlation coefficients from given time series and signals. We illustrate its use by analyzing paleoclimate data in the context of the Milankovitch theory of glacial cycles.

Keywords: Time series, periodic phenomena, Fourier series, discrete Fourier transform (DFT), fast Fourier transform (FFT), power spectrum, covariance, (auto)correlation, Fourier integral, Milankovitch theory.

11.1 ▪ Fourier Analysis

The Fourier transform is one of the most powerful tools for extracting *information* from a *time series* or *signal*—a sequence of data (a record) collected by sampling a variable at more or less regular intervals over a certain length of time. If the signal represents a periodic phenomenon, or if it is the superposition of several periodic phenomena with different periods, Fourier analysis enables us to find the periods and relative strengths of the individual components. This data analysis problem—extracting information from a signal—is common in climate science, as well as in other sciences, and is especially hard when the signal has been corrupted, for example with noise (additive or multiplicative random contributions), jitter (random variations in the observations), or components that are clearly not periodic such as trends or jumps.

The basic ideas of Fourier analysis go back to the early 19th century, to the investigations of heat flow problems by the French mathematician JEAN BAPTISTE JOSEPH FOURIER (1768–1830). The ideas were first presented to the French "Académie des Sciences" in 1811 and subsequently published (in part) in Fourier's celebrated paper *Théorie analytique de chaleur*, which appeared in 1822 [23, 24]. Fourier's ideas turned out to be fundamental; it is fair to say that they have made their way into every branch of mathematics and found applications in every scientific discipline.

Figure 11.1 sets the tone for the discussion. The graphs represent four signals. The signal in the top left panel is a sum of five sines and cosines with different frequencies (all integer multiples of $2\pi/128$), amplitudes, and phases, sampled at 128 points labeled $0, 1, \ldots, 127$, plus a constant term; the signal in the top right panel is the same signal per-

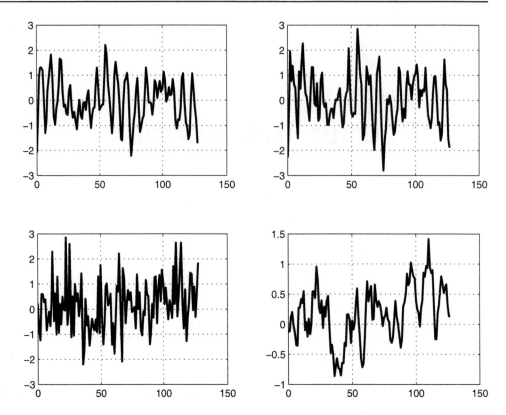

Figure 11.1. *Four different time signals: a superposition of five trigonometric functions (top left), the same five trigonometric functions plus noise (top right), pure noise (bottom left), and pure noise averaged (bottom right).*

turbed with random noise at every point. The signal in the bottom left panel is pure random noise sampled at 128 points; the signal in the bottom right panel is the same random noise signal averaged over every four consecutive observation points. It is not easy to see with the naked eye what is going on, let alone to make precise statements about the frequencies, amplitudes, noise levels, correlation lengths, etc. The data analysis techniques to be presented in this chapter are capable of doing all this at remarkably little computational expense.

11.2 ▪ Trigonometric Interpolation

Suppose we are given a sequence $\{x_j : j = 0, \ldots, N-1\}$ of N data which have been obtained by sampling a function $x : t \mapsto x(t)$ at times $t = 0, 1, \ldots, N-1$. The function x is unknown, but the hypothesis is that it is a superposition of sines and cosines with periods N/k, $k = 1, 2, \ldots, N-1$. To verify this hypothesis, we need to solve a *trigonometric interpolation* problem, writing x as a linear combination of trigonometric functions and determining the coefficients such that the value of x matches the data x_j at $t = j$ for each $j = 0, 1, \ldots, N-1$.

It is actually more convenient to formulate the trigonometric interpolation problem in terms of complex exponentials, using Euler's formula, $e^{it} = \cos t + i \sin t$, and writing $x : t \mapsto x(t) = \sum_{k=0}^{N-1} c_k e^{ik\omega_N t}$, where $\omega_N = 2\pi/N$. Evaluating this expression at $t =$

sum of sines/cos
w./ period N/k

$0, 1, \ldots, N-1$, we obtain a system of N equations for the unknown coefficients c_k,

$$X(j) = \quad x_j = \sum_{k=0}^{N-1} c_k e^{i(jk\omega_N)}, \quad j = 0, \ldots, N-1. \tag{11.1}$$

The coefficients c_k are known as the *Fourier coefficients* of x. Since $e^{i(jk\omega_N)} = (e^{i(j\omega_N)})^k$, we also have the identity $x_j = p(z_j)$ for $j = 0, 1, \ldots, N-1$, where $p : z \mapsto p(z) = \sum_{k=0}^{N-1} c_k z^k$ and $z_j = e^{ij\omega_N}$ and thus $|z_j| = 1$. Thus, the problem can also be viewed as a polynomial interpolation problem on the unit circle in the complex plane.

11.3 ▪ Discrete Fourier Transform

If we consider the data $x_0, \ldots, x_{N-1}$ and the Fourier coefficients $c_0, \ldots, c_{N-1}$ as the entries of column vectors x and c, respectively, we can summarize the system (11.1) in matrix form,[4]

$$F_N = \begin{bmatrix} -x_0- \\ \vdots \\ -x_{N-1}- \end{bmatrix} \qquad x = F_N c, \tag{11.2}$$

$$\uparrow N-1 \times N-1$$

where

$$F_N = \left((z_N)^{jk} \right)_{j,k=0}^{N-1}, \quad z_N = e^{i\omega_N} = e^{2\pi i/N}. \tag{11.3}$$

The matrix F_N is called a *Fourier matrix*. The points z_N^j $(j = 0, \ldots, N-1)$ lie at the vertices of a regular N-gon inscribed in the unit circle in the complex plane ($|z_N^j| = 1$), and $z_N^N = 1$.

The matrix F_N is symmetric, $F_N = F_N^T$, but not Hermitian, $F_N^* \neq \overline{F}_N^T$. Complex conjugation maps $z_N = e^{i\omega_N}$ to $\overline{z}_N = e^{-i\omega_N} = (z_N)^{-1}$, so the conjugate transpose F_N^* is obtained from F_N by taking the inverse element by element,

$$F_N^* = \left((z_N)^{-jk} \right)_{j,k=0}^{N-1}. \tag{11.4}$$

It follows that

$$(F_N^* F_N)_{kl} = \sum_{j=0}^{N-1} z_N^{-kj} z_N^{jl} = \sum_{j=0}^{N-1} z_N^{j(l-k)} = \frac{z_N^{(l-k)N} - 1}{z_N^{l-k} - 1} = 0, \quad k \neq l,$$

$$\uparrow \text{B/c } z_N^N = 1$$

$$(F_N^* F_N)_{kk} = \sum_{j=0}^{N-1} z_N^{-kj} z_N^{jk} = \sum_{j=0}^{N-1} 1 = N,$$

and, similarly, $(F_N F_N^*)_{kl} = 0$ if $k \neq l$ and $(F_N F_N^*)_{kk} = N$, so

$$F_N^* F_N = F_N F_N^* = N I_N. \tag{11.5}$$

Hence, the Fourier matrix F_N is invertible; its inverse is $F_N^{-1} = (1/N)F_N^*$. Consequently, x and c are different but equivalent representations of the same signal. They are related by the equations

$$\star \quad x = F_N c, \quad c = \frac{1}{N} F_N^* x. \tag{11.6}$$

Multiplication with the matrix F_N^* is known as the *discrete Fourier transform* (DFT). $\star$ MATLAB computes the complex vector $F_N^* x = \tilde{c}$, with the function `fft` and x from $\tilde{c}$ with the function `ifft`. The computation of the vector of Fourier coefficients then

---

[4]In this chapter we follow the standard notation of Fourier analysis and employ regular typeface for vectors and matrices.

$$C_j = \frac{1}{N}\left(\sum_{k=1}^{N-1} \left((z_N)^{jk} \right) \cdot x_k \right)$$

$$C_0 = \frac{1}{N}\left(\sum_{k=1}^{N-1} (z_N)^{0 \cdot k} \cdot x_k \right) = \frac{1}{N} \sum_{k=1}^{N-1} x_k$$

requires a further scaling, $c = (1/N)\tilde{c}$. In this way, the computation of the vector c from the signal x requires about N^2 multiplications.

The vector x has N real components, while the vector c has N complex components and therefore $2N$ real components; hence, there must be a redundancy in the components of c. Clearly, $c_0 = (1/N)\sum_{j=0}^{N-1} x_j$ is real. Furthermore, a straightforward computation shows that $c_k = \overline{c}_{N-k}$ for $k = 1,\ldots,N-1$, so the components c_k with $k > \frac{1}{2}N$ do not contain any additional information about the signal x. If one rewrites the complex exponentials as linear combinations of sines and cosines, one sees that only terms with $k \leq \frac{1}{2}N$ are needed; details are left to the reader.

There is a beautiful relation between a signal and its DFT. Recall that x^* is the complex conjugate transpose of a vector x.

Theorem 11.1 (Parseval's theorem). *Let x and y be two signals, each with N components, and let $c = (1/N)F_N^* x$ and $d = (1/N)F_N^* y$ be their DFTs (divided by N). Then*

$$x^* y = N c^* d. \tag{11.7}$$

Proof. The theorem follows from a direct computation, using Eq. (11.5). □

11.4 ▪ Fast Fourier Transform

The computation of the DFT requires about N^2 multiplications. We now show how this computation can be done much faster if N is a power of 2.

Assume for the moment that N is even, say $N = 2M$. As before, we write $z_N = e^{i\omega_N}$. To compute $c = F_N x$, we first split the vector x into a component with even indices, $x' = (x_0, x_2, \ldots, x_{N-2})^T$, and a component with odd indices, $x'' = (x_1, x_3, \ldots, x_{N-1})^T$, and compute $c' = F_M x'$ and $c'' = F_M x''$ (both vectors of length M). The vector c is split into halves, one half for "low" frequencies, c_j for $j = 0,\ldots,M-1$, the other for "high" frequencies, c_j for $j = M,\ldots,N-1$. Then the low- and high-frequency components of c can be computed from c' and c'' without any additional matrix operations, according to the *Lanczos–Danielson* formulas,

$$\tilde{c}_j'' = z_N^j c_j''; \quad c_j = c_j' + \tilde{c}_j'', \, c_{j+M} = c_j' - \tilde{c}_j'', \quad j = 0,1,\ldots,M-1. \tag{11.8}$$

These formulas are proved by splitting the sums that define the components c_j into sums of even- and odd-indexed terms and reinterpreting each of them as Fourier transforms. Figure 11.2 shows the data flow for this step. Computing first c' and c'' and then c from Eq. (11.8) requires $2M^2 + M = \frac{1}{2}N^2 + \frac{1}{2}N$ complex multiplications, plus $M = \frac{1}{2}N$ complex multiplications to find the so-called *twiddle factors* z_N^j. If M is also even, the same trick can be repeated to compute each of x' and x'' from two quarter-size components, and if N is a power of 2, $N = 2^n$, say, the process can be continued all the way down to 1×1 Fourier

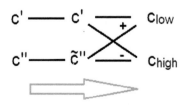

Figure 11.2. *Data flow for a single FFT step, with low- and high-frequency components of c.*

transforms. In that case, the procedure does not require any matrix multiplications at all; $c = F_N x$ can be computed with $2^{n-1} n = \frac{1}{2} N \log_2 N$ complex multiplications.

This method for computing the DFT is known as the *fast Fourier transform* (FFT). It reduces the effort to compute the DFT of a vector of length N from N^2 to $\frac{1}{2} N \log_2 N$ complex multiplications. For example, if $N = 10^6 \approx 2^{20}$, the number of complex multiplications is reduced five orders of magnitude, from 10^{12} to 10^7.

The above results were obtained under the assumption that the signal was sampled at unit time intervals. They are readily generalized to accommodate any fixed positive sampling interval. In that case, write the real-valued function $x : t \mapsto x(t)$ as a linear combination of N complex exponentials with periods that are fractions of T,

$$x(t) \sim \sum_{k=0}^{N-1} c_k e^{2\pi i k t / T}. \tag{11.9}$$

Matching the value $x(t_j) = x(jT/N)$ of x at t_j with a given value x_j for $j = 0, 1, \ldots, N-1$ results in a system that is identical to the system (11.1), so it can be solved with the DFT or FFT as before.

The first known reference to an FFT algorithm can be found in some posthumous work by the German mathematician CARL FRIEDRICH GAUSS (1777–1855), which was published (in Latin) in 1866. Around the turn of the century, a related method reappeared in a book by Runge and König [95], who used sines and cosines instead of complex exponentials. Danielson and Lanczos published the expressions (11.8), together with a fast algorithm, in 1942 [18]; they were aware of the operation count for this method. In spite of these early discoveries and rediscoveries by well-known researchers, the method did not become well known until 1965, when Tukey gave the essence of the algorithm and Cooley wrote the code to make it work [16]. Today, the FFT is probably the most widely used algorithm in scientific computing. Whenever a computational problem can be cast in a form where an FFT can be applied, a substantial breakthrough has been achieved.

11.5 ▪ Power Spectrum

Since the Fourier transform is invertible, the Fourier coefficients contain all the information about the original signal. For a signal x that is T-periodic and has been sampled at times $t_j = jT/N$, a nonzero coefficient c_k indicates the presence of a component with frequency $2\pi k/T$, and the squared amplitude $|c_k|^2$ is a measure of the strength of the component or its *power*. The vector of the $|c_k|^2$ is called the *power spectrum* and is often plotted against the frequencies $2\pi k/T$. The FFT provides a fast way to compute the power spectrum.

The power spectrum is always symmetrical about the frequency $\frac{1}{2} N \omega_N = \pi$. Moreover, Parseval's Theorem 11.1 tells us that $\sum_k |x_k|^2 = N \sum_k |c_k|^2$. A "spike" in the power spectrum at a certain frequency indicates that there is a prominent periodic component with that frequency in the signal.

Figure 11.3 shows the power spectra for the four signals in Figure 11.1 (same ordering) for frequencies between 0 and π ($k = 0, \ldots, \frac{1}{2} N$). The graph in the top left panel clearly shows the five different frequency components. The two strongest components have frequencies $2\pi k/N$ with $k = 15$ and $k = 17$. There are also components with $k = 3, 8, 36$, which are visible even after noise has been added. The power spectrum of a purely random signal is essentially purely random, and averaging over several adjacent observations suppresses higher frequencies in the power spectrum.

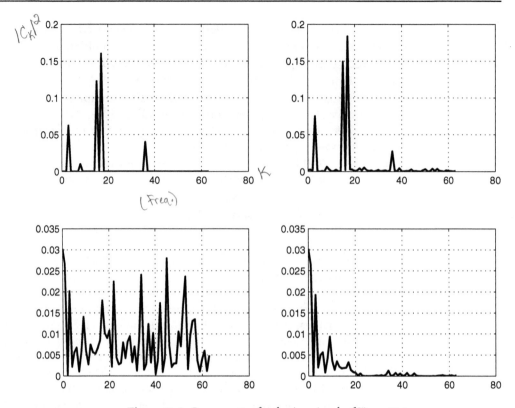

Figure 11.3. *Power spectra for the time signals of Figure* 11.1.

11.6 ▪ Correlation and Autocorrelation

The degree of similarity of two signals is measured by their *covariance* or *correlation*. In the following definition we assume that we have two (nonconstant) signals of length N, $x = \{x_j : j = 0,\dots,N-1\}$ and $y = \{y_j : j = 0,\dots,N-1\}$, and that both x and y have been extended periodically with period N. We use the notation $\overline{x} = (1/N)\sum_{k=0}^{N-1} x_k$ for the mean of x and similarly for y.

Definition 11.1. (i) *The* covariance *of two signals x and y is*

$$\mathrm{cov}(x,y) = \frac{1}{N}\sum_{k=0}^{N-1}(x_k - \overline{x})(y_k - \overline{y}). \tag{11.10}$$

The covariance of a signal with itself is called the variance, $\mathrm{var}(x) = \mathrm{cov}(x,x)$.
(ii) *The* correlation *of two signals x and y is the normalized covariance,*

$$r(x,y) = \frac{\mathrm{cov}(x,y)}{\sqrt{\mathrm{var}(x)\mathrm{var}(y)}}. \tag{11.11}$$

The definition of covariance is slightly different from the definition usually employed in statistics where, instead of $1/N$, the factor $1/(N-1)$ is used in Eq. (11.10). We use the factor $1/N$ for notational simplicity. Note that $\mathrm{cov}(x,y) = (1/N)x^T y - \overline{xy}$ and $\mathrm{var}(x) = (1/N)\|x\|_2^2 - \overline{x}^2$, where $\|\cdot\|_2$ is the Euclidean norm.

It follows from the Cauchy–Schwarz inequality that the correlation is always between -1 and 1. If $r(x,y) = 1$ $(r(x,y) = -1)$, then x and y are related by an affine transformation, $y = \alpha x + \beta$ with $\alpha \geq 0$ $(\alpha \leq 0)$.

The definition of covariance and correlation can be generalized to include a *lag* of length j $(j = \pm 1, \pm 2, \ldots)$,

$$\text{cov}_j(x,y) = \frac{1}{N}\sum_{k=0}^{N-1}(x_{k+j} - \overline{x})(y_k - \overline{y}), \quad r_j(x,y) = \frac{\text{cov}_j(x,y)}{\sqrt{\text{var}(x)\text{var}(y)}}.$$

The correlation with lag j measures the similarity of two signals, one of which has been shifted j units of time. Effectively, if $r_j(x,y)$ is close to 1 for some positive j, then the signal x "echoes" the signal y with a lag of j time units.

For periodically extended signals, covariance and correlation with lag j include $N-j$ products of terms from the two signals that are j time units apart and also j products of terms that are $N-j$ time units apart. For large N and modest j, the effect from including these distant-in-time components is negligible.

The covariance (with or without lag) is symmetric, $\text{cov}_j(x,y) = \text{cov}_j(y,x)$. Furthermore, $\text{cov}_{j+N}(x,y) = \text{cov}_j(x,y)$ for all j. The autocorrelation depends only on the absolute value of the lag, $r_j(x,x) = r_{-j}(x,x)$.

The definitions of covariance, correlation, and autocorrelation in statistics and in signal processing are somewhat different. In particular, in the signal processing community, the terms $\overline{x}$ and $\overline{y}$ in the definition (11.10) of the covariance are usually omitted, and the resulting expression is called *cross correlation*. The definitions (11.10) and (11.11) are close to the corresponding definitions from statistics, with the exception of the factor in Eq. (11.10).

All these quantities can be computed efficiently from the DFTs and the power spectra of the two signals, due to Parseval's Theorem 11.1. Let $c = (1/N)F_N^* x$ and $d = (1/N)F_N^* y$. For $j = -N+1, \ldots, N-1$, let $c_{(j)}$ be the vector with components $c_k z_N^{jk} = c_k e^{i(jk\omega_N)}$ $(k = 0, \ldots, N-1)$ and let $\|c\|_*^2 = \sum_{j=1}^{N-1}|c_j|^2$. Then

$$\text{cov}_j(x,y) = c_{(j)}^* d - c_0 d_0 = \sum_{k=1}^{N-1} e^{-i(jk\omega_N)}\overline{c_k}d_k, \tag{11.12}$$

$$r_j(x,y) = \frac{c_{(j)}^* d - c_0 d_0}{\|c\|_* \|d\|_*}, \tag{11.13}$$

$$r_j(x,x) = \frac{\sum_{k=1}^{N-1} e^{-i(jk\omega_N)}|c_k|^2}{\|c\|_*^2}. \tag{11.14}$$

Formula (11.12) says that the DFT of the vector $q = (0, \overline{c}_1 d_1, \ldots, \overline{c}_{N-1}d_{N-1})^T$ results in the vector of covariances of x and y for lags $j = 0, 1, \ldots, N-1$ (that is, if $g = F_N^* q$, then $g_j = \text{cov}_j(x,y)$ for all j). Furthermore, the DFT of the power spectrum (minus its "dc" component c_0) $p = (0, |c_1|^2, \ldots, |c_{N-1}|^2)^T$ is the vector of covariances of x with itself—that is, if $e = F_N^* p$, then $e_j = \text{cov}_j(x,x)$ for all j.

Note that both e and g are real vectors, although they are DFTs. The reason is that $\overline{q}_k = q_{N-k}$ and $p_k = p_{N-k}$ for $k = 1, \ldots, N-1$. In actual computations, these two DFTs are computed as complex vectors; their imaginary parts are on the order of the machine precision and can be ignored.

Figure 11.4 shows the autocorrelations of the four signals of Figure 11.1 for positive lags up to 20. If the lag is 0, the autocorrelation is always 1. The two graphs in the top

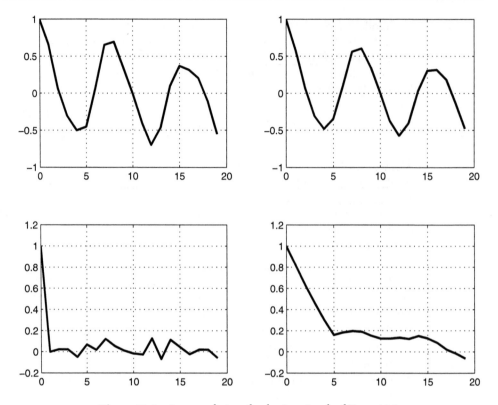

Figure 11.4. *Autocorrelations for the time signals of Figure 11.1.*

row show autocorrelations close to 1 for a lag of 1, which reflects the fact that the signals are smooth functions sampled at equidistant times. The fairly large autocorrelations for lags between 7 and 8 come from the two strong components with frequencies $2\pi k/N$ with $k = 15$ and $k = 17$, which result in periods near 8. After a lag of about 15, the signal again looks like a faint echo of itself. For the purely random signal (bottom left), the autocorrelations are close to zero for all positive lags. In the bottom right graph, they decrease steadily to zero for lags up to 5, reflecting the width of the "window" over which the random signal has been averaged.

11.7 ▪ Fourier Series and Fourier Integrals

It is mathematically interesting and very useful for applications to investigate the limit of the DFT as N goes to infinity. This limit can be taken in various ways. The case where $N \to \infty$ while the sampling interval is kept constant is rarely encountered in applications and will not be considered here.

   More interesting is the case where the data come from sampling a function x which is defined on a fixed interval $[0, T]$, where the length of the sampling interval is T/N. The kth component of the DFT of the signal can be interpreted as a Riemann sum, so we have the approximation

$$c_k = \sum_{j=0}^{N-1} x\left(j\frac{T}{N}\right) e^{-2\pi i(jk)/N} \approx \frac{1}{T}\int_0^T x(t) e^{-2\pi i(k/T)t}\, dt. \qquad (11.15)$$

In the limit $N \to \infty$, the number of coefficients increases indefinitely, and we obtain, at least formally,

$$x(t) = \sum_{k=0}^{\infty} c_k e^{2\pi i(k/T)t}, \quad c_k = \frac{1}{T}\int_0^T x(t)e^{-2\pi i(k/T)t}\,dt. \tag{11.16}$$

The infinite series is called the *Fourier series* of x, and the coefficients $c_0, c_1, \ldots$ are the *Fourier coefficients* of x. The Fourier coefficient c_k is associated with the frequency $2\pi k/T$ and is a factor for functions with periods T/k in Eq. (11.16). The power spectrum values $|c_k|^2$ of the function x can be computed from the c_k for these discrete frequencies. They form a convergent infinite series.

We emphasize that none of the previous arguments justifies the validity of the formulas in Eq. (11.16). Their rigorous justification is the objective of Fourier analysis, for which we refer the reader to the literature; see, for example, [19, 110].

Going back to Eq. (11.15), we consider one more limiting case, where $T, N \to \infty$, and $T/N \to 0$. As in Eq. (11.15), we interpret the DFT of the signal as a Riemann sum, so for large N we obtain, at least formally, Eq. (11.15). Now, assume that the Fourier coefficients, which depend on T, satisfy the identity $c_k = (1/T)\hat{x}(k/T)$ for some function $\hat{x}$, at least approximately for large T. The factors $1/T$ on both side of the resulting approximate identity cancel, and Eq. (11.15) yields an expression for $\hat{x}(s)$ at $s = k/T$ in terms of x, namely $\hat{x}(s) \approx \int_0^T x(t)e^{-2\pi i t s}\,dt$ for $s = k/T$. In the limit as $T \to \infty$ we obtain the relation

$$\hat{x}(s) = \int_0^\infty x(t)e^{-2\pi i t s}\,dt. \tag{11.17}$$

On the other hand, $x(t) \approx (1/T)\sum_{k=0}^{N-1}\hat{x}(k/T)e^{2\pi i k t/N}$ for $t = j/T$, and in the limit as $N \to \infty$ we obtain

$$x(t) = \int_0^\infty \hat{x}(s)e^{2\pi i t s}\,ds. \tag{11.18}$$

The integrals in Eqs. (11.17) and (11.18) can be extended if x is extended trivially by assigning the value $x(t) = 0$ for all $t < 0$. In that case, we obtain the formulas

$$x(t) = \int_{-\infty}^\infty \hat{x}(s)e^{2\pi i t s}\,ds, \quad \hat{x}(s) = \int_{-\infty}^\infty x(t)e^{-2\pi i s t}\,dt. \tag{11.19}$$

The map $F : x \mapsto \hat{x}$ is called the *Fourier transform*.

In conclusion, the theory of Fourier transforms is one of the most powerful tools in mathematics. Its applications go far beyond signal processing and data analysis. For example, the celebrated Uncertainty Principle in quantum physics can be stated as a relation between functions and their Fourier transforms, and Nobel-prize winning work in chemistry has been enabled by the application of Fourier transforms in crystallography.

11.8 • Milankovitch's Theory of Glacial Cycles

In 1941, the Serbian mathematician MILUTIN MILANKOVITCH (sometimes spelled MILANKOVIĆ or MILANKOVICH) (1879–1958) suggested that past glacial cycles might be correlated to cyclical changes in the insolation—the amount of solar energy that reaches Earth from the Sun [68]. This theory is known as the *Milankovitch theory* of glacial cycles and is an integral part of *paleoclimatology*—the study of prehistoric climates.

Changes in insolation are due to variations in the Earth's orbit and can be computed accurately by integrating the equations of celestial mechanics. Such integrations have been done with increasing precision, most recently by Laskar and collaborators [59, 57], who used a symplectic integration scheme to integrate the equations of motion of the full solar system comprising all nine planets including Pluto. According to the authors, the numerical solution is good for paleoclimate studies for periods up to 50 million years, and even over longer periods if only the most stable features of the solution are used. The solution is available on the Web [56], together with a set of routines for the computation of the insolation quantities as described in [58]. Our goal is to compare the results of these computations to observed paleoclimate records using Fourier transform techniques.

11.8.1 • Orbital Parameters

Milankovitch argued that three parameters are of particular importance for the characterization of the Earth's orbit around the Sun: (i) the *eccentricity* of the elliptical orbit; (ii) the *obliquity*—that is, the tilt of the Earth's equatorial plane relative to its orbital plane; and (iii) the *precession* of the Earth's spin axis around the normal to the orbital plane (Figure 11.5). Each parameter causes the insolation to vary cyclically with time, so a Fourier analysis should be able to separate out the various contributions.

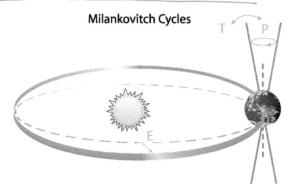

Figure 11.5. *Schematic of the Earth's orbital parameters that drive the variation of the insolation: eccentricity E, tilt (obliquity) T, precession P. Reprinted with permission from IPCC.*

(i) Eccentricity. Figure 11.6 shows a time series of the eccentricity of the Earth's orbit for the past 4.5 Myr in the top plot and the power spectrum of the eccentricity anomaly (deviation from the average) in the bottom plot. The eccentricity, whose current value is about 0.017, has varied from almost zero (a nearly circular orbit) to about 0.06. The power spectrum shows a periodic component with a period of about 400 Kyr and two additional periodic components, a larger one with about a 95 Kyr cycle and a smaller 125 Kyr cycle, which together average out to a period of about 100 Kyr.

(ii) Obliquity. Changes in the obliquity have an important effect on the Earth's climate, because they affect the intensity of the solar input near the poles. Higher values of the obliquity result in more solar energy hitting the North Pole during the northern summer and the South Pole during the northern winter.

Figure 11.7 shows the time series of the obliquity of the Earth's rotation axis for the past 4.5 Myr. The current value is about 23.5°, but the obliquity has varied between about 22° and 24.5°. The time series (top plot) shows a dominant frequency with a modulating amplitude. The power spectrum of the anomaly (bottom plot) shows that the dominant frequency has a period of about 41 Kyr.

(iii) Precession. The axis of the Earth's rotation changes due to small differences in the gravitational force on the Earth's equatorial bulge. As with a spinning top, the axis

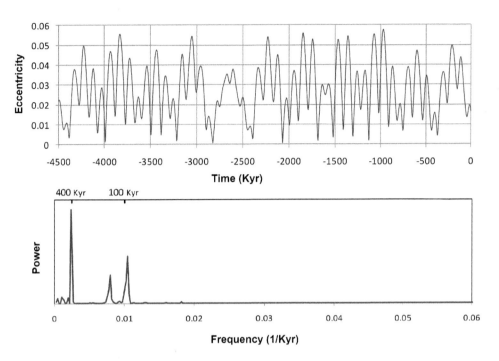

Figure 11.6. *Time series (top) and power spectrum (bottom) of the Earth's eccentricity over the past* 4.5 *Myr* [65].

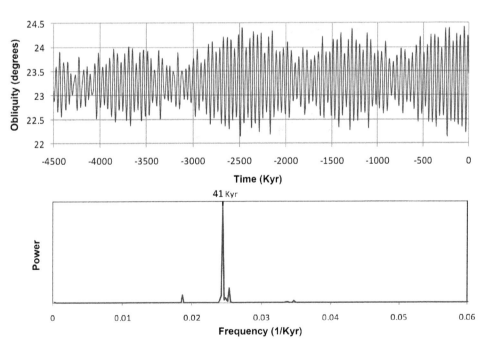

Figure 11.7. *Time series (top) and power spectrum (bottom) of the Earth's obliquity over the past* 4.5 *Myr* [65].

precesses about the vector perpendicular to the orbital plane. The appropriate precession variable for climate is the longitudinal angle between the rotation vector and the major axis of the Earth's elliptical orbit. This angle determines where the seasons occur along the ellipse and affects the relative insolation during winter and summer months.

Figure 11.8 shows the time series of the *precession index*, which is the product of the eccentricity and the sine of the precession angle, for the past 4.5 Myr. The amplitude modulation is the effect of multiplying by the eccentricity. The underlying precession cycle has a period of about 23 Kyr, although it is actually composed of three dominant periods.

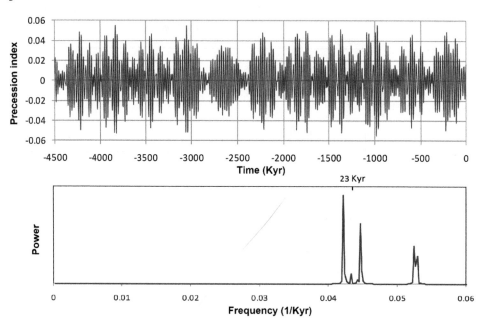

Figure 11.8. *Time series (top) and power spectrum (bottom) of the Earth's precession index over the past 4.5 Myr* [65].

11.8.2 • Daily Insolation at 65°N

The three parameters—eccentricity, obliquity, precession—together affect the Earth's weather and climate. Traditionally, their effects are combined into one variable, the *average daily insolation* at 65° North latitude at the summer solstice, denoted by Q^{65}. This quantity is a function of the three Milankovitch variables and hence varies with time. Its time series and power spectrum for the last 4.5 Myr are shown in Figure 11.9. There is a dominant frequency around 23 Kyr, coming from three closely clustered spikes in the power spectrum, and another frequency component around 41 Kyr.

11.8.3 • Climate Data

The theoretical results obtained for the Milankovitch cycles can be tested against temperature data from the paleoclimate record. In the 1970s, Hayes et al. used data from ocean sediment core samples to relate the Milankovitch cycles to the climate of the last 468,000 years [38]. One of their conclusions was that "… climatic variance of these records is concentrated in three discrete spectral peaks at periods of 23,000, 42,000, and approximately

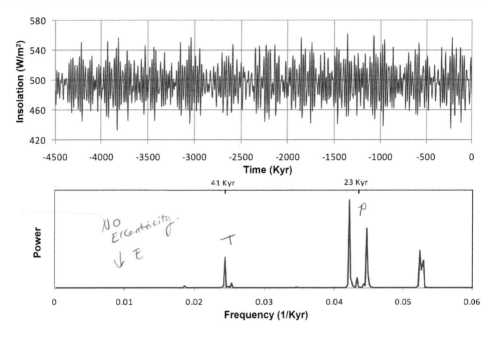

Figure 11.9. *Time series (top) and power spectrum (bottom) of the average daily insolation at 65° North at summer solstice (Q⁶⁵) [65].*

Q⁶⁵ orbital parameter var.

100,000 years." This study was repeated recently by Zachos et al., who used much more extensive data [123]. The reconstructed temperature profile and the corresponding power spectrum are shown in Figure 11.10. Again, we see periods of 100, 41, and 23 Kyr.

The best data we have for temperatures during the ice age cycles come from the analysis of isotope ratios of air trapped in pockets in the polar ice. The ratio $\delta^{18}O = {}^{18}O/{}^{16}O$ of oxygen isotopes is a good proxy for global mean temperature; higher ^{18}O concentrations correspond to colder climates. In the 1990s, Petit et al. studied data from the Vostok ice core to reconstruct a temperature profile for the past 420 Kyr [83]. Although the record is only half a million years long, it allows for fairly precise dating from the progressive layering process that laid down the ice. A spectral analysis shows cycles with periods of 100, 43, 24, and 19 Kyr, in reasonable agreement with previous findings and with the calculated periods of the Milankovitch cycles.

oxygen isotope

$\delta^{18}O = $ *Temp Proxy*

(Global mean temp)

11.8.4 ▪ Critique

At this point we might conclude that Milankovitch's idea was correct and that ice ages are indeed correlated with orbital variations. There are, however, some serious caveats. Changes in the oxygen isotope ratio reflect the combined effect of changes in global ice volume and temperature at the time of deposition of the material, and the two effects cannot be separated easily. Furthermore, the cycles do not change the total energy received by the Earth if this is averaged over the course of a year. An increase in eccentricity, or obliquity, means the insolation is larger during part of the year and smaller during the rest of the year, with very little net effect on the total energy received at any latitude over a year. However, a change in eccentricity or obliquity could make the seasonal cycle more severe and thus change the extent of the ice caps. A possible scenario for the onset of ice ages would then be that minima in high-latitude insolation during the summer en-

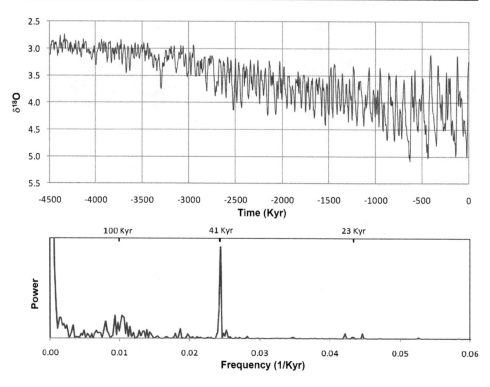

Figure 11.10. *Time series (top) and power spectrum (bottom) of the Earth's climate record for the past 4.5 Myr* [65].

able winter snowfall to persist throughout the year and thus accumulate to build glacial ice sheets. Similarly, times with especially intense high-latitude summer insolation could trigger a deglaciation. Clearly, additional detailed modeling would be needed to account for these effects.

Even allowing for the scenario described in the previous paragraph, we would expect an asymmetric climate change, where the ice cap over one of the poles increases while the cap over the other decreases. Yet, the entire globe cooled during the ice ages and warmed during periods of deglaciation.

An even more disturbing observation arises when one considers not just the periods of the cycles but also their relative strengths. Note in Figure 11.10 the dominance of the frequency corresponding to a period of 41 Kyr, and note the resemblance to the spectrum of the obliquity shown in Figure 11.7. Note also the suite of frequencies associated with periods around 100 Kyr, corresponding to the spectrum associated with eccentricity shown in Figure 11.6. We also see a small suite of frequencies around 23 Kyr, presumably a small contribution from precession. The substantial power at very low frequencies is probably due to the long-term trends seen in the data, corresponding to factors other than insolation. The bottom line is that, in the data, the relative contributions are ordered as obliquity, followed by eccentricity, followed by precession, while, for Q^{65}, the order is the reverse: precession, followed by obliquity, followed by eccentricity. The dominance of precession in the forcing term Q^{65} does not show up in the data (the "100,000-year problem"). Eccentricity

The lesson learned here is that actual climate dynamics are very complex, involving much more than insolation and certainly much more than insolation distilled down to a single quantity, Q^{65}. Feedback mechanisms are at work that are hard to model or explain. On the other hand, an analysis of the existing signals shows that astronomical factors most likely play a role in the Earth's long-term climate.

11.9 ▪ Exercises

Recall that, in this chapter, N-column vectors are written as $x = (x_0, x_1, \ldots, x_{N-1})^T$. The $N \times N$ Fourier matrix is denoted by F_N and is defined in Eq. (11.3).

1. Find all entries of F_N^2.

2. Let r_j be the autocorrelation with lag j ($j = 0, 1, \ldots$) of a signal of length N, where N is large.

   (i) Is it possible that $r_1 = 1$ and $r_2 \approx 0$? Explain your answer. If your answer is "yes," show how an example could be constructed.

   (ii) Is it possible that $r_1 \approx 0$ and $r_2 = 1$? Explain your answer. If your answer is "yes," show how an example could be constructed.

   (iii) If $r_1 = -0.9$, what do you expect r_2 to be? Explain your answer.

3. Let $x = (x_0, \ldots, x_{N-1})^T$ be a column vector, and let $c = F_N x$ be its DFT. Also assume that $N = rM$ for some integers r and M with $1 < r < N$. Find the DFT of the following vectors in terms of c:

   (i) $u = (u_0, u_1, \ldots, u_{N-1})^T = (x_{N-1}, x_{N-2}, \ldots, x_0)^T$ (time reversal);

   (ii) $v = (v_0, v_1, \ldots, v_{N-1})^T = (x_k, x_{k+1}, \ldots, x_{N-1}, x_0, \ldots, x_{k-1})^T$ for any $k \in \{1, \ldots, N-1\}$ (shift);

   (iii) $w = (w_0, w_1, \ldots, w_{M-1})^T = (x_0, x_{r-1}, x_{2r-1}, \ldots, x_{N-1})^T$, where $N = rM$ (down-sampling).

4. Let $x_0 \in (0, 1)$, and let $N > 1$ be an integer. Define the remaining components of the N-vector x recursively by the relation $x_i = 4x_{i-1}(1 - x_{i-1})$ for $i = 1, \ldots, N-1$. Clearly, there is an association between each x_i and x_{i+k} for any $k \in \{1, \ldots, N-1\}$, since x_{i+k} can be computed from x_i. Generate such a sequence with MATLAB using a randomly chosen x_0 from a uniform distribution over $(0, 1)$, and compute its power spectrum and autocorrelation.

   (i) Does the power spectrum indicate any periodic components of this sequence?

   (ii) How much of the association between the original sequence $\{x_i : i = 0, \ldots, N-1\}$ and the shifted sequence $\{x_{i+k} : i = 0, \ldots, N-1\}$ is detected by the autocorrelation function?

   (iii) What does this mean for detecting a nonlinear association between a signal and its shifted version, using the Fourier transform?

5. For this exercise you will need the eccentricity data for the Earth's orbit that were computed by Laskar et al. [57]. The data are available at [56]. Use the file `La2010a_ecc3L.dat`, which contains eccentricities from 250 Myr before the present until now, in steps of 1 Kyr.

   (i) Download the data. Use a computer to plot the eccentricity for several intervals of length 2 Myr, starting at various times in the past.

(ii) Compute the power spectrum of the eccentricity from 2 Myr before the present until now. Check that the power spectrum is symmetric. Repeat this computation after subtracting the mean of the eccentricities. Explain in detail how the two power spectra are related.

From now on, always remove the mean of the signal before computing the power spectrum.

(iii) We wish to examine the power spectrum for periodic components with periods between 20 Kyr and 500 Kyr. Plot a suitable segment of the power spectrum and identify the three most prominent components. Compare their periods and magnitudes to those in Figure 11.6.

6. Continue with the eccentricity data from [56], which you downloaded for the previous exercise. The goal of this exercise is to examine how the power spectrum of the eccentricities changes over time.

Remember: Remove the mean of the signal before computing the power spectrum.

(i) Graph the power spectrum of the eccentricity for the intervals $[-10, -8]$, $[-8, -6]$, $[-6, -4]$, $[-4, -2]$, and $[-2, 0]$ Myr before the present in the same plot.

(ii) Which features in the power spectra remain the same? Which features change?

(iii) Are the power spectra for time intervals of length 2 Myr different from the "deep" past, say 100 Myr or 200 Myr before the present? Explore this question graphically and summarize your observations.

7. It is shown in [65] that the average annual insolation at latitude θ equals $s_\beta(\theta)Q$, where $Q = \frac{1}{4}S_0$ (S_0 is the solar constant) and $s_\beta(\theta)$ is given by an *elliptic integral*,

$$s_\beta(\theta) = \frac{2}{\pi^2} \int_0^{2\pi} \sqrt{1 - (\cos\theta \sin\beta \cos\gamma - \sin\theta \cos\beta)^2} \, d\gamma, \quad \theta \in (-\tfrac{1}{2}\pi, \tfrac{1}{2}\pi).$$

(11.20)

Here, β is the obliquity of the Earth's axis of rotation. The Earth's orbit is assumed to be circular (zero eccentricity).

(i) Evaluate this integral in closed form for $\beta = 0$ and for $\theta = \frac{1}{2}\pi$. Interpret the results.

(ii) The function s_β is even in θ. Verify this analytically.

(iii) Let $y = \sin\theta \in (-1, 1)$. Use a computer to evaluate and plot s_β as a function of y for the current value of Earth's obliquity, $\beta = 23.4°$, on the interval $(-1, 1)$.

(iv) Use a least-squares approach to approximate s_β as a function of $y = \sin\theta$ with a quadratic function $y \mapsto 1 + s_2 P_2(y)$, where P_2 is the Legendre polynomial of degree 2, $P_2(y) = \frac{1}{2}(3y^2 - 1)$. How does your value of s_2 compare with the value $s_2 = -0.477$ suggested by North [77] (cf. Eq. (12.35))?

(v) Use a least-squares approach to approximate s_β as a function of $y = \sin\theta$ with a function $y \mapsto 1 + s_2 P_2(y) + s_4 P_4(y)$, where P_2 is as above and P_4 is the Legendre polynomial of degree 4, $P_4(y) = \frac{1}{8}(35y^4 - 30y^2 + 3)$. Compare the maximum error of the approximations that use Legendre polynomials of degrees 2 and 4.

8. Continue with the function s_β that is defined by the integral in Eq. (11.20). Use a computer to set up a way to plot s_β for various values of β. Then identify values of β for which

   (i) this function is decreasing on $(0, \frac{1}{2}\pi)$,

   (ii) this function is increasing on $(0, \frac{1}{2}\pi)$, and

   (iii) there is a $\theta_0 \in (0, \frac{1}{2}\pi)$ such that s_β decreases on $(0, \theta_0)$ and increases on $(\theta_0, \frac{1}{2}\pi)$.

   What would be the climate on planets that have these obliquities?

Chapter 12

Zonal Energy Budget

In this chapter we develop an energy balance model (EBM) for the temperature averaged over circular zones along lines of constant latitude, with diffusion-driven transport across latitudes. Using a spectral method involving Legendre polynomials, we show how the model gives rise to an infinite-dimensional dynamical system. A few dimensions suffice to obtain a temperature profile that matches the current state of the Earth's climate system.

Keywords: Energy balance, zonal average, ice line, Legendre polynomials, eigenvalue problem, diagonalization, spectral method, orthogonality, normalization, completeness, equilibrium temperature profile.

12.1 ▪ Zonal Energy Balance Model

Global energy balance models (EBMs) of the type discussed in Chapter 2 cannot account for the fact that lower latitudes receive on average more energy from the Sun than higher latitudes, nor for the fact that the polar regions may be covered with snow and ice and therefore have a higher albedo than the equatorial region. To account for these latitudinal variations, we must consider EBMs for circular zones of constant latitude around the globe, so-called *zonal* EBMs. Because zonal EBMs involve one spatial dimension, they are also referred to as "one-dimensional" EBMs. Zonal EBMs were introduced in the 1960s independently by Budyko [10] and Sellers [102] and have been studied extensively by North and coworkers in the 1970s [77, 78].

The model we have in mind is again that of a spherical climate system, ignoring differences in the atmosphere's composition, differences among continents and oceans, topography, and all other local features. The system varies with latitude but is uniform along lines of constant latitude. The polar regions, on both the Northern and Southern Hemispheres, may be covered with snow and ice and have a higher cloud cover and thus a greater albedo than the equatorial region, which is free of snow and ice and covered with land and open water. An *ice line* separates the polar ice caps from the lower ice-free latitudes. On each hemisphere, the ice line coincides with a circle of constant latitude. The state of the climate system is described by the *zonal mean surface temperature*—that is, the surface temperature averaged over longitude. At any time t, this is a function of latitude,
$$T(t,\cdot): \theta \mapsto T(t,\theta) \text{ for } \theta \in (-\tfrac{1}{2}\pi, \tfrac{1}{2}\pi).$$

Latitude

To derive an energy balance equation, we consider a zonal strip which encircles the globe between latitudes θ and $\theta+\Delta\theta$; see Figure 12.1. The energy content of a column of ocean/atmosphere or soil/atmosphere above the strip changes over time due to incoming solar radiation and outgoing longwave radiation, as before. But thermal energy is also being redistributed and transported across latitudes. The energy balance in the zonal strip therefore takes the form

$$\frac{\partial}{\partial t} \int_{\theta}^{\theta+\Delta\theta} C(\eta) T(t,\eta) \cos\eta \, d\eta = \int_{\theta}^{\theta+\Delta\theta} (E_{\text{in}} - E_{\text{out}}) \cos\eta \, d\eta - \Phi. \qquad (12.1)$$

Here, Φ represents the net loss of energy per unit length per unit time due to the transfer of thermal energy. The other symbols have the same meaning as before: C is the heat capacity (W yr m^{-2} deg^{-1}), and E_{in} and E_{out} are the average amounts of solar energy flowing into and out of a unit area of the Earth's surface per unit time. The weights ($\cos\eta$) reflect the fact that each latitude contributes in proportion to its circumference. The circumference is proportional to the radius which, in turn, is proportional to the cosine of the latitude. Leaving out common factors, we therefore assign the weight $\cos\eta$ to the contribution from the circle at latitude η.

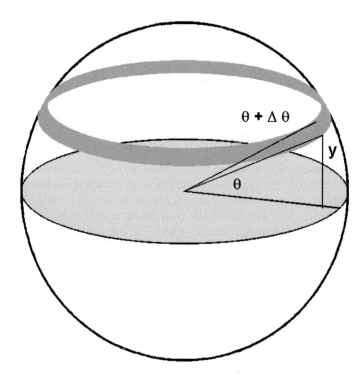

Figure 12.1. *Circular strip between latitudes θ and $\theta+\Delta\theta$.*

Eq. (12.1) is a generic zonal EBM. Details depend on the specific forms of E_{in}, E_{out}, and Φ. Although the heat capacity most likely varies with latitude (as well as with the medium under consideration), we assume henceforth that it is constant and equal to its global average value C everywhere.

12.1.1 ▪ Incoming and Outgoing Radiation

To find an expression for E_{in}, we follow the same arguments as in the global case, Eq. (2.3). Recall that the amount of solar energy intercepted by the Earth per unit area per unit time is $Q = \frac{1}{4}S_0$, where S_0 is the solar constant (Wm^{-2}).

On average, the equatorial region receives more energy than the polar regions. To account for this nonuniformity, we introduce a latitudinal energy distribution function s: $\theta \mapsto s(\theta)$. An explicit expression for $s(\theta)$ in terms of the orbital parameters can be found from celestial mechanics. The precession can be ignored, since its effect averaged over the yearly cycle is zero. The Earth's orbit around the Sun is almost circular, so the eccentricity is small. If we assume that it is zero, then $s(\theta)$ depends only on the obliquity—that is, the angle between the Earth's equatorial plane and the plane of its orbital motion around the Sun. The dependence is *parametrically*, in the sense that if the obliquity changes, then $s(\theta)$ changes at every latitude θ. To indicate this dependence, we temporarily include β as a subscript, writing $s_\beta(\theta)$ rather than $s(\theta)$. It is shown in [65] that s_β is given by an *elliptic integral*,

$$s_\beta(\theta) = \frac{2}{\pi^2} \int_0^{2\pi} \sqrt{1 - (\cos\theta \sin\beta \cos\phi - \sin\theta \cos\beta)^2}\, d\phi, \quad \theta \in (-\tfrac{1}{2}\pi, \tfrac{1}{2}\pi). \quad (12.2)$$

Note that s is symmetric about the equator, $s_\beta(\theta) = s_\beta(-\theta)$ for all $\theta \in (-\tfrac{1}{2}\pi, \tfrac{1}{2}\pi)$. The distribution satisfies the normalization condition

$$\frac{1}{2} \int_{\pi/2}^{\pi/2} s_\beta(\theta) \cos\theta\, d\theta = 1. \quad (12.3)$$

In the limiting case of zero obliquity, s_0 is monotonically decreasing to zero at the poles. For any nonzero value of β, s_β is positive at the poles, indicating that the poles receive some of the incoming solar radiation. Currently, the Earth's obliquity is $\beta = 23.4°$. According to Eq. (12.2), s_β with $\beta = 23.4°$ decreases monotonically from a maximum value of approximately 1.221 at the equator to a minimum value of approximately 0.505 at the poles. A good approximation for $\beta = 23.4°$ is

$$\tilde{s}(\theta) = 0.523 + 0.716 \cos^2\theta, \quad \theta \in (-\tfrac{1}{2}\pi, \tfrac{1}{2}\pi). \quad (12.4)$$

Figure 12.2 shows the graphs of s_β for $\beta = 23.4°$ and its approximation $\tilde{s}$ on the interval $(0, \tfrac{1}{2}\pi)$.

Since snow and ice reflect more energy back into the stratosphere and thus have a higher albedo than land and open water, we allow for a latitude-dependent albedo. The model we have in mind is that of a piecewise continuous function, with jump discontinuities at the ice lines. This means that if the location of an ice line changes, then the functional form of the albedo and therefore the value of the albedo at every latitude change. In other words, the albedo depends *parametrically* on the location of the ice lines. To indicate this dependence, we introduce θ_N and θ_S, the latitude of the ice line on the Northern and Southern Hemispheres, respectively, and let θ_c denote the pair (θ_N, θ_S). Then the albedo α is a function of θ which depends parametrically on θ_c; to indicate this dependence, we include θ_c as a second argument.

Putting it all together we see that the generalization of Eq. (2.3) to the zonal case is

$$E_{in} = (1 - \alpha(\theta, \theta_c))s_\beta(\theta)Q, \quad Q = \tfrac{1}{4}S_0. \quad (12.5)$$

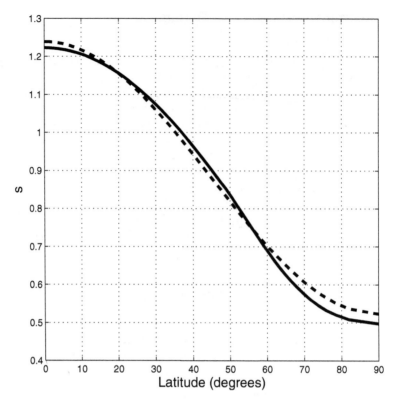

Figure 12.2. *Graphs of the latitudinal energy distribution function s_β for $\beta = 23.4°$ (Eq. (12.2), solid line) and the approximation $\tilde{s}$ (Eq. (12.4), dashed line) on the interval $(0, \frac{1}{2}\pi)$.*

We adopt Budyko's linear model, Eq. (2.12), for the outgoing energy E_{out},

T = Global mean
    surface Temp.

$$E_{\mathrm{out}} = A + BT, \tag{12.6}$$

where A and B are constants. (Recall that, in the context of Budyko's model, temperatures are defined in degrees C.)

12.1.2 ▪ Latitudinal Energy Transfer

The explicit form of Φ depends on the processes involved in the energy exchange among latitudes. Physically, these processes are driven by subscale phenomena like seasonally varying wind patterns and ocean gyres. Following North and coworkers [77, 78], we model these phenomena as a *diffusion process*, where thermal energy is transported like a passive scalar from warm to cool areas, with a flux density that is proportional to the negative local temperature gradient. The proportionality constant is the *thermal diffusion coefficient*, D, which measures the intensity of the energy exchange mechanism (strengths of wind patterns and ocean gyres). The dimension of D is $\mathrm{Wm^{-2}deg^{-1}}$. To keep the model as simple as possible, we take D to be constant, equal to its global average value.

The zonal strip between latitudes θ and $\theta + \Delta\theta$ gains energy due to a flux into the strip at the lower latitude θ and loses energy due to a flux out of the strip at the higher latitude $\theta + \Delta\theta$, so

$$\Phi = -D\cos\eta\frac{\partial T}{\partial\eta}(t,\eta)\bigg|_{\eta=\theta}^{\eta=\theta+\Delta\theta} = -D\int_{\theta}^{\theta+\Delta\theta}\frac{\partial}{\partial\eta}\cos\eta\frac{\partial T}{\partial\eta}(t,\eta)d\eta.$$

Thus, the EBM (12.1) becomes

$$C\frac{\partial}{\partial t}\int_{\theta}^{\theta+\Delta\theta}T(t,\eta)\cos\eta\,d\eta = \int_{\theta}^{\theta+\Delta\theta}\left[(E_{\text{in}}-E_{\text{out}})\cos\eta + D\frac{\partial}{\partial\eta}\cos\eta\frac{\partial T}{\partial\eta}\right]d\eta.$$

After dividing by $\Delta\theta$ and taking the limit $\Delta\theta \to 0$, we obtain the partial differential equation (PDE)

$$C\frac{\partial T}{\partial t}\cos\theta = (E_{\text{in}}-E_{\text{out}})\cos\theta + D\frac{\partial}{\partial\theta}\left(\cos\theta\frac{\partial T}{\partial\theta}\right). \tag{12.7}$$

Note that Eq. (12.7) becomes singular at the poles, so an extra condition must be imposed to guarantee that the equation remains valid there. A sufficient condition that is physically reasonable is that $\cos\theta\,(\partial T/\partial\theta)$ tend to zero as $\theta \to \pm\frac{1}{2}\pi$.

12.1.3 ▪ Changing Variables

At this point it is convenient to change variables from θ to $y = \sin\theta$ $(-1 < y < 1)$. Thus, negative values of y correspond to points on the Southern Hemisphere and positive values to points on the Northern Hemisphere; the points $y = 1$ and $y = -1$ are singular points corresponding to the North and South Poles, respectively.

With this change of variables, a function f of θ becomes a function $\tilde{f}$ of y, according to the identities $f(\theta) = f(\sin^{-1}y) = \tilde{f}(y)$. Strictly speaking, we should distinguish between f and $\tilde{f}$, but with a slight abuse of notation we shall use the same symbol to denote the function before and after the change of variables. In other words, we omit the tilde and denote the function $\tilde{f}$ again by the symbol f. It is usually clear from the context to which function the symbol f refers.

After the change of variables, Eq. (12.7) becomes

$$C\frac{\partial T}{\partial t} = E_{\text{in}} - E_{\text{out}} + D\frac{\partial}{\partial y}(1-y^2)\frac{\partial T}{\partial y}, \tag{12.8}$$

where

$$E_{\text{in}} = (1-\alpha(y,y_c))s(y)Q, \quad Q = \tfrac{1}{4}S_0, \tag{12.9}$$

and

$$E_{\text{out}} = A + BT. \tag{12.10}$$

Since the obliquity does not play a role in the subsequent analysis, we have dropped the subscript β from the distribution function s.

Eq. (12.8) is a PDE, since it involves partial derivatives of T both with respect to t and y. In fact, it is a PDE from the class of *reaction-diffusion equations*, for which there is a rich mathematical literature.

The global mean surface temperature T_0 is obtained by integrating T over all latitudes, $T_0(t) = \frac{1}{2}\int_{-1}^{1}T(t,y)\,dy$. Integration of Eq. (12.8) over all latitudes y yields the governing equation,

$$C\frac{dT_0}{dt} = (1-\alpha_0(y_c))Q - (A+BT_0). \tag{12.11}$$

This is a global energy balance equation of the type discussed in Chapter 2, with albedo $\alpha_0(y_c) = \frac{1}{2}\int_{-1}^{1}\alpha(y,y_c)s(y)\,dy$.

Before proceeding to a discussion of Eq. (12.8), we mention that Budyko [10] and Sellers [102] independently proposed an energy-transfer model where the energy exchange among latitudes is driven by the difference between the zonal temperature and the global average temperature. In that case, the zonal EBM (12.8) has a term $-k(T(t,y)-T_0(t))$ instead of the second-order differential term. A discussion of this model can be found in [114]; the corresponding EBM was analyzed recently by Widiasih [119].

12.2 ▪ Legendre Polynomials

Our first objective is to eliminate the partial derivatives of T with respect to y from Eq. (12.8) by expressing them in terms of T itself. As long as we are focusing on the partial derivatives with respect to y, t is just a parameter. Ignoring t, we therefore focus on the differential expression

$$\mathscr{L}u(y) = -\big((1-y^2)u'(y)\big)', \quad y \in (-1,1). \tag{12.12}$$

Here, $u : y \mapsto u(y)$ is any function defined and twice differentiable on the interval $(-1,1)$, which satisfies the condition $\lim_{y\to\pm1}(1-y^2)u'(y) = 0$. The prime $'$ denotes differentiation with respect to y. Our objective is to express $\mathscr{L}u$ in terms of u. The simplest way to accomplish this is to assume that $\mathscr{L}u$ is some multiple of u,

$$\mathscr{L}u = \lambda u. \tag{12.13}$$

This equation is known as the *eigenvalue problem* for the differential expression $\mathscr{L}$, by analogy with the eigenvalue problem for matrices. If, for some (possibly complex) constant λ, Eq. (12.13) admits a nontrivial solution u, then λ is called an *eigenvalue* of the differential expression $\mathscr{L}$ and u the *eigenfunction* associated with λ. The process of finding eigenvalues and eigenfunctions of a differential expression or similar operation goes by the name of *spectral analysis*, and a method that uses eigenvalues and eigenfunctions to replace a complicated operation (like differentiation) by the operation of multiplication by a scalar is known as a *spectral method*.

A moment's reflection shows that if u is an even (odd) polynomial function in y, then $\mathscr{L}u$ is also an even (odd) polynomial function with the same degree. We therefore look for eigenfunctions that are polynomials, using the letter P instead of u. Two eigenfunctions of $\mathscr{L}$ are readily found, namely $P_0(y) = 1$ and $P_1(y) = y$, with eigenvalues $\lambda_0 = 0$ and $\lambda_1 = 2$, respectively. In fact, there are countably many eigenfunctions; they are the *Legendre polynomials* [17, Section II.8], which are denoted by P_n, n being the degree of the polynomial,

$$\mathscr{L}P_n(y) = -\big((1-y^2)P_n'(y)\big)' = n(n+1)P_n(y), \quad n = 0, 1, \dots. \tag{12.14}$$

The equation shows that the eigenvalue associated with P_n is $\lambda_n = n(n+1)$. The Legendre polynomials are commonly normalized such that $P_n(1) = 1$ for all n. The P_n are given explicitly by the formula

$$P_n(y) = \sum_{i=0}^{p} \frac{(-1)^i(2n-2i)!}{2^n i!(n-i)!(n-2i)!}\, y^{n-2i}, \quad n = 0, 1, \dots. \tag{12.15}$$

The index i runs from 0 to p, where $p = \frac{1}{2}n$ if n is even and $p = \frac{1}{2}(n-1)$ if n is odd. The first few Legendre polynomials are

$$P_0(y) = 1, \quad P_2(y) = \frac{1}{2}(3y^2 - 1), \quad P_4(y) = \frac{1}{8}(35y^4 - 30y^2 + 3),$$
$$P_1(y) = y, \quad P_3(y) = \frac{1}{2}(5y^3 - 3y), \quad P_5(y) = \frac{1}{8}(63y^5 - 70y^3 + 15y). \tag{12.16}$$

Figure 12.3 shows the graph of P_n on $(-1, 1)$ for $n = 0, \ldots, 5$. The figure shows that P_n becomes more and more oscillatory as n increases; P_n has exactly n zeros, and the zeros of P_n and P_{n+1} are interlaced.

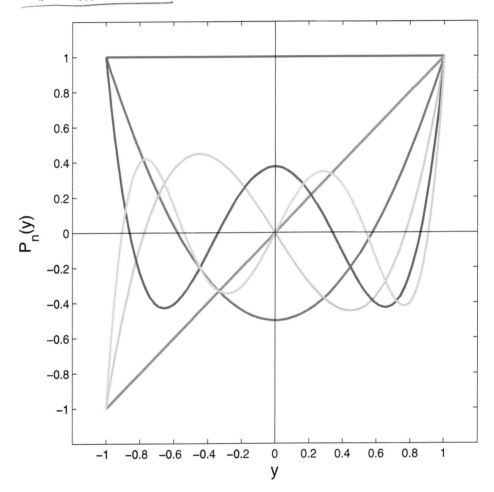

Figure 12.3. *Legendre polynomials P_n for $n = 0, \ldots, 5$ for $-1 \leq y \leq 1$.*

If we were to arrange the Legendre polynomials in an infinitely long column vector and (formally) apply the operation $\mathscr{L}$ to it, the result would be a matrix with entries λ_n on the diagonal and zeros everywhere else. This explains why the process of finding the eigenfunctions and eigenvectors of a differential expression is also known as *diagonalization*.

The most important property of Legendre polynomials is that they are *complete* or, more accurately, form a *basis* for functions defined on $(-1, 1)$. That is, they are linearly independent and any function $u : y \mapsto u(y)$ that is defined on $(-1, 1)$ and satisfies some very general regularity conditions can be expanded in a series,

$$u(y) = \sum_{n=0}^{\infty} u_n P_n(y). \tag{12.17}$$

In this sense, Legendre polynomials are like the trigonometric functions $\sin(n\pi x)$ and $\cos(n\pi x)$ for $n = 0, 1, \ldots$, which also form a basis for functions defined on $(-1, 1)$. Like the trigonometric functions, the Legendre polynomials satisfy an *orthogonality condition*,

$$\int_{-1}^{1} P_m(y)P_n(y)\,dy = 0 \text{ if } m \neq n. \tag{12.18}$$

In particular, by taking $m = 0$, we have $\int_{-1}^{1} P_n(y)\,dy = 0$ for $n = 1, 2, \ldots$. If $m = n$, there is the normalization

$$\int_{-1}^{1} (P_n(y))^2\,dy = \frac{2}{2n+1}, \quad n = 0, 1, \ldots. \tag{12.19}$$

By multiplying both sides of Eq. (12.17) with $P_n(y)$, integrating over $(-1, 1)$, and interchanging the integral and the infinite sum (which is permitted for sufficiently regular functions u), we obtain $\int_{-1}^{1} u(y)P_n(y)dy = (2/(2n+1))u_n$, so the coefficients u_n in the expansion (12.17) are given by

$$u_n = \frac{2n+1}{2} \int_{-1}^{1} u(y)P_n(y)\,dy, \quad n = 0, 1, \ldots. \tag{12.20}$$

The convergence of the series (12.17) depends on the smoothness of u. For example, if u is differentiable, the series converges pointwise in the sense that the partial sums $u^{(N)}(y) = \sum_{n=0}^{N} u_n P_n(y)$ satisfy the condition $|u^{(N)}(y) - u(y)| \to 0$ as $N \to \infty$ at every $y \in (-1, 1)$. If u is not smooth, for example if there are jumps, the convergence is no longer pointwise, but the series may still converge in some weaker sense, for example in the L^2 or root-mean-square (rms) sense,

$$\|u^{(N)} - u\| = \left(\frac{1}{2} \int_{-1}^{1} |u^{(N)}(y) - u(y)|^2\,dy \right)^{1/2} \to 0 \text{ as } N \to \infty.$$

The difference is illustrated in Figure 12.4. The left panel shows the graphs of the approximations $u^{(1)}$ and $u^{(3)}$ for the smooth function $u : y \mapsto u(y) = e^y$ on $(-1, 1)$ and the right panel the graphs of the approximations $u^{(3)}$ and $u^{(5)}$ for the discontinuous function $u : y \mapsto u(y) = \text{sign}(y - \frac{1}{3})$ on $(-1, 1)$. For the former, a few terms suffice to get a good approximation everywhere; for the latter, the pointwise approximation is poor even at large values of N.

12.3 ▪ Spectral Method

We now return to Eq. (12.8) and apply the *spectral method*, using the Legendre polynomials as the basis functions and expanding the surface temperature T as in Eq. (12.17),

$$T(t, y) = \sum_{n=0}^{\infty} T_n(t)P_n(y), \tag{12.21}$$

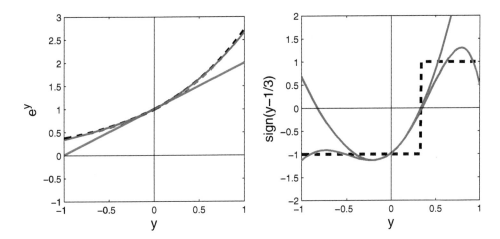

Figure 12.4. *Left: Legendre approximations* $u^{(1)}$ *(red) and* $u^{(3)}$ *(blue) of* $u : y \mapsto u(y) = e^y$ *on* $(-1, 1)$. *Right: Legendre approximations* $u^{(3)}$ *(blue) and* $u^{(5)}$ *(magenta) of* $u : y \mapsto u(y) = \mathrm{sign}(y - \frac{1}{3})$ *on* $(-1, 1)$.

where

$$T_n(t) = \frac{2n+1}{2} \int_{-1}^{1} T(t,y) P_n(y) \, dy, \quad n = 0, 1, \ldots. \tag{12.22}$$

Our goal is to translate Eq. (12.8) into a set of ODEs for the functions T_n.

Upon substitution of the Legendre expansion (12.21) for T, the term $\partial T / \partial t$ becomes the partial derivative of an infinite sum. Interchanging the order of differentiation and summation is certainly allowed if the sum is finite. But if the sum is infinite, the interchange may not be legitimate, unless T is sufficiently smooth. We assume that this is the case, so the following identity is true:

$$\frac{\partial T}{\partial t}(t,y) = \sum_{n=0}^{\infty} \frac{dT_n}{dt}(t) P_n(y). \tag{12.23}$$

The Legendre expansions of the incoming and outgoing energy can be computed directly. We multiply the expression Eq. (12.9) for E_{in} with $P_n(y)$ and integrate,

$$\int_{-1}^{1} E_{\mathrm{in}}(y) P_n(y) \, dy = \left[\int_{-1}^{1} s(y) P_n(y) \, dy - \int_{-1}^{1} \alpha(y, y_c) s(y) P_n(y) \, dy \right] Q.$$

Apart from a numerical factor, the integrals are the coefficients in the Legendre expansions of the functions $y \mapsto s(y)$ and $y \mapsto \alpha(y, y_c) s(y)$,

$$s(y) = \sum_{n=0}^{\infty} s_n P_n(y), \quad s_n = \frac{2n+1}{2} \int_{-1}^{1} s(y) P_n(y) \, dy, \tag{12.24}$$

$$\alpha(y, y_c) s(y) = \sum_{n=0}^{\infty} \overline{\alpha}_n(y_c) P_n(y), \quad \overline{\alpha}_n(y_c) = \frac{2n+1}{2} \int_{-1}^{1} \alpha(y, y_c) s(y) P_n(y) \, dy. \tag{12.25}$$

Therefore,

$$E_{\mathrm{in}}(y) = \sum_{n=0}^{\infty} (s_n - \overline{\alpha}_n(y_c)) Q P_n(y). \tag{12.26}$$

For the outgoing energy we have

$$E_{\text{out}} = AP_0(y) + \sum_{n=0}^{\infty} BT_n(t)P_n(y). \tag{12.27}$$

Next, we turn to the energy transport term. Given the definition (12.12) of the differential expression $\mathscr{L}$ and the fact that the Legendre polynomials P_n are eigenfunctions of $\mathscr{L}$ which satisfy Eq. (12.14), we obtain the identity

$$-\frac{\partial}{\partial y}\left((1-y^2)\frac{\partial}{\partial y}T(t,y)\right) = \mathscr{L}T(t,y) = \sum_{n=0}^{\infty} n(n+1)T_n(t)P_n(y). \tag{12.28}$$

Here we have interchanged the order of differentiation (this time with respect to y) and summation, which is justified provided T is sufficiently smooth with respect to y.

Collecting terms, we see that the spectral method leads to an equivalent formulation of Eq. (12.8),

$$C\sum_{n=0}^{\infty} \frac{dT_n}{dt} P_n(y) = \sum_{n=0}^{\infty} [(s_n - \overline{\alpha}_n(y_c))Q - (A\delta_{n0} + BT_n) - n(n+1)DT_n]P_n(y). \tag{12.29}$$

Here, δ_{mn} is the *Kronecker delta*, $\delta_{mn} = 0$ if $m \neq n$, $\delta_{mn} = 1$ if $m = n$. Since the Legendre polynomials are linearly independent, it must be the case that the coefficients of each P_n in the left- and right-hand sides of the equation are equal,

$$C\frac{dT_n}{dt} = (s_n - \overline{\alpha}_n(y_c))Q - (A\delta_{n0} + BT_n) - n(n+1)DT_n, \quad n = 0, 1, \dots. \tag{12.30}$$

Thus, the zonal EBM (12.8) for T is equivalent with an infinite system of ODEs for the expansion coefficients T_n. Note that the differential equation for T_0 is identical with Eq. (12.11).

12.4 · Equilibrium Solutions

Equilibrium solutions satisfy the equations

$$(s_n - \overline{\alpha}_n(y_c))Q - (A\delta_{n0} + BT_n) - n(n+1)DT_n = 0, \quad n = 0, 1, \dots. \tag{12.31}$$

These equations are readily solved,

$$T_n^*(y_c) = \frac{(s_n - \overline{\alpha}_n(y_c))Q - A\delta_{n0}}{B + n(n+1)D}, \quad n = 0, 1, \dots. \tag{12.32}$$

The resulting equilibrium temperature profile is

$$T^*(y, y_c) = \sum_{n=0}^{\infty} P_n(y)\frac{(s_n - \overline{\alpha}_n(y_c))Q - A\delta_{n0}}{B + n(n+1)D}. \tag{12.33}$$

In particular, the global mean surface temperature is

$$T_0^*(y_c) = \frac{(1 - \overline{\alpha}_0(y_c))Q - A}{B}. \tag{12.34}$$

This last expression matches the solution of the global EBM, Eq. (2.14).

12.5 ▪ Temperature Profile

To obtain an actual equilibrium temperature profile, we need to specify the incoming energy distribution and the variation of the albedo with latitude. In addition, we must find a value of the diffusion coefficient D. (The constants A, B, and C are determined from experimental observations.)

A formula for the incoming energy distribution is given in Eq. (12.2). The expression depends parametrically on β, the obliquity of the Earth's orbit. At the current value $\beta = 23.4°$, the distribution is very close to the function $\tilde{s}$ defined in Eq. (12.4); see Figure 12.2. In terms of y, this function becomes $\tilde{s}(y) = 0.523 + 0.716(1 - y^2)$ or, in terms of Legendre polynomials, $\tilde{s}(y) = 1 + s_2 P_2(y)$ with $s_2 = -0.477$. We are interested in climate states that are close to the current one. Therefore we take s to be a quadratic polynomial function, given by the general two-term Legendre expansion

$$s(y) = 1 + s_2 P_2(y). \tag{12.35}$$

With this expression for s, the Legendre coefficients $\bar{\alpha}_n$ ($n = 0, 1, \ldots$) are given by the integral

$$\bar{\alpha}_n(y_c) = \frac{2n+1}{2} \int_{-1}^{1} \alpha(y, y_c)(s_0 + s_2 P_2(y)) P_n(y)\, dy. \tag{12.36}$$

To simplify the problem, we now assume that the albedo satisfies the symmetry condition

$$\alpha(y, y_c) = \alpha(-y, y_c), \quad y \in (-1, 1), \tag{12.37}$$

and that the ice lines are located symmetrically with respect to the equator, $y_N = -y_S$. Then the entire climate system is symmetric with respect to the equator, we need only consider the Northern Hemisphere ($y > 0$), y_c stands for the single scalar parameter y_N, and all the odd-indexed Legendre coefficients $\bar{\alpha}_n$ are zero. We emphasize that this symmetry assumption is introduced strictly for convenience; the following analysis can be extended to the nonsymmetric case.

We choose the simplest possible model for the albedo. Below the ice line, the albedo is equal to the average value for soil and water, α_f (subscript f for fluid); above the ice line it is equal to the average value for snow and ice, α_i (subscript i for ice), where $\alpha_i > \alpha_f$; at the ice line, we allow for a jump discontinuity in the albedo. Thus

$$\alpha(y, y_c) = \begin{cases} \alpha_f, & y \in (0, y_c), \\ \alpha_i, & y \in (y_c, 1). \end{cases} \tag{12.38}$$

Given the explicit expressions for s and α, we can compute the coefficients $\bar{\alpha}_n$ from Eq. (12.36),

$$\begin{aligned}
\bar{\alpha}_{2n}&(y_c) \\
&= (4n+1) \int_0^1 \alpha(y, y_c)(1 + s_2 P_2(y)) P_{2n}(y)\, dy \\
&= (4n+1) \left[\alpha_i \int_0^1 (1 + s_2 P_2(y)) P_{2n}(y)\, dy - (\alpha_i - \alpha_f) \int_0^{y_c} (1 + s_2 P_2(y)) P_{2n}(y)\, dy \right] \\
&= \alpha_i(\delta_{n0} + s_2 \delta_{n1}) - (\alpha_i - \alpha_f)(4n+1) \int_0^{y_c} (1 + s_2 P_2(y)) P_{2n}(y)\, dy, \quad n = 0, 1, \ldots.
\end{aligned}$$

$$\tag{12.39}$$

The first two coefficients are

$$\overline{\alpha}_0(y_c) = \alpha_i - (\alpha_i - \alpha_f)y_c\left(1 - \tfrac{1}{2}s_2(1 - y_c^2)\right),$$
$$\overline{\alpha}_2(y_c) = \alpha_i s_2 + \tfrac{5}{2}(\alpha_i - \alpha_f)y_c\left[1 - y_c^2 - s_2\left(1 - 2y_c^2 + \tfrac{9}{5}y_c^4\right)\right].$$

(12.40)

In general, since the product $P_2 P_{2n}$ is a polynomial of degree $2n + 2$, $\overline{\alpha}_{2n}$ is a polynomial in y_c of degree $2n + 3$.

The expression (12.33) for the equilibrium temperature reduces to

$$T^*(y, y_c) = \frac{Q}{B}\sum_{n=0}^{\infty} P_{2n}(y)\frac{s_{2n} - \overline{\alpha}_{2n}(y_c) - (A/Q)\delta_{n0}}{1 + 2n(2n + 1)(D/B)},$$

(12.41)

where it is understood that $s_{2n} = 0$ for $n = 2, 3, \ldots$. The global average surface temperature T_0^* is still given by Eq. (12.34), with the coefficient $\overline{\alpha}_0(y_c)$ given by Eq. (12.40).

We now check that, for a reasonable choice of the parameter values, the equilibrium temperature matches the current state of the Earth's climate system. The parameter values

$$A = 203, \quad B = 2.09, \quad Q = 343, \quad s_2 = -0.477, \quad \alpha_f = 0.31, \quad \alpha_i = 0.62 \quad (12.42)$$

are those used in [78]. We determine the location of the ice line by the condition that the global mean temperature equals 14.6°C for the Northern Hemisphere and 13.6°C for the Southern Hemisphere [47]. With $\overline{\alpha}_0$ given by Eq. (12.40), the expression for T_0^* becomes

$$T_0^*(y_c) = \frac{Q}{B}\left[1 - \alpha_i - \frac{A}{Q} + (\alpha_i - \alpha_f)y_c\left(1 - \tfrac{1}{2}s_2(1 - y_c^2)\right)\right].$$

(12.43)

By choosing the location of the ice line at

$$y_c = 0.9405 \quad (\theta = 70°),$$

(12.44)

we find that $T_0^*(y_c) = 14.4°C$, which is quite reasonable. To make the result consistent with the requirement that near the ice line the temperature should equal the critical temperature $-10°C$ (the value used in [78]), we choose

$$D/B = 0.208.$$

(12.45)

Using a few terms in the expansion (12.41), we find the temperature at the ice line, $T^*(y_c, y_c) = -9.85°C$.

We can now compute the temperature profile from Eq. (12.41). The result is shown in Figure 12.5 for all latitudes between $-90°S$ and $90°N$. The graph was obtained using two and six terms in the infinite sum in Eq. (12.41). They illustrate the rapid convergence of the series. Also shown are observed average temperatures at various latitudes. These were computed from the data that accompany [47]. In particular, the entire mean temperature profile for the Northern Hemisphere, including the actual location of the latitude at which $T = -10°C$, is reproduced with remarkable accuracy. For the Southern Hemisphere, the location where $T = -10°C$ is predicted to be too far south, and the very cold temperature averages of Antarctica are not reproduced accurately.

12.6 • Assessment

The results in the previous section show that the mean temperature distribution on the Northern Hemisphere can be faithfully reproduced from just a few parameters. On the

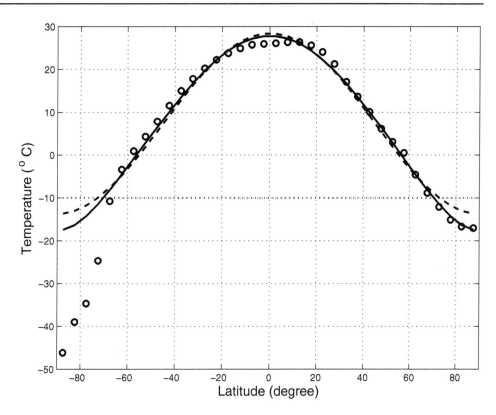

Figure 12.5. *Temperature profile obtained with two terms (dashed) and six terms (solid) in the expansion (12.41), together with observed mean latitudinal temperatures.*

other hand, the symmetric model that was used here shows a large systematic error at high southern latitudes. It is plausible that an asymmetric model will do a better job, since it would allow for a southern ice line that is closer to the equator.

More generally, this is a good opportunity to discuss various errors in the model and its mathematical discussion. Clearly, there are *model errors* that come from simplifications and neglect or incomplete understanding of important physical phenomena. There are some obvious geometric simplifications in this model, for example the symmetry assumption, the assumption of longitude independence, and the neglect of the distributions of oceans and continents across latitudes. Most parameterizations are likely to introduce additional errors. For example, it is possible that a single diffusion constant is not adequate to describe energy exchange processes across latitudes and that a latitude-dependent diffusion coefficient must be used. To identify and assess such errors, a detailed understanding of physical processes and perhaps detailed data can be helpful. For example, the mean latitudinal temperatures plotted in Figure 12.5 contradict the symmetry assumption that was made throughout. The mean latitudinal temperature data were obtained from monthly absolute temperature data for a $5° \times 5°$ grid obtained from the data sources in [47]. These data do indeed show longitude-dependent temperature variations.

There are also *truncation errors* present in the current model. These are errors that come from mathematical simplifications such as linearization and the use of finitely many terms in an infinite series. The expression (12.10) for the outgoing energy is an example, as is the use of only a few terms of the infinite series (12.41) for the equilibrium temperature distribution. The magnitude of such errors can be assessed by mathematical techniques.

The data themselves have *measurement errors* of a statistical nature. These types of errors are discussed elsewhere, and there are many techniques known to assess and control them.

Finally, there are *computation errors* that come from rounding, solving equations inaccurately, and so on. For example, the computed temperature at the ice line in our model does not agree perfectly with the target temperature of $-10°C$. These errors are easiest to control and reduce. In the present discussion, they contribute relatively little to the overall error structure.

12.7 ▪ Exercises

1. Show in detail that the change of variables $y = \sin\theta$ transforms Eq. (12.7) into Eq. (12.8).

2. Consider a spherical planet of radius R which is surrounded by an atmosphere. In the atmosphere, a trace gas is being transported by a velocity field and continually being formed and destroyed by chemical processes. The density of the gas per unit area, ρ, depends on time t and latitude θ, while the velocity field and the terms describing the formation and destruction of the gas depend only on latitude θ. The flux of the gas per unit length across the latitudinal circle at θ is given by $-\alpha\rho(t,\theta)$, where the wind speed α is positive and a flux is considered to be positive if it is directed northward. The net gain due to the formation and destruction of the gas per unit area is $\beta\sin\theta$, where $\beta > 0$.

   (i) In which direction is the gas being transported? Where is the flux strongest? Where is the gas being produced? Where is it being destroyed? Give your answer in geographical terms ("North Pole," "Southern Hemisphere," etc.).

   (ii) Derive the equation for the density ρ,

   $$R\cos\theta\frac{\partial\rho}{\partial t}(t,\theta)+\frac{\partial}{\partial\theta}(\alpha\rho(t,\theta)\cos\theta)=\beta R\sin\theta\cos\theta.$$

   (iii) Rewrite the equation for ρ in terms of the variable $y = \sin\theta \in (-1,1)$.

   (iv) Given α, β, and R, find the unique stationary (time-independent) solution ρ^* that is finite for all latitudes.

   (v) Discuss in physical terms how the solution found under (iv) depends on the strength α of the velocity field, the magnitude β of the production term, and the size R of the planet.

3. In this and the following exercises we show how to set up a balance equation like Eq. (12.11) for transport processes in arbitrary regions $\Omega \subset \mathbb{R}^2$.

   Let $\mathbf{q} : (t_0, t_1) \times \Omega \to \mathbb{R}^2$ be a time-dependent vector field which describes the flow of some physical quantity in Ω, and let γ be a fixed curve in $\mathbb{R}^2$ which is parameterized by a one-to-one C^1 function $\mathbf{x} : [a, b] \to \mathbb{R}^2$. The amount of the physical quantity that is transported across γ per unit time from left to right as the curve is traversed from $\mathbf{x}(a)$ to $\mathbf{x}(b)$ is called the *flux* of F across γ and is given by the integral

   $$\int_\gamma \mathbf{q}(\mathbf{x})\cdot J\,d\mathbf{x} = \int_a^b \mathbf{q}(\mathbf{x}(t))^T J\,\dot{\mathbf{x}}(t)\,dt, \qquad (12.46)$$

   where $J = \begin{pmatrix} 0 & 1 \\ -1 & 0 \end{pmatrix}$ is the matrix for a clockwise rotation by $90°$.

(i) Let $\rho : (t_0, t_1) \times \Omega \to \mathbb{R}$ be the density of the quantity that is transported by the vector field $\mathbf{q}$. It depends on position and time. Consider a rectangular patch $P = [x_1, x_1 + h_1] \times [x_2, x_2 + h_2] \subset \Omega$. Express the rate of change of the quantity contained in P,

$$\frac{d}{dt} \iint_P \rho(t, \mathbf{x}) \, d\mathbf{x},$$

in terms of flux integrals of $\mathbf{q}$ over the line segments that make up the boundary of P.

(ii) By letting h_1 and h_2 tend to 0, deduce that

$$\frac{\partial \rho}{\partial t}(\mathbf{x}, t) + \nabla \cdot \mathbf{q}(\mathbf{x}, t) = 0. \qquad (12.47)$$

Here, $\nabla \cdot$ is the two-dimensional divergence operator, $\nabla \cdot \mathbf{q} = \partial q_1 / \partial x_1 + \partial q_2 / \partial x_2$.

4. Continue with the setup of the previous exercise. Assume that the vector field $\mathbf{q}$ is given by *Fourier's Law*,

$$\mathbf{q}(t, \mathbf{x}) = -\varkappa \nabla \rho(t, \mathbf{x}), \qquad (12.48)$$

where $\varkappa$ is a positive constant with suitable dimensions.

(i) Explain in physical terms the meaning of Fourier's Law, Eq. (12.48).

(ii) Deduce the PDE for the quantity ρ,

$$\frac{\partial \rho}{\partial t}(t, \mathbf{x}) - \varkappa \Delta \rho(t, \mathbf{x}) = 0. \qquad (12.49)$$

Here, Δ is the two-dimensional *Laplace operator*, $\Delta u = \partial^2 u / \partial x_1^2 + \partial^2 u / \partial x_2^2$.

(iii) Replace Eq. (12.48) by the more general assumption

$$\mathbf{q}(t, \mathbf{x}) = -\varkappa \nabla \rho(t, \mathbf{x}) + \mathbf{h}(t, \mathbf{x}), \qquad (12.50)$$

where $\mathbf{h}$ is an arbitrary sufficiently smooth vector field. Deduce a PDE similar to Eq. (12.49).

5. Recall the differential expression $\mathscr{L}$, $\mathscr{L} u(y) = -((1 - y^2)u'(y))'$ (from the definition (12.12)) for any real-valued function $u : y \mapsto u(y)$ defined on $(-1, 1)$. Prove the following statements:

(i) If u is C^2 on $[-1, 1]$ and $\mathscr{L} u(y) = 0$ for all $y \in [-1, 1]$, then u is constant.

(ii) If u is continuous and bounded on $[-1, 1]$ and C^2 on $(-1, 1)$, and $\mathscr{L} u(y) = 0$ for all $y \in [-1, 1]$, then u is constant.

(iii) Find a nonconstant C^2-function u for which $\mathscr{L} u(y) = 0$ for all $y \in (-1, 1)$. Note: (ii) implies that u cannot be bounded on $[-1, 1]$.

6. Find coefficients $a_0, a_1, \ldots, a_5$ such that $y^5 = \sum_{j=0}^5 a_j P_j(y)$, where P_j is the Legendre polynomial of degree j, in two ways:

(i) by comparing coefficients, and

(ii) by using Eq. (12.20).

7. Prove *Rodrigues's formula*,

$$P_n(y) = \frac{1}{2^n n!} \frac{d^n}{dy^n} (y^2 - 1)^n, \quad n = 0, 1, \ldots, \tag{12.51}$$

by showing that

(i) $Q_n : y \mapsto (d^n/dy^n)(y^2 - 1)^n$ is an eigenfunction of $\mathscr{L}$, and

(ii) $Q_n/(2^n n!)$ is identical with the Legendre polynomial P_n defined in Eq. (12.15).

8. Use Eq. (12.51) to prove the orthogonality of the Legendre polynomials, Eq. (12.18),

$$\int_{-1}^{1} P_m(y) P_n(y) \, dy = 0 \text{ if } m \neq n. \tag{12.52}$$

Hint: Show that $y \mapsto (d^k/dy^k)(y^2 - 1)^n$ vanishes at $y = \pm 1$ for $k = 0, 1, \ldots, n - 1$; then integrate by parts.

9. Show that

$$\int_{-1}^{1} (P_n(y))^2 \, dy = \frac{2}{2n + 1}. \tag{12.53}$$

10. Show the following recurrence relations for the Legendre polynomials P_n:

$$(2n + 1) y P_n(y) = (n + 1) P_{n+1}(y) + n P_{n-1}(y), \quad n = 1, 2, \ldots, \tag{12.54}$$

$$(2n + 1) P_n(y) = P'_{n+1}(y) - P'_{n-1}(y), \quad n = 1, 2, \ldots, \tag{12.55}$$

with the conventions $P_{-1}(y) = 0$ and $P_0(y) = 1$.

11. The location of the ice line, y_c, is a parameter in the expression (12.33) for the equilibrium temperature profile T^*. Assume that A, B, Q, s_2, α_f, and α_i have the values given in Eq. (12.42). Explore the following extreme cases:

(i) Ice line at the equator, $y_c = 0$. Show that if $\frac{1}{10}B \leq D \leq B$, then $T^*(y) < -10$ for all $y \in (-1, 1)$ (Snowball Earth). What is the range of values for D for which $T^*(y) < -10$ for all $y \in (-1, 1)$? Note: This result shows that the present model allows for a Snowball Earth under the assumption that the parameters in Eq. (12.42) are unchanged. (This assumption may very well be unrealistic.)

(ii) Ice line at the North Pole, $y_c = 1$. Is it possible to find D such that $T^*(y) > -10$ for all $y \in (-1, 1)$ (ice-free Earth)? Note: This condition is necessary to make the model consistent.

12. In this exercise, we explore the model for temperature transport proposed by Budyko [10] and Sellers [102], in which the energy exchange across latitudes is driven by the difference between the zonal temperature and the global average temperature. Then the zonal EBM, Eq. (12.8), has a term $-k(T(t, y) - T_0(t))$ instead of the second-order differential term, where T_0 is the global mean temperature, $T_0(t) = \frac{1}{2} \int_{-1}^{1} T(t, y) \, dy$. All other assumptions are to remain the same.

(i) Set up a differential equation similar to Eq. (12.8). What is the analogue of Eq. (12.11) in this case?

(ii) Set up differential equations for the Legendre coefficients T_n, similar to Eq. (12.30).

(iii) Try to find an equilibrium profile for which the northern ice line is at approximately 70° latitude and the global mean temperature is approximately 14.4°C. Plot it and compare it to the temperature profile given in Figure 12.5.

Chapter 13
Atmosphere and Climate

In this chapter we introduce the fundamental variables describing the atmosphere, namely pressure, temperature, and moisture content, and discuss the global atmospheric circulation pattern.

Keywords: Atmospheric pressure, hydrostatic balance, ideal gas law, lapse rate, atmospheric circulation, Hadley cell, Ferrel cell, polar cell.

13.1 ▪ Earth's Atmosphere

The atmosphere and ocean are major players in the Earth's climate system. Although the density and therefore the heat capacity of air are several orders of magnitude smaller than those of ocean water, the time scale for ocean circulation is several orders of magnitude greater than for atmospheric circulation. It turns out that the differences balance each other out, at least in order of magnitude; consequently, the atmosphere and ocean are roughly of equal importance for maintaining the global energy balance.

The much smaller scale, both in space and in time, of the phenomena in the Earth's atmosphere compared to the phenomena in its oceans makes it considerably more difficult to model the atmosphere at the system level. There is no analogue of box models at the scale of the atmosphere. The atmosphere's response to forcing is extremely variable on a time scale of hours or days and over regions that are much smaller than a typical ocean basin. Realistic models must take these variations into account, so it makes more sense to take a physics-based approach to modeling the Earth's atmosphere. We will do so in Chapter 14, where we derive the governing equations for mass density and hydrodynamic velocity from the conservation laws of physics. Mass density and hydrodynamic velocity are the fundamental variables describing the state of the atmosphere. Additional variables may come into play, depending on the level of detail. For example, temperature (a measure of internal energy of the atmosphere) and temperature gradients may be taken into account as possible drivers of transport mechanisms, and moisture content, mass fractions, and species velocities may enter into a model of the atmosphere in conjunction with water vapor and chemical species (greenhouse gases). In general, the atmospheric circulation is driven by pressure gradients and temperature differences across the globe. We devote brief sections to these variables before we discuss the general circulation pattern.

13.2 ▪ Pressure

The *pressure* at any point in the Earth's atmosphere is simply the downward force exerted at that point by the air above it. According to Newton's law, it is therefore the product of the mass of the column of air above it and g, the acceleration of gravity at the point. The SI unit of pressure is the *pascal*, abbreviated *Pa* and equal to $1\,\mathrm{Nm}^{-2}$ (newton per square meter). Other commonly used units are *hectopascal*, abbreviated *hPa*, equal to $100\,Pa$, and the *bar*, equal to $10^5\,Pa$. The change in pressure dp as one rises over a vertical distance dz is

$$dp = -\rho g\,dz, \tag{13.1}$$

where ρ is the *density* (mass per unit volume) of the air at this altitude. This is the *hydrostatic balance* equation. The minus sign indicates that pressure decreases with increasing height.

The *ideal gas law* connects the pressure p, volume V, and temperature T of 1 mole of air,

$$pV = RT. \tag{13.2}$$

Here R is the gas constant, $R = 8.314\,\mathrm{JK}^{-1}\mathrm{mol}^{-1}$. Since $\rho = M/V$, where M is the mass of 1 mole, it follows that $\rho = Mp/RT$. If we substitute this expression in the hydrostatic balance equation (13.1), we find the relation

$$\frac{dp}{p} = -\frac{dz}{H}, \tag{13.3}$$

where $H = RT/gM$. By integrating this relation, for example from $z = 0$, we obtain the pressure p as a function of height z,

$$p(z) = p(0)\exp\left(-\int_0^z \frac{dz}{H}\right). \tag{13.4}$$

Note that H depends on T, so H varies throughout the atmosphere. In practice, H is often taken to be constant. Figure 13.1 (left panel) shows the atmospheric pressure as a function of altitude. The figure shows observations made from a sounding in Charleston, South Carolina, in January 2013.

13.3 ▪ Temperature

The mean temperature of the Earth's atmosphere declines with height up to about 10 km above the surface, at which point it begins to increase slowly up to about 50 km. Above 50 km, radiative cooling into space becomes the dominant process and the temperature decreases again. The lower region of decreasing temperature (below 10 km altitude) is known as the *troposphere*, the region of slowly increasing temperature between about 10 km and 50 km altitude as the *stratosphere*, and the region above 50 km with decreasing temperature as the *mesosphere*. The transition layer between troposphere and stratosphere is known as the *tropopause*. Figure 13.1 (middle panel) shows the observed atmospheric temperature (solid curve) as a function of altitude up to an altitude of about 30 km, the middle of the stratosphere.

Sometimes, the *potential temperature* θ is used instead of the temperature T. The potential temperature of a parcel of fluid at pressure p is the temperature that the parcel would acquire if it were adiabatically brought to a standard reference pressure p_0 (usually

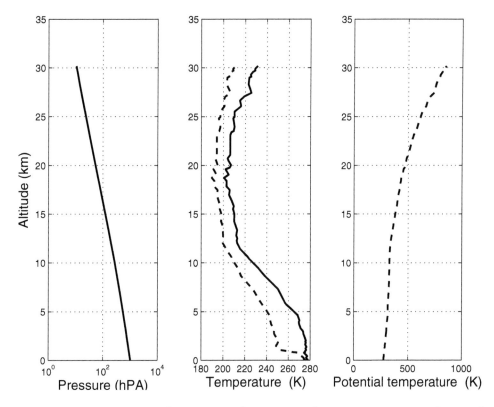

Figure 13.1. *Vertical pressure, actual temperature, dew point temperature, and potential temperature, observed in Charleston, South Carolina, on January 18, 2013.*

1000 hPa). It is defined by the infinitesimal relation

$$\frac{d\theta}{\theta} = \frac{dT}{T} - \frac{R}{c_p}\frac{dp}{p},$$ (13.5)

where R is the gas constant and c_p the specific heat at constant pressure. Potential temperature is a more important quantity dynamically than the actual temperature because it is not affected by the physical lifting or sinking associated with flow over obstacles or large-scale atmospheric turbulence. The concept applies to any stratified fluid. In oceanic conditions, the difference between θ and T is small (less than 0.5 K), so θ and T are often identified. However, in the atmosphere the potential and actual temperatures are quite different. Figure 13.1 (right panel) shows the potential temperature, computed from measurements obtained during the sounding, as a function of altitude. The integrated form of Eq. (13.5),

$$\theta = T\,(p_0/p)^{R/c_p}\,,$$ (13.6)

is known as *Poisson's equation*.

The rate at which the mean vertical temperature T decreases with height z is called the *lapse rate*. This parameter is of critical importance in meteorology, as it determines how high a parcel of air will rise until its water vapors begin to condense and form clouds and, once clouds are formed, whether the air will continue to rise and form rain or thunderclouds.

The actual change of temperature with altitude for the stationary atmosphere (i.e., the temperature gradient) is called the *environmental lapse rate*,

$$L = -dT/dz. \tag{13.7}$$

The value of L in the troposphere is typically around 6 to 7 deg/km.

The environmental lapse rate is distinct from the *adiabatic lapse rate*, which refers to the change in temperature of a mass of air as it moves upwards. Here we must distinguish between dry or unsaturated air and saturated air. Unsaturated air has less than 100% relative humidity—that is, its actual temperature is higher than its dew point. The temperature of saturated air is exactly at the dew point.

The *dry adiabatic lapse rate* is the rate of temperature decrease with height for a parcel of dry or unsaturated air rising under adiabatic conditions. (The term *adiabatic* means that no heat transfer occurs into or out of the parcel.) Dry air has low thermal conductivity, and the bodies of air involved are very large, so transfer of heat by conduction is negligibly small. As a parcel of dry air rises (for instance, by convection), it expands because the pressure is lower at higher altitudes, and as it expands, it pushes on the air around it, doing work. Since the parcel does work but gains no heat, it loses internal energy; as a consequence, its temperature decreases. The rate of decrease is the dry adiabatic lapse rate, Γ. Assuming that the atmosphere is hydrostatically balanced, one finds the expression $\Gamma = g/c_p$ from the laws of thermodynamics, with c_p the specific heat at constant pressure. The value of Γ is approximately 9.7 deg/km.

When the air is saturated with water vapor (at its dew point), the *moist adiabatic lapse rate* or *saturated adiabatic lapse rate* applies. This lapse rate varies strongly with temperature; a typical value is around 5 deg/km. The reason for the difference between the dry and moist adiabatic lapse rate values is that latent heat is released when water condenses, thus decreasing the rate of temperature drop as altitude increases. (This heat release process is an important source of energy in the development of thunderstorms.)

As an unsaturated parcel of air of given temperature, altitude, and moisture content rises, it cools at the dry adiabatic lapse rate. Figure 13.1 (middle panel) shows the dewpoint temperature (dashed curve) as a function of height. The graph shows that the dewpoint temperature was substantially below the actual temperature throughout the sounding, indicating that this was probably a clear January morning over Charleston, South Carolina. In general, the dew point also drops (as a result of decreasing air pressure) but often much more slowly, typically about 2 deg/km. In such a situation, the temperature of the unsaturated parcel will eventually catch up with the dew point for the given moisture content. At that point, the water vapor starts condensing and, as the parcel of air rises further, it cools at the slower moist adiabatic lapse rate.

Stability prevails when the moist adiabatic lapse rate exceeds the environmental lapse rate, since the rising air will cool faster than the surrounding air and lose buoyancy. (This often happens in the early morning, when the air near the ground has cooled overnight.) Cloud formation in stable air is unlikely. On the other hand, if the dry adiabatic lapse rate is less than the environmental lapse rate, a parcel of air will gain buoyancy as it rises and the whole configuration is unstable. (This often happens in the afternoon over many land masses. In these conditions, the likelihood of cumulus clouds, showers, or even thunderstorms is increased.)

13.4 ▪ Atmospheric Circulation

The Earth's climate system maintains its energy balance by permanently circulating heat, momentum, moisture, and chemical elements around the globe. Both the ocean and the atmosphere contribute to this process, in roughly equal amounts, although on very different time scales. Typically, the atmosphere cycles on a yearly scale, while the ocean cycle is of the order of one hundred years or more. The system is self-adjusting, so circulation patterns are dynamic and evolving continuously over time.

The large-scale structure of the atmospheric circulation varies from year to year, but the basic climatological structure is stable over longer periods of time. It has both a latitudinal and a longitudinal component. The latitudinal circulation is a consequence of the fact that incident solar radiation per unit area is highest at the equator and decreases as the latitude increases, reaching its minimum at the poles. Longitudinal circulation, on the other hand, comes about because water has a higher specific heat capacity than land and thereby absorbs and releases more heat, but the temperature changes less than on land.

The latitudinal circulation is dominated by three closed loops (cells): the Hadley cell, the Ferrel cell, and the polar cell, as illustrated in Figure 13.2. The *Hadley cell* begins at the equator with warm, moist air lifted aloft in equatorial low-pressure areas (the Intertropical Convergence Zone, ITCZ) to the tropopause and carried poleward. At about 30 degrees latitude (both north and south), the air descends in a high pressure area. Some of the descending air travels along the surface toward the equator, thus closing the loop of the Hadley cell. The Coriolis force drives these returning air masses toward the west, creating the trade winds.

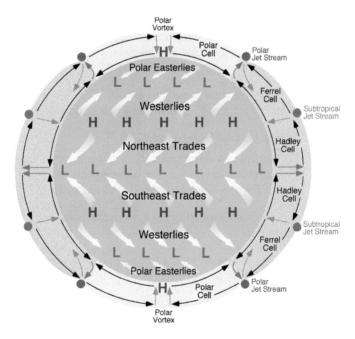

Figure 13.2. *Schematic of the atmospheric circulation pattern. Reprinted with permission from Michael Pidwirny.*

The *polar cell* is likewise a simple system. Though cool and dry relative to equatorial air, air masses at the 60th parallel are still sufficiently warm and moist to undergo convection and drive a thermal loop. Air circulates within the troposphere, limited vertically by the tropopause at about 8 km. Warm air rises at lower latitudes and moves poleward through the upper troposphere at both the North and South Poles. When the air reaches the polar areas, it has cooled considerably and descends as a cold, dry high-pressure area, moving away from the pole along the surface but twisting westward as a result of the Coriolis force to produce the polar easterlies.

The Hadley cell and the polar cell are similar in that they are a direct consequence of surface temperatures. Their thermal characteristics override the effects of weather in their domain. The sheer volume of energy the Hadley cell transports, and the depth of the heat sink that is the polar cell, ensure that the effects of transient weather phenomena are not only not felt by the system as a whole, but—except under unusual circumstances—are not even permitted to form.

The *Ferrel cell* is a secondary circulation feature, dependent for its existence upon the Hadley cell and the polar cell. It behaves much as an atmospheric ball bearing between the Hadley cell and the polar cell, and comes about as a result of the eddy circulations (the high- and low-pressure areas) of the midlatitudes. For this reason it is sometimes known as the "mixing zone." At its southern extent (in the Northern Hemisphere), it overrides the Hadley cell, and at its northern extent, it overrides the polar cell.

In the context of climate, the dominant longitudinal circulation pattern is one where warm air rises over the equatorial, continental, and western Pacific Ocean regions; flows eastward or westward, depending on its location, when it reaches the tropopause; and subsides in the Atlantic and Indian Oceans, and in the eastern Pacific.

13.5 ▪ Exercises

1. The behavior of the temperature lapse rate for the profile shown in Figure 13.1 differs from the general description given in this chapter. Describe the main differences and the features that are in common.

2. The environmental lapse rate L is in general smaller in magnitude than the dry adiabatic lapse rate Γ. Explain why this is to be expected, giving physical reasons.

3. (i) Derive Eq. (13.4) from Eq. (13.3).

   (ii) Assuming a temperature T_0 at altitude 0 and a lapse rate L up to an altitude h_1 and constant temperature between altitudes h_1 and h_2, derive closed formulas for the pressure distribution between altitudes 0 and h_2.

4. Derive Poisson's equation (13.6) from the differential equation (13.5) by integrating over altitude.

5. Some of the global wind patterns that are depicted in Figure 13.2 have regional names. Identify at least three of them by name.

6. The Monsoon winds over the northern Indian Ocean do not quite fit into the schematic given in Figure 13.2. Explain what is meant by this and why this is the case.

Chapter 14

Hydrodynamics

Our main goal in this chapter is to derive the fundamental equations of hydrodynamics—the continuity equation and the equation of motion (Navier–Stokes equation)—from the conservation laws of mass and momentum. Before doing so, we discuss the Coriolis effect, which plays an important role in the dynamics of the atmosphere and ocean. We also discuss two of the more common approximations of the hydrodynamic equations, namely the shallow water equations and the Boussinesq equations.

Keywords: Coriolis effect, conservation of mass, continuity equation, conservation of momentum, equation of motion, Navier–Stokes equation, shallow water equations, Boussinesq approximation.

14.1 ▪ Coriolis Effect

The Earth's atmosphere and ocean are essentially layers of fluid that are gravitationally bound to the surface of a rapidly rotating sphere. Because we follow their motion from a position on the surface of the Earth, it is important that we account for the *Coriolis effect* when we set up the equations of motion. The Coriolis effect exists only when one uses a rotating frame of reference; it does not exist in an inertial frame (that is, a frame of reference that is fixed to the stars). While the effect is quite small for objects moving across a rotating turntable, it is extremely important when the spatial scales are large, as is the case in the Earth's atmosphere and ocean.

14.1.1 ▪ Intuitive Argument

The Coriolis effect is illustrated in Figure 14.1. The left panel shows the effect in the case of a rotating plane that is spinning in the clockwise direction, as indicated by the arrow around the origin. The trajectory of an object that moves along a straight line in an inertial frame of reference appears to be curved to the left to an observer who moves with the rotating plane. The observer concludes that there is a "virtual" force acting on the object, and since the motion of a slow-moving object (thin line) appears to be less curved than the motion of a fast-moving object (thick line), the observer also concludes that the magnitude of the force must depend on the velocity of the object.

The right panel shows the same phenomenon on the surface of the Earth. The Earth is spinning around its north-south axis in a counterclockwise direction as viewed from above

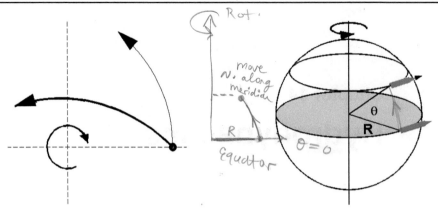

Axis of Rot.

move N. along meridian

R

Equator

θ

R

θ = 0

Figure 14.1. *Coriolis effect; linear movement as it appears in a rotating plane (left) and poleward movement of an object on a rotating sphere (right).*

with angular velocity Ω; $\Omega = 2\pi \, \text{day}^{-1} = 2\pi/86400 \, \text{sec}^{-1}$. Suppose an object of mass m is located in a stationary position (relative to the surface of the Earth) on the equator. Its distance to the Earth's axis of rotation is equal to the radius R of the Earth, so in an inertial frame of reference the object has a linear velocity $R\Omega$ in the eastward direction. This velocity is indicated by the red arrow tangent to the equator. Now suppose that the object begins to move northward along a meridian on the surface of the globe, with velocity v and without friction. As soon as it leaves the equator, its distance to the Earth's axis of rotation decreases like $\cos\theta$, where θ is latitude, so the eastward velocity that the object carries with it ($R\Omega$) exceeds the eastward velocity of the underlying surface of the Earth ($r\Omega$, where $r = R\cos\theta$) by $u = (R - r)\Omega$. The velocity $r\Omega$ is indicated by the red arrow tangent to the latitudinal circle at latitude θ and the excess velocity u by the thin black arrow that extends $r\Omega$. An observer on Earth "sees" only the excess velocity and concludes that a force must be acting on the object which causes it to accelerate in the eastward direction. This force is the so-called *Coriolis force*. The argument shows that the Coriolis force can be explained completely in terms of rotational dynamics, so it is not really a force but rather a *pseudoforce*. It is therefore more appropriate to refer to the *Coriolis effect*.

Coord. Syst. which is not accelerating (Constant Velocity)

R const. object velocity $R\Omega$ earth " $r\Omega$ $r = R\cos\theta$ ↓ as θ ↑

14.1.2 ▪ Rotating Coordinates

To describe the Coriolis effect mathematically, we consider a sphere in $\mathbb{R}^3$ that is rotating around its north-south axis with angular frequency Ω. The rotation is described by a family of rotation matrices,

$\Omega = 0$ recover org. x,y,z coord.

$$R(t) = \begin{pmatrix} \cos\Omega t & -\sin\Omega t & 0 \\ \sin\Omega t & \cos\Omega t & 0 \\ 0 & 0 & 1 \end{pmatrix}, \quad t \in \mathbb{R}. \qquad (14.1)$$

← Rot. about z-axis z coord. unchanged!

The trajectory of a point $\mathbf{z}_0 \in \mathbb{R}^3$ under the rotation is given by the function $\mathbf{z} : t \mapsto \mathbf{z}(t) = R(t)\mathbf{z}_0$, which satisfies the equation

$$\dot{\mathbf{z}} = \mathbf{\Omega} \times \mathbf{z}, \qquad (14.2)$$

where $\mathbf{\Omega} = \Omega(0, 0, 1)^T$ is the rotation vector. The dot ˙ denotes differentiation with respect to t.

Now consider a single point mass m that is in motion. In an inertial frame of reference, its trajectory is given by the function $\mathbf{y} : t \mapsto \mathbf{y}(t)$, while in a frame of reference that moves with the rotating sphere it is given by the function $\mathbf{x} : t \mapsto \mathbf{x}(t)$. The two descriptions are related at any time t through the rotation matrix,

$$\mathbf{y}(t) = R(t)\mathbf{x}(t). \tag{14.3}$$

Differentiating both sides of this relation once with respect to t, using the product rule together with Eq. (14.2), we obtain the relation

$$\dot{\mathbf{y}}(t) = \Omega \times \mathbf{y}(t) + R(t)\dot{\mathbf{x}}(t),$$

and differentiating once more, we obtain the relation

$$\ddot{\mathbf{y}}(t) = \Omega \times \dot{\mathbf{y}}(t) + \Omega \times (R(t)\dot{\mathbf{x}}(t)) + R(t)\ddot{\mathbf{x}}(t)$$
$$= \Omega \times (\Omega \times \mathbf{y}(t)) + 2\Omega \times (R(t)\dot{\mathbf{x}}(t)) + R(t)\ddot{\mathbf{x}}(t).$$

Since t is arbitrary, we may take $t = 0$. The quantity $\ddot{\mathbf{y}}(0)$ on the left-hand side corresponds to the object's acceleration in the inertial coordinate system at $t = 0$, which we denote by $\mathbf{a}_F$, so $\ddot{\mathbf{y}}(0) = \mathbf{a}_F$. The matrix $R(0)$ is the identity matrix. According to Eq. (14.2), $y(0) = x(0)$, so the terms on the right-hand side all refer to the rotating frame of reference. We use the notation $\mathbf{x}_0 = \mathbf{x}(0)$ for the position, $\mathbf{v} = \dot{\mathbf{x}}(0)$ for the velocity (v for its magnitude), and $\mathbf{a}_R = \ddot{\mathbf{x}}(0)$ for the acceleration in the rotating frame of reference. The term $\Omega \times (\Omega \times \mathbf{x}_0)$ corresponds to a centrifugal acceleration; with a minus sign, it becomes a centripetal acceleration, which we denote by $\mathbf{a}_{cp}$, so $\mathbf{a}_{cp} = -\Omega \times (\Omega \times \mathbf{x}_0)$. Solving for $\mathbf{a}_R$, we thus obtain a decomposition of the acceleration as seen in the rotating coordinate system, $\mathbf{a}_R$, into the actual acceleration $\mathbf{a}_F$, the centripetal acceleration $\mathbf{a}_{cp}$, and a third term which is perpendicular to both Ω and $\mathbf{v}$,

$$\mathbf{a}_R = \mathbf{a}_F + \mathbf{a}_{cp} - 2\Omega \times \mathbf{v}. \tag{14.4}$$

This third term is the *Coriolis acceleration* $\mathbf{a}_C$,

$$\mathbf{a}_C = -2\Omega \times \mathbf{v}. \tag{14.5}$$

After multiplying with the mass m of the moving object, we obtain a similar relation among forces. (The centripetal and Coriolis "force" are not really forces but pseudo-forces.)

For the Earth's rotation, the centripetal force is very small. The centripetal acceleration at the equator is only about 0.3% of the gravitational acceleration and less at higher latitudes. It can therefore be ignored. The *Coriolis force* is

$$\mathbf{F}_C = -2m\Omega \times \mathbf{v}. \tag{14.6}$$

14.1.3 ▪ Coriolis Force

We now assume that the point $\mathbf{x}_0$ is located on the surface of the rotating sphere, $\mathbf{x}_0 = R\mathbf{n}$, where R is the radius of the sphere and $\mathbf{n}$ a unit vector. Since we are interested in the motion of atmosphere and ocean around the globe, we assume that $\mathbf{v}$ is tangent to the sphere, $\mathbf{v} \cdot \mathbf{n} = 0$. For the same reason, we are mainly interested in the tangential component of the Coriolis force,

$$\mathbf{F}_{\text{tan}} = \mathbf{F}_C - (\mathbf{F}_C \cdot \mathbf{n})\mathbf{n}. \tag{14.7}$$

At the equator, both the rotation vector and $\mathbf{v}$ are tangent to the surface, so $\mathbf{F}_C$ points straight up or straight down and $\mathbf{F}_{\text{tan}} = 0$. Also, $\mathbf{F}_{\text{tan}}$ is perpendicular to $\mathbf{v}$ everywhere and points to the right of $\mathbf{v}$ on the Northern Hemisphere (east if $\mathbf{v}$ points north) and to the left of $\mathbf{v}$ on the Southern Hemisphere (east if $\mathbf{v}$ points south). To find the magnitude of $\mathbf{F}_{\text{tan}}$, let $\mathbf{x}_0$ be at latitude $\theta \in (-\frac{1}{2}\pi, \frac{1}{2}\pi)$. Without loss of generality, we may assume that $\mathbf{x}_0$ is at longitude 0, so $\mathbf{n} = (\sin\theta, 0, \cos\theta)$. It follows from Eq. (14.7) that the magnitude of $\mathbf{F}_{\text{tan}}$ is

$$F_{\text{tan}} = 2m\,\Omega|\sin\theta|v. \tag{14.8}$$

The quantity

$$f = 2\Omega \sin\theta \tag{14.9}$$

is known as the *Coriolis parameter*.

14.1.4 • β-Plane Approximation

For many flows in the atmosphere and ocean, the meridional gradient of the Coriolis parameter is the quantity of interest, more so than the value of the Coriolis parameter itself. We use Eq. (14.9) to derive the relevant expression.

Consider the configuration of Figure 14.2. A tangent plane is attached to the surface of the Earth at the point P at latitude θ_0 and a local Cartesian coordinate system is set up, with the y-direction pointing poleward, the x-direction pointing along a latitudinal circle, and the z-direction pointing straight up. The y-coordinate of P is y_0. (The x- and z-coordinates are irrelevant.) At P, the value of the Coriolis parameter is $f_0 = 2\Omega \sin\theta_0$. As P moves to A' along the meridian, the latitude increases from θ_0 to θ, so the value of the Coriolis parameter changes from f_0 to $f = 2\Omega \sin\theta$. Expand in a Taylor expansion around θ_0,

$$f = 2\Omega(\sin\theta_0 + (\theta - \theta_0)\cos\theta_0 + \cdots) = f_0 + 2\Omega(\theta - \theta_0)\cos\theta_0 + \cdots.$$

Now follow the projection of this same motion in the tangent plane. As P moves to A', the projection moves from P to A. The y-coordinate of A in the tangent plane is y. (The x- and z-coordinates are again irrelevant.) If the angle $\theta - \theta_0$ is small, a good approximation is $y - y_0 \approx R(\theta - \theta_0)$, so

$$f \approx f_0 + \beta(y - y_0), \tag{14.10}$$

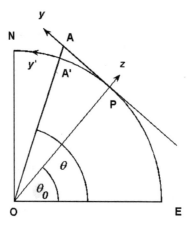

Figure 14.2. *The β-plane approximation.*

where $\beta = (2\Omega\cos\theta_0)/R$. This is the so-called *β-plane approximation*, which captures the main effect of the Earth's spherical geometry on large-scale motions in the atmosphere and ocean. The constant β is known as the *Rossby parameter* (not to be confused with the Rossby number, which is a dimensionless number used in describing fluid flow). While the formal requirement for the validity of the β-plane approximation is $|y - y_0| \ll R$, the approximation is often a useful starting point for motions with scales comparable with the Earth's radius because of the simplicity of the resultant equations in Cartesian coordinates compared with spherical coordinates. We will use it in Chapter 16, when we discuss equatorial Kelvin and Rossby waves in connection with ENSO.

14.2 ▪ State Variables

In a physics-based approach to climate modeling, the Earth's atmosphere and ocean are regarded as fluids—continuous media that are characterized by their mass *density ρ*, a scalar quantity; hydrodynamic *velocity* **v**, a vector quantity; and *temperature T*, a scalar quantity. Additional state variables are moisture content (in the case of the atmosphere) and salinity (in the case of the ocean). If details of chemistry and biology are included, there may be other variables like species concentrations or mass fractions as well.

All these quantities vary through space and in time, so a climate model consists of a set of differential equations for the *prognostic* variables and a set of functional relations for the *diagnostic* variables. Among the diagnostic variables is the *pressure p*, which is determined by the density, temperature, and mass fractions through an *equation of state*.

The equations for the prognostic variables follow in principle from the laws of physics and chemistry. For example, the continuity equation, which governs the evolution of the mass density, is a translation in mathematical terms of the law of conservation of mass; the equation of motion is essentially Newton's law; and since the temperature is determined by the internal energy (heat content) of the fluid, its evolution follows from the law of conservation of internal energy. These equations are not independent; each state variable influences directly or indirectly the evolution of the other variables, so a mathematical climate model is indeed a *coupled* system of (usually nonlinear) differential equations.

Figure 14.3 gives a schematic view of the various couplings. Some couplings manifest themselves explicitly in the mathematical equations (indicated in black), while other cou-

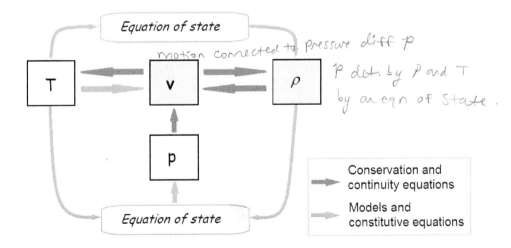

Figure 14.3. *Schematic of the interdependence of the hydrodynamic equations.*

plings appear indirectly in the equations through so-called constitutive relations or are dictated by physics (indicated in blue). The continuity equation and the equation of motion (Navier–Stokes equation) play a central role in the system. They govern the evolution of mass density and hydrodynamic velocity. The motion of air and water is intimately connected to pressure differences, while the pressure itself is determined by mass density and temperature through the equation of state. The evolution of temperature depends on the fluid motion and in turn affects the fluid motion through various transport processes; for example, viscosity, which is defined by internal friction, is strongly temperature dependent.

Needless to say, discussing the details of this system would lead us well beyond the scope of the present text. Nevertheless, to give the reader an impression of what goes into a climate model, we will focus in this chapter on the equations for mass density and hydrodynamic velocity.

14.3 ▪ Continuity Equation

The *continuity equation* is the governing equation for the total mass density. It is a direct consequence of the mass conservation law.

Consider an arbitrary but fixed volume element Ω in the interior of the Earth's atmosphere or ocean. The mass density is given by ρ, which is a function of position $\mathbf{x}$ and time t, so the total mass inside Ω at time t is given by the volume integral $\int_\Omega \rho(\mathbf{x}, t) \, dV$. The derivative of this integral with respect to t gives the increase per unit time—that is, the rate of change—of the total mass in Ω.

If there are no sources or sinks inside Ω, the total mass changes only because matter flows into and out of Ω through the boundary $\partial\Omega$ of Ω. The flow velocity is $\mathbf{v}$; like ρ, $\mathbf{v}$ varies with $\mathbf{x}$ and t. The mass flux is a vector, namely the product of the mass density ρ and the flow velocity $\mathbf{v}$. Assuming that $\partial\Omega$ is smooth, so the outward unit normal vector $\mathbf{n}$ is well defined at every point of $\partial\Omega$, we obtain the net loss of mass per unit time due to the flow into and out of Ω by integrating the normal component of the mass flux over the surface $\partial\Omega$, $\int_{\partial\Omega}(\rho\mathbf{v})\cdot\mathbf{n}\,dA$. ($dA$ denotes an infinitesimal surface element of $\partial\Omega$.) If $\mathbf{v}$ and $\mathbf{n}$ point into the same hemisphere, mass is lost; otherwise, mass is gained. Mass conservation is expressed by the equation

$$\frac{d}{dt}\int_\Omega \rho \, dV = -\int_{\partial\Omega}(\rho\mathbf{v})\cdot\mathbf{n}\,dA.$$

Applying Gauss's theorem, we transform the integral over the boundary $\partial\Omega$ to an integral over the volume Ω,

$$\frac{d}{dt}\int_\Omega \rho \, dV = -\int_\Omega \nabla\cdot(\rho\mathbf{v})\,dV.$$

Assuming that ρ is a smooth function, we bring the operation of differentiation under the integral sign,

$$\int_\Omega\left[\frac{\partial\rho}{\partial t}+\nabla\cdot(\rho\mathbf{v})\right]dV = 0.$$

Since Ω is arbitrary, it follows that

$$\frac{\partial\rho}{\partial t}+\nabla\cdot(\rho\mathbf{v}) = 0. \tag{14.11}$$

This equation is known as the *continuity equation*. The equation holds in a fixed frame of reference and is referred to as the Eulerian form of the equation.

We can transform the equation into an equation that holds in a frame of reference that moves with the fluid. Using the identity $\nabla \cdot (\rho \mathbf{v}) = \rho \nabla \cdot \mathbf{v} + \mathbf{v} \cdot \nabla \rho$, we see that Eq. (14.11) is equivalent to

$$\frac{D\rho}{Dt} = -\rho \nabla \cdot \mathbf{v}, \tag{14.12}$$

where the symbol D/Dt is the *Lagrangian derivative* or *material derivative*,

$$\ast \quad \frac{D}{Dt} = \frac{\partial}{\partial t} + \mathbf{v} \cdot \nabla. \qquad \frac{DP}{Dt} = \frac{\partial P}{\partial t} + \overline{V} \cdot \nabla P$$

The Lagrangian derivative gives the rate of change in a frame of reference that moves with the fluid. It establishes a direct connection between the equation of motion and the continuity equation, as indicated by the gray arrows in Figure 14.3. Equation (14.12) is referred to as the Lagrangian form of the continuity equation.

14.4 ▪ Equation of Motion

Next, we derive the equation for the hydrodynamic velocity $\mathbf{v}$. The equation is essentially Newton's law in a rotating frame of reference.

Again, consider an arbitrary but fixed volume element Ω in the interior of the Earth's atmosphere or ocean. The conservation law for linear momentum is

$$\frac{d}{dt} \int_\Omega \rho \mathbf{v} \, dV = - \int_{\partial \Omega} (\rho \mathbf{v}) \mathbf{v} \cdot \mathbf{n} \, dA + \int_\Omega \rho \mathbf{b} \, dV + \int_{\partial \Omega} \mathbb{T} \cdot \mathbf{n} \, dA.$$

Here, $\mathbf{b}$ is the body force density per unit mass acting on the fluid and $\mathbb{T}$ is the Cauchy stress tensor, both to be discussed below.

Applying Gauss's theorem again and rearranging terms, we obtain

$$\frac{d}{dt} \int_\Omega \rho \mathbf{v} \, dV + \overbrace{\int_\Omega \nabla \cdot (\rho \mathbf{v} \mathbf{v}) \, dV}^{Gauss} - \int_\Omega \rho \mathbf{b} \, dV - \overbrace{\int_\Omega \nabla \cdot \mathbb{T} \, dV}^{Gauss} = 0.$$

Assuming $\rho \mathbf{v}$ is smooth, we bring the operation of differentiation under the integral sign,

$$\int_\Omega \left[\frac{\partial}{\partial t} (\rho \mathbf{v}) + \nabla \cdot (\rho \mathbf{v} \mathbf{v}) - \rho \mathbf{b} - \nabla \cdot \mathbb{T} \right] dV = 0.$$

Since Ω is arbitrary, it follows that

$$\frac{\partial}{\partial t} (\rho \mathbf{v}) + \nabla \cdot (\rho \mathbf{v} \mathbf{v}) - \rho \mathbf{b} - \nabla \cdot \mathbb{T} = 0. \tag{14.13}$$

To obtain the equation of motion, we need more details about $\mathbf{b}$ and $\mathbb{T}$.

The body forces acting on the fluid are due to the Coriolis effect and gravity, so

$$\mathbf{b} = -2\Omega \times \mathbf{v} + \mathbf{g}, \tag{14.14}$$

where $\mathbf{g}$ is the acceleration due to gravity.

Air and water are deformable media. In a deformable medium, particles act across imaginary internal boundaries as a reaction to external forces. The resulting force per unit area is called a "stress." Cauchy showed that the stress at any point is completely defined by the nine components σ_{ij} of a second-order symmetric tensor, which is known as the (Cauchy) *stress tensor*,

$$\mathbb{T} = \begin{pmatrix} \sigma_{xx} & \sigma_{xy} & \sigma_{xz} \\ \sigma_{yx} & \sigma_{yy} & \sigma_{yz} \\ \sigma_{zx} & \sigma_{zy} & \sigma_{zz} \end{pmatrix} \equiv \begin{pmatrix} \sigma_x & \tau_{xy} & \tau_{xz} \\ \tau_{yx} & \sigma_y & \tau_{yz} \\ \tau_{zx} & \tau_{zy} & \sigma_z \end{pmatrix}. \tag{14.15}$$

Here, $\sigma_{\alpha\beta}$ ($\alpha, \beta = x, y$, or z) is the force per unit area in the α-direction exerted on a plane surface perpendicular to the β-direction; the vector with components $\sigma_{\alpha x}$, $\sigma_{\alpha y}$, and $\sigma_{\alpha z}$ is the resultant force in the α-direction. In the last representation, the σ are the *normal stresses*, and the τ are the *shear stresses*. The stress tensor obeys the tensor transformation rules under a change of coordinates.

If the medium is in hydrostatic equilibrium, the shear stresses are zero, the normal stresses are equal, and $\mathbb{T} = \sigma \mathbb{I}$, where $\mathbb{I}$ is the unit tensor of rank two (which is represented by the 3×3 identity array). The *hydrostatic pressure* p is the negative value of the normal stress, $p = -\sigma$.

In general, the *pressure* is equal to the negative of the mean normal stress,

$$p = -\tfrac{1}{3}\left(\sigma_x + \sigma_y + \sigma_z\right). \tag{14.16}$$

The remainder

$$\mathbb{S} = \mathbb{T} + p\,\mathbb{I} \tag{14.17}$$

is a symmetric traceless tensor of rank two. It represents the effects of friction and is known as the *stress deviator tensor*.

Putting it all together, we obtain the Eulerian form of the *equation of motion*,

$$\frac{\partial}{\partial t}(\rho\mathbf{v}) + \nabla\cdot(\rho\mathbf{vv}) + 2\rho\,\mathbf{\Omega}\times\mathbf{v} - \rho\,\mathbf{g} + \nabla p - \nabla\cdot\mathbb{S} = 0. \tag{14.18}$$

The equivalent Lagrangian form is

$$\frac{D\mathbf{v}}{Dt} = -2\mathbf{\Omega}\times\mathbf{v} + \mathbf{g} - \frac{1}{\rho}(\nabla p - \nabla\cdot\mathbb{S}). \tag{14.19}$$

Note that only the gradient of the pressure enters into the equation. The effect is that air (or water) moves from areas of high pressure to areas of low pressure. The pressure itself is given by the equation of state.

Equation (14.18) becomes the classical Navier–Stokes equation of fluid dynamics if the stress deviator tensor $\mathbb{S}$ is a linear function of the velocity gradients. Since $\mathbb{S}$ is traceless and symmetric, the proper definition is

$$\mathbb{S} = 2\nu\left(\mathbb{E} - \tfrac{1}{3}(\nabla\cdot\mathbf{v})\mathbb{I}\right), \tag{14.20}$$

where $\mathbb{I}$ is the unit tensor and $\mathbb{E}$ the (symmetric) rate-of-strain tensor,

$$\mathbb{E} = \tfrac{1}{2}\left(\nabla\mathbf{v} + (\nabla\mathbf{v})^{\mathsf{T}}\right). \tag{14.21}$$

The trace of $\mathbb{E}$, which is $\nabla\cdot\mathbf{v}$, represents the rate of expansion of the flow. The proportionality constant ν in the definition (14.20) is the *kinematic viscosity*. Note that the viscosity ν is introduced in a purely ad hoc manner. It is simply the coefficient in a presumed linear relationship between strains and velocity gradients; it is not derived from the laws of physics, and the viscosity term differs therefore fundamentally from the other terms in the equation of motion. The viscosity coefficient is strongly temperature dependent; thus, it couples the Navier–Stokes equation indirectly to the temperature equation, as indicated by the blue arrow in Figure 14.3.

14.5 ▪ Other Prognostic Variables

Equations for other prognostic variables like temperature, moisture content, salinity, species concentrations, and so on are derived in a similar manner from the laws of physics and chemistry. For example, the *temperature equation* relates the rate of change of internal energy to the temperature and its spatial gradients. It includes terms for convective and radiative heat transfer and heat release due to evaporation and condensation. Several terms involve transport coefficients analogous to the viscosity coefficient in the Navier–Stokes equation. For example, according to Fourier's law of heat conduction, the conductive heat flux is linearly proportional to the temperature gradient; the proportionality constant is the coefficient of *thermal conductivity*. This coefficient is similar to the viscosity coefficient in the Navier–Stokes equation, in the sense that it is introduced in a purely ad hoc manner. The thermal conductivity is strongly dependent on the moisture content, so it provides an indirect coupling between the temperature equation and the equation governing the moisture content in the atmosphere (not indicated in Figure 14.3). Other transport coefficients are the *thermal diffusion* coefficient, which correlates the diffusive heat flux with temperature gradients, and the coefficient of *mass diffusion*, usually simply referred to as the *diffusion coefficient*, which correlates mass diffusion and density gradients.

These transport coefficients are convenient devices to connect fluxes to gradients; they are not derived from first principles and are therefore subject to considerable uncertainty. They are part of the parameterization process, to represent subscale phenomena that are not captured otherwise by the model. It is possible to derive expressions for the transport coefficients in terms of intermolecular potentials. This is the main topic of investigation in the kinetic theory of gases. The basic equation is the *Boltzmann equation*, which models the behavior of a gas at the molecular level. In practice, however, the various transport coefficients are more readily obtained from experimental data, since they can be quite sensitive to conditions like temperature and pressure.

14.6 ▪ Equation of State

The continuity equation and the equation of motion specify the time derivatives of the fluid density and velocity, so they must be integrated forward in time to yield information about the variables at a later time. For this reason, density and velocity are *prognostic variables*. The same is true for temperature, moisture content, and species concentrations. The pressure, on the other hand, is a *diagnostic variable*. No integration in time is needed. The pressure follows from the so-called *equation of state* once the prognostic variables are known,

$$p = f(\rho, T, \ldots), \qquad (14.22)$$

where the dots indicate possibly other state variables. The ideal gas law, Eq. (13.2), is the special case for a monatomic gas. The special position of the equation of state is highlighted by the blue box in the schematic of Figure 14.3.

14.7 ▪ Coupling Ocean and Atmosphere

In a climate model, the equations for ocean and atmosphere must be coupled through boundary conditions, which express the physical fact that the fluxes of momentum, internal energy, mass fractions, and so on are continuous across the interface. For instance, the net vertical heat flux leaving the atmosphere at the air-sea interface must be taken up by the advective and diffusive fluxes of heat in the ocean. The processes by which this happens can be quite complicated, and the formulation of the boundary conditions requires intimate familiarity with the physical processes involved. In practice, these processes are most often captured by parameterization.

14.8 ▪ Need for Approximations

The differential equations for the prognostic variables, complemented with an equation of state and appropriate initial and boundary conditions, give a more or less complete description of the Earth's atmosphere-ocean system. Suppose we wish to solve these equations numerically over a given region, or over the entire globe. Typically, this is done by dividing the spatial region into many small subregions of similar shape and assuming that the motion of interest is very simple in each subregion; for example, it is independent of position and constant for a short time interval. If a spatial region of $1,000 \times 1,000$ km with atmospheric height of 10 km is divided into cube-shaped subregions of 1 km on the side, then there will be 10^7 such subregions. None of these subregions will be small enough to capture atmospheric or topographic features such as clouds and mountains. If the motion of air is assumed to be constant in each subregion for one minute at a time, then the information for each subregion must be updated approximately $1.5 \cdot 10^3$ times for a simulation of a single day, and approximately $6 \cdot 10^5$ times for a simulation that extends over a year. Clearly, such a simulation will stretch the capabilities of the most powerful of today's supercomputers—and the result will most likely be subject to significant uncertainty.

The difficulty stems in large part from the fact that we are dealing with three spatial dimensions, in addition to time. Therefore, various approaches have been developed to reduce the dimensionality. Averaging over altitude will result in a model with two spatial dimensions, and averaging also over longitude will reduce the number of spatial dimensions further to one. One such model, which has only one spatial dimension (latitude), has been discussed in Chapter 12. Averaging over all spatial dimensions leads to global models of the type discussed in Chapter 2, where the entire planet is treated as if it were a single point.

Typically, averaging a PDE such as a conservation law does not lead to an equation that is still a conservation law. Intuitively this is clear, since conservation laws are derived under the assumption that interactions occur only between neighboring regions, for times that are close. Usually, it is better and cleaner to derive a lower-dimensional model from scratch, since then the model assumptions become clear. In the following section we will discuss two approximations, namely the shallow water equations and the Boussinesq equations.

14.9 ▪ Shallow Water Equations

The atmosphere and ocean are essentially thin layers of fluid, so we should be able to make a distinction between what happens in the "horizontal" directions (in the tangent plane) and in the "vertical" direction (pointing away from the center of the Earth). This is the idea behind the *shallow water equations*, which is the simplest set of equations governing the evolution of an incompressible fluid in response to gravitational and rotational accelerations. In this section we show how the shallow water equations can be obtained from the continuity equation and the equation of motion,

$$\frac{D\rho}{Dt} + \rho \nabla \cdot \mathbf{v} = 0, \tag{14.23}$$

$$\rho \left(\frac{D\mathbf{v}}{Dt} + 2\mathbf{\Omega} \times \mathbf{v} \right) = \rho \mathbf{g} - \nabla p; \tag{14.24}$$

cf. Eqs. (14.12) and (14.19). We ignore the effects of friction ($\mathbb{S} = 0$).

Denote the coordinate in the vertical direction by z. Gravitational forces are directed toward the center of the Earth, so $\mathbf{g} = -g\mathbf{e}_z$, where $\mathbf{e}_z$ is the unit vector in the direction of increasing z. The velocity vector $\mathbf{v}$ decomposes into its component $\mathbf{u}$ in the tangent plane and its vertical component w,

$$\mathbf{v} = (\mathbf{u}, 0) \oplus (0, w). \tag{14.25}$$

In a hydrostatically balanced system, the density and pressure are related by the differential expression $dp = -\rho g\, dz$. Therefore, if we denote the density and pressure of the system at equilibrium by ρ_r and p_r, respectively, then ρ_r and p_r are functions of z such that $dp_r/dz = -\rho_r g$. To separate the equilibrium and nonequilibrium contributions, we decompose the density and pressure,

$$\rho = \rho_r(z) + \rho', \quad p = p_r(z) + p'. \tag{14.26}$$

Here, ρ' and p' are functions of $\mathbf{x}$ and t. Thus, the equation of motion, Eq. (14.24), reduces to

$$\rho \left(\frac{D\mathbf{v}}{Dt} + 2\mathbf{\Omega} \times \mathbf{v} \right) = -\rho' g\mathbf{e}_z - \nabla p'. \tag{14.27}$$

So far, no approximations have been made, and Eqs. (14.23) and (14.27) hold pointwise. The obtain the shallow water equations, we need to make two assumptions:

- The fluid is *hydrostatically balanced*, $\rho' = 0$.

- The fluid is *incompressible*, $\nabla \cdot \mathbf{v} = 0$.

The first assumption implies that $\rho = \rho_r$; in particular, ρ is constant in time. The density is thus transformed from a prognostic variable to a diagnostic variable.

Since water is nearly incompressible, the second assumption explains, at least in part, the reference to "water" in "shallow water equations." The adjective "shallow" refers to the fact that the effects of vertical shear of the horizontal velocity are ignored. Such effects can occur, for example, if horizontal momentum is advected vertically. Ignoring these effects is justified as long as the horizontal length scale is much greater than the vertical length scale—that is, as long as the basin is shallow.

With the two assumptions above, the continuity equation, Eq. (14.23), reduces to $D\rho_r/Dt = 0$. But ρ_r is not only constant in time; it also varies spatially only in the vertical direction, so the continuity equation reduces further to $d\rho_r/dz = 0$. In other words, ρ_r is constant and, by extension, ρ is constant.

14.9.1 ▪ Continuity Equation

Consider a vertical fluid column with unit cross-section. The column is bounded below by the Earth's surface, which is fixed in time and located at $z = h_s$, and bounded above by a "free" surface of constant or nearly constant pressure. The location of the free surface may vary with time and is given by $z = h_f(t)$ at time t. The difference $h = h_f - h_s$ is the height of the fluid column. Both h_s and h_f may depend on the base point in the horizontal plane.

In terms of the velocities $\mathbf{u}$ and w, the incompressibility condition is

$$\nabla \cdot \mathbf{u} + \frac{\partial w}{\partial z} = 0, \tag{14.28}$$

where $\nabla \cdot \mathbf{u}$ is the horizontal divergence. Since the horizontal velocity is independent of z, we can integrate Eq. (14.28) with respect to z from $z = h_s$ to $z = h_f$,

$$(h_f - h_s)\nabla \cdot \mathbf{u} + w_f - w_s = 0, \tag{14.29}$$

where $w_f = w|_{z=h_f}$ and $w_s = w|_{z=h_s}$. No fluid crosses the free surface, so the vertical velocity at a point on the free surface is equal to the total derivative of h_f at that point,

$$w_f = \frac{Dh_f}{Dt} = \frac{\partial h_f}{\partial t} + \mathbf{u} \cdot \nabla h_f. \tag{14.30}$$

Similarly, no fluid crosses the lower boundary, so the fluid motion there must follow the topography,

$$w_s = \frac{Dh_s}{Dt} = \frac{\partial h_s}{\partial t} + \mathbf{u} \cdot \nabla h_s. \tag{14.31}$$

(Ignore the fact that $\partial h_s / \partial t = 0$.) Subtracting Eq. (14.31) from Eq. (14.30) and combining the resulting expression with Eq. (14.29), we obtain the equation

$$\frac{\partial}{\partial t}(h_f - h_s) + \nabla \cdot ((h_f - h_s)\mathbf{u}) = 0. \tag{14.32}$$

With $h = h_f - h_s$, the equation takes the form

$$\frac{\partial h}{\partial t} + \nabla \cdot (h\mathbf{u}) = 0, \tag{14.33}$$

which is the *continuity equation* for the shallow water system.

14.9.2 ▪ Equation of Motion

Next, we turn to the equation of motion, Eq. (14.27). The horizontal component decouples from the vertical component if all forces arising from the meridional component of the angular velocity are eliminated. At latitude θ, the angular velocity is $\mathbf{\Omega} = (0, \Omega \cos\theta, \Omega \sin\theta)$, so the approximation implies that we set $\Omega \cos\theta = 0$. As a result, the effect of the Coriolis forces is confined to the horizontal directions,

$$2\mathbf{\Omega} \times \mathbf{v} = f(J\mathbf{u}, 0), \tag{14.34}$$

where $f = 2\Omega \sin\theta$ is the Coriolis parameter and $J = \begin{pmatrix} 0 & -1 \\ 1 & 0 \end{pmatrix}$ describes a clockwise rotation by 90° and hence $J^2 = -I$, the negative of the 2×2 identity matrix I.

The equation of motion in the horizontal plane thus becomes

$$\frac{D\mathbf{u}}{Dt} = -fJ\mathbf{u} - \frac{1}{\rho}\nabla p'. \tag{14.35}$$

Integrating the hydrostatic equation from some arbitrary height z within the fluid up to the free surface, we obtain the identity $p_f - p(z) = -\rho g(h_f - z)$, where $p_f = p|_{z=h_f}$; hence,

$$p(z) = \rho g(h_f - z) + p_f, \quad z \in (h_s, h_f). \tag{14.36}$$

Since $p = -\rho g z + p'$, it follows that $p' = \rho g h_f + p_f$. Assuming that p_f is horizontally uniform, we take the horizontal gradient and obtain the expression

$$\nabla p' = \rho g \nabla h_f. \tag{14.37}$$

This expression shows that the horizontal pressure gradient is driven by the gradient of the free surface. Equation (14.35) now takes the form

$$\frac{D\mathbf{u}}{Dt} + fJ\mathbf{u} = -g \nabla h_f. \tag{14.38}$$

This is the Lagrangian form of the *equation of motion* for the shallow water system. Using the vector identity $(\mathbf{u} \cdot \nabla)\mathbf{u} = \nabla(\frac{1}{2}\mathbf{u} \cdot \mathbf{u}) + (\nabla \times \mathbf{u}) \times \mathbf{u}$, we obtain the Eulerian form of this equation,

$$\frac{\partial \mathbf{u}}{\partial t} + (\zeta + f)J\mathbf{u} = -\nabla(g h_f + E), \tag{14.39}$$

where $E = \frac{1}{2}\mathbf{u} \cdot \mathbf{u}$ is the *kinetic energy* per unit mass and $\zeta = \mathbf{e}_z \cdot (\nabla \times \mathbf{u})$ is the *relative vorticity*.

Since we have assumed no vertical shear, the equations apply at any height within the fluid, but the range of heights in question is presumably small, since the fluid is "shallow."

The coupled system of equations (14.33) and (14.38) is referred to as the *shallow water equations*. When expressed in spherical coordinates of the Earth, with the radius taken as constant in the metric coefficients, they are called *primitive equations*. Because of the hydrostatic approximation, the shallow water equations do not describe vertical convection. Vertical convection occurs when the buoyancy increases, which happens in the ocean by cooling or evaporation and in the atmosphere by heating the air. These processes are commonly included via complex parameterization schemes.

14.10 ▪ Further Approximations

Further approximations can be obtained from the shallow water equations if some of the forces underlying the flow are close to being balanced. For example, the *geostrophic approximation* applies when the forces due to the pressure gradient and the Coriolis effect are approximately in balance; in the notation of Eq. (14.35), this is when $fJ\mathbf{u} \approx -\nabla(p/\rho)$. The use of the geostrophic approximation is indeed quite common, since the major currents of the world's oceans, such as the Gulf Stream and the Antarctic Circumpolar Current (ACC), are all approximately in geostrophic balance.

The assumption of a balanced flow is often quite accurate and is useful in improving the qualitative understanding and interpretation of atmospheric and oceanic motion. Its formal justification is usually given through a *scaling analysis*, where the strength of each term in the governing equations is measured in terms of some characteristic lengths L and H in the tangent plane and vertical direction, respectively, and some characteristic velocity U in the tangent plane. The relative strengths of the various terms can then be expressed in terms of dimensionless parameters, the most important of which are the *Rossby number*, $\text{Ro} = U/Lf$, and the *Ekman numbers*, $\text{Ek} = \nu_h/(2\Omega L^2)$ or $\nu_v/(2\Omega H^2)$.

The Rossby number measures the strength of the inertial forces, $\mathbf{v} \cdot \nabla \mathbf{v} \sim U^2/L$, relative to the Coriolis force, $\mathbf{\Omega} \times \mathbf{v} \sim f U$. When the Rossby number is small, the system is strongly affected by the Coriolis force, the effects of planetary rotation are large, and the net acceleration is comparably small. In this case, the use of the geostrophic approximation may be appropriate. On the other hand, when the Rossby number is large (either because f is small or because L is small or for large velocities), inertial and centrifugal forces dominate, and the effects of planetary rotation are unimportant and can be neglected.

The Ekman number measures the strength of the viscous forces, $\nabla \cdot (\nu \nabla \mathbf{u}) \sim \nu U/L^2$, relative to the Coriolis force, $\mathbf{\Omega} \times \mathbf{v} \sim f U$. When the Ekman number is small, disturbances are able to propagate before decaying owing to frictional effects. The Ekman number gives the order of magnitude of the thickness of an *Ekman layer*, a boundary layer in which viscous diffusion is balanced by Coriolis effects rather than the usual convective inertia.

Additional dimensionless parameters to consider are the *planetary scale ratio*, L/R, and the *aspect ratio*, H/R; here, R is the radius of the Earth. The *quasi-geostrophic theory* results if all these parameters are small and of the same order of magnitude [81].

14.11 ▪ Boussinesq Equations

The mass density of ocean water depends on temperature and salinity but varies only little from its average value across the globe. We can make use of this observation and derive a mathematical model of the ocean where the vertical coordinate is eliminated from the flow equations but some of the influences of the vertical structure of the flow are retained.

Suppose the fluid density ranges over an interval of width ρ_d around an average value ρ_0, where the ratio ρ_d/ρ_0 is negligible but the product $g\rho_d/\rho_0$ is not. The expression

$$g' = g \frac{\rho - \rho_0}{\rho_0} \tag{14.40}$$

is known as the *reduced gravity*.

Using the notation of Section 14.9, we replace the density ρ in the inertial terms by a constant ρ_0. In addition, we ignore the contribution p' to the pressure in Eq. (14.27). The resulting equations are known as the *Boussinesq equations*,

$$\frac{D\mathbf{u}}{Dt} = -f J\mathbf{u} - \nabla\left(\frac{p}{\rho_0}\right), \tag{14.41}$$

$$\frac{\partial}{\partial z}\left(\frac{p}{\rho_0}\right) = -g \frac{\rho}{\rho_0}, \tag{14.42}$$

$$\nabla \cdot \mathbf{u} + \frac{\partial w}{\partial z} = 0. \tag{14.43}$$

The horizontal equation of motion, Eq. (14.41), may have additional diffusive flux terms representing subscale phenomena. The quantity in the right-hand side of Eq. (14.42) is known as the *buoyancy*,

$$b = -g \frac{\rho}{\rho_0}. \tag{14.44}$$

Note: There are several mathematical models which are loosely referred to as Boussinesq equations. Many are variants of the equations first proposed by Boussinesq, so it would be more appropriate to call them *Boussinesq-type equations*.

14.11.1 ▪ Waves

The Boussinesq equations can be used to study the propagation of waves in the ocean. We simplify the equations by rescaling the pressure, replacing p/ρ_0 by p. Next, we focus on the linear terms by bringing all nonlinear terms to the right-hand side,

$$\frac{\partial u}{\partial t} - fv + \frac{\partial p}{\partial x} = \mathscr{F}^u, \tag{14.45}$$

$$\frac{\partial v}{\partial t} + fu + \frac{\partial p}{\partial y} = \mathscr{F}^v, \tag{14.46}$$

$$\frac{\partial b}{\partial t} + N^2 w = \mathscr{G}_b, \tag{14.47}$$

$$\frac{\partial p}{\partial z} - b = 0, \tag{14.48}$$

$$\nabla \cdot \mathbf{u} + \frac{\partial w}{\partial z} = 0. \tag{14.49}$$

The terms on the right-hand side contain the nonlinearities due to advection (terms derived from $\mathbf{u} \cdot \nabla \mathbf{u}$ and $w(\partial \mathbf{u}/\partial z)$) and the diffusive fluxes (if present). The quantity N is known as the *Brunt–Väisälä frequency* or *buoyancy frequency*; it is a function of z and is the only remnant of the stratification of the reference density field. In terms of the gradients of the potential temperature and salinity,

$$N^2(z) = g\left[\alpha\frac{d\theta_r}{dz} - \beta\frac{dS_r}{dz}\right], \tag{14.50}$$

where α is the thermal expansion coefficient and β the saline contraction coefficient.

The buoyancy frequency quantifies the importance of stability and is a fundamental variable in the dynamics of stratified flows. In simplest terms, the buoyancy frequency is the vertical frequency excited by a vertical displacement of a fluid parcel. It is the maximum frequency of internal waves in the ocean; typical values are a few cycles per hour.

14.12 ▪ Exercises

1. (i) To an observer on the Earth's surface, the orbit of a satellite looks like a curved path. For example, unless the orbit is circumpolar, it reaches a maximum southern latitude and then turns north again. Is this phenomenon due to the Coriolis effect?

    (ii) Explain why there is no Coriolis effect on the surface of a cylinder that rotates around its axis of symmetry.

    (iii) Find the magnitude of the Coriolis force for an object that is moving on a rotating plane in terms of Ω, $\mathbf{v}$, and any other relevant quantities.

2. Let $\rho_0 : \mathbb{R}^3 \to [0, \infty)$ be a smooth mass density function with $\int_{\mathbb{R}^3} \rho_0(\mathbf{x})\,d\mathbf{x} < \infty$.

    (i) Suppose the flow velocity is constant, $\mathbf{v}(\mathbf{x}, t) = \mathbf{a}$ for some $\mathbf{a} \in \mathbb{R}^3$. Find the solution ρ of the Eulerian continuity equation, Eq (14.11), that satisfies the condition $\rho(\mathbf{x}, 0) = \rho_0(\mathbf{x})$ for all $\mathbf{x} \in \mathbb{R}^3$ and $t > 0$. What is the material derivative in this case? What is the Lagrangian form of the continuity equation?

(ii) Suppose the flow velocity is constant in space but varies with time, $\mathbf{v}(\mathbf{x}, t) = \mathbf{a}(t)$ for some smooth function $\mathbf{a} : t \mapsto \mathbf{a}(t) \in \mathbb{R}^3$. Find the solution ρ of the Eulerian continuity equation, Eq (14.11), that satisfies the condition $\rho(\mathbf{x}, 0) = \rho_0(\mathbf{x})$ for all $\mathbf{x} \in \mathbb{R}^3$ and $t > 0$. What is the material derivative in this case? What is the Lagrangian form of the continuity equation?

(iii) Suppose the flow velocity is constant in time but varies through space, $\mathbf{v}(\mathbf{x}, t) = \gamma \mathbf{x}$ for some $\gamma \in \mathbb{R}$. Find the solution of the Eulerian continuity equation, Eq (14.11), that satisfies the condition $\rho(\mathbf{x}, 0) = \rho_0(\mathbf{x})$ for all $\mathbf{x} \in \mathbb{R}^3$ and $t > 0$. Give a physical interpretation of the flow in the case of positive and negative γ. Hint: Look for a solution in the form $\rho(\mathbf{x}, t) = e^{\alpha t} \rho_0(e^{\beta t} \mathbf{x})$ for suitable α and β.

3. Consider an *incompressible flow* in a spatial domain $\Omega \subset \mathbb{R}^3$. In an incompressible flow, the volume of a small parcel of matter does not change as it is transported by the flow, so $\nabla \cdot \mathbf{v} = 0$ for all $\mathbf{x} \in \Omega$ and $t > 0$. Suppose that $\rho(\mathbf{x}, 0) = C$ for some positive constant C. Show that the solution ρ of the Eulerian continuity equation, Eq (14.11), satisfies the condition $\rho(\mathbf{x}, t) = C$ for all $\mathbf{x} \in \Omega$ and $t > 0$.

4. Consider the Navier–Stokes equation of fluid mechanics, Eq. (14.18), in a domain $\Omega = (0, 1) \times \mathbb{R}^2$. Ignore the Coriolis and gravitational forces, $\mathbf{\Omega} = 0$ and $\mathbf{g} = 0$, and let the stress deviator tensor $\mathbb{S}$ be given by Eq. (14.20) for some value for the kinematic viscosity ν. Suppose that the velocity field has the form $\mathbf{v}(\mathbf{x}, t) = \mathbf{v}(x_1, x_2, x_3, t) = (0, u(x_1), v(x_1))^T$ for all $t > 0$. Show that if the density is constant and equal to one everywhere, $\rho(\mathbf{x}, t) = 1$ for all $\mathbf{x} \in \Omega$ and $t > 0$, then there is a function p which does not depend on t, such that the Navier–Stokes equation is satisfied. What is the general form of $\mathbf{v}$ if the pressure p is linear in all its variables, $p(\mathbf{x}, t) = \mathbf{d} \cdot \mathbf{x}$ for some fixed vector $\mathbf{d} \in \mathbb{R}^3$?

5. Rewrite Eqs. (14.23), (14.24), and (14.27) in their Eulerian forms.

6. Let $\mathbf{v} : \mathbb{R}^3 \times (0, T) \to \mathbb{R}^3$ be a smooth vector field, and let $g, h : \mathbb{R}^3 \times (0, T) \to \mathbb{R}$ be smooth functions. Let D/Dt denote the material derivative.

(i) Prove the product rule,

$$\frac{D}{Dt}(gh) = g\frac{D}{Dt}h + h\frac{D}{Dt}g.$$

(ii) Find a formula for $D^2 h / Dt^2 = (D/Dt)(Dh/Dt)$.

7. The classical Navier–Stokes equation of fluid dynamics governs the behavior of an incompressible viscous fluid. For a fluid in a spatial domain $\Omega \subset \mathbb{R}^3$ with body forces $\mathbf{b}$, constant density ρ, and kinematic viscosity $\nu > 0$, the equation is usually written in the form

$$\frac{\partial \mathbf{v}}{\partial t} + (\mathbf{v} \cdot \nabla)\mathbf{v} + \nabla p - \nu \Delta \mathbf{v} = \mathbf{b}, \quad \nabla \cdot \mathbf{v} = 0. \tag{14.51}$$

Deduce this equation from Eqs. (14.18)–(14.20), assuming that the Coriolis force can be ignored, $\mathbf{\Omega} = 0$, and that the gravity term $\mathbf{g}$ can be replaced by the general body force term $\mathbf{b}$.

Note: The Clay Mathematics Institute (CMI) of Cambridge, Massachusetts, will award a Millennium Prize of $1,000,000 to anyone who establishes a complete mathematical theory for the Navier–Stokes equation.

8. Let $\mathbf{v} : \mathbb{R}^3 \times (0, T) \to \mathbb{R}^3$ be a smooth solution of the Navier–Stokes equation (14.51) with $\mathbf{b} = 0$, for which $\int_{R^3} |\mathbf{v}(\mathbf{x}, t)|^2 d\mathbf{x} < \infty$ for all $t > 0$. Prove the *energy estimate*,

$$\int_{R^3} |\mathbf{v}(\mathbf{x}, t_1)|^2 \, d\mathbf{x} \leq \int_{R^3} |\mathbf{v}(\mathbf{x}, t_0)|^2 \, d\mathbf{x}, \quad 0 < t_0 < t_1 < T < \infty.$$

Hint: Use the differential equation to show that $d/dt \int_{R^3} |\mathbf{v}(\mathbf{x}, t)|^2 \, d\mathbf{x} \leq 0$.

9. Consider the Navier–Stokes equation (14.51) in a pipe domain $\Omega = \Omega_0 \times \mathbb{R} \subset \mathbb{R}^3$, where $\Omega_0 \subset \mathbb{R}^2$. Assume that $\mathbf{b} = (0, 0, b_3)$, where b_3 depends only on x_1 and x_2. Then it is plausible to look for a solution of the form $\mathbf{v} = (0, 0, u)$, where u also depends only on x_1 and x_2. Show that $\nabla \cdot \mathbf{v} = 0$ holds automatically, and write down the equation that u must satisfy.

10. Derive the Eulerian form of the shallow water equations, Eq. (14.39), from the Lagrangian form, Eq. (14.38).

11. Consider the continuity equation and the equations of motion for the shallow water wave approximations, Eq. (14.33) and Eq. (14.39).

   (i) Write down the equations for the special case where all quantities depend only on x and t.

   (ii) Write down the equations for the special case where the second component of $\mathbf{u}$ vanishes identically Then show that if the Coriolis parameter f does not vanish in the spatial domain of interest (that is, if the Equator does not pass through this domain), the equations reduce to a single PDE for h.

   (iii) Show that, unless the Coriolis parameter f vanishes identically, there are no nontrivial solutions for which the second component of $\mathbf{u}$ vanishes identically, and all other quantities depend only on x and t.

12. Consider the continuity equation and the equations of motion for the shallow water wave approximations, Eqs. (14.33) and (14.39). Assume that there is no rotation, so $f = 0$, and that $\mathbf{u} = \varepsilon \mathbf{u}'$, $h = H + \varepsilon h'$, where H is the average height of the fluid column and ε is a small number.

   (i) Linearize the equations by substituting these expressions into the continuity equation and the equations of motion and retaining only terms of order ε. Deduce the single PDE

   $$\frac{\partial^2 h'}{\partial t^2} = gH \Delta h',$$

   where $\Delta = \partial^2/\partial x^2 + \partial^2/\partial y^2$ is the two-dimensional Laplacian differential operator.

   (ii) Substitute a trial solution $h'(x, y, t) = e^{i(kx + \ell y - \omega t)}$ to obtain the *dispersion relation*,

   $$\omega^2 = gH(k^2 + \ell^2).$$

The waves for this dispersion relation are called *gravity waves*, since the restoring force for the wave motion is gravity. For one-dimensional motion ($\ell = 0$), their phase speed $\omega/k = \pm\sqrt{gH}$ is constant and depends only on the depth of the fluid column and on gravity. All waves travel at the same phase speed, and the waves therefore are nondispersive. It is thought that tsunamis are reasonably well described by this linearization.

13. Consider again the shallow water wave equations, Eqs. (14.33) and (14.39). Assume that the Coriolis parameter is a constant, $f = f_0$.

(i) Linearize the equations as in the previous exercise. Deduce the single PDE for the scaled depth anomaly h',

$$\frac{\partial}{\partial t}\left(\frac{\partial^2 h'}{\partial t^2} + f_0^2 h' - gH\Delta h'\right) = 0.$$

(ii) Substitute a trial solution $h'(x, y, t) = e^{i(kx + \ell y - \omega t)}$ to obtain the *dispersion relation*,

$$\omega = 0 \quad \text{or} \quad \omega^2 = f_0^2 + gH\left(k^2 + \ell^2\right).$$

(iii) For the case of one-dimensional motion ($\ell = 0$), find the phase speed ω/k and the group velocity $\partial\omega/\partial k$. Show that the group velocity decreases with wave length.

The waves with nontrivial dispersion relation, $\omega \neq 0$, are called *Poincaré waves* or *inertio-gravity waves*, since they are influenced by a combination of rotation and gravity.

14. Compute the Rossby numbers at 45° latitude for

(i) water flowing out of a bath tub, moving at a speed $U \approx 1\,\text{m/s}$ over a distance $L \approx 0.1\,\text{m}$;

(ii) a golf ball in flight, moving at a speed $U \approx 50\,\text{m/s}$ over a distance $L \approx 200\,\text{m}$;

(iii) a cannonball, moving at a speed $U \approx 300\,\text{m/s}$ over a distance $L \approx 3000\,\text{m}$; and

(iv) a ballistic rocket, moving at a speed $U \approx 5000\,\text{m/s}$ over a distance $L \approx 1000\,\text{km}$.

Note: The first investigation into what became known as the Coriolis effect was done in the mid-1600s to understand the motion of cannonballs. Also note that the Coriolis effect is not responsible for the direction of water flowing out of a bath tub.

Chapter 15

Climate Models

In this chapter we show how a general climate model gives rise to an infinite-dimensional dynamical system. We then demonstrate a dimension-reduction procedure that reduces the Boussinesq equations of hydrodynamics to the three-dimensional dynamical system of the Lorenz model. We conclude the chapter with a discussion of some general aspects of abstract climate models.

Keywords: Abstract climate model, infinite-dimensional dynamical system, dimension reduction, Lorenz model, critical points, stochasticity.

15.1 ▪ Climate Models as Dynamical Systems

The hydrodynamic equations constitute the core of any general circulation model (GCM). They describe the evolution of the coupled atmosphere-ocean system with a certain level of detail, but a complete climate model needs more than atmosphere and ocean modules. A radiation model is needed to calculate the short- and long-wave radiation fields in the atmosphere from the incoming solar flux at the top of the atmosphere. A sea-ice module is needed to simulate the freezing, melting, and storage of frozen water as well as the transport of sea ice. The building and decay of ice sheets must be included, as well as modules for terrestrial and marine ecosystems, atmospheric chemistry, and biogeochemical tracers. On top of all this, these modules must be coupled together through physically realistic interface conditions. The challenge for mathematics is how to approach such a complicated model at the conceptual level.

Essentially, a climate model is a system of coupled PDEs which relate the time rate of change of the state variables to the current state of the physical system. If the system is autonomous, it is possible to associate a dynamical system with such a model by reinterpreting the variables in a functional sense. This idea was pioneered within the mathematical community by Henry in the 1980s [39].

Let u be the vector of all the prognostic state variables in a climate model—density, the components of the hydrodynamic velocity, temperature, salinity, and so on. The state variables vary throughout space and with time, so u is a function of the spatial coordinates (x) and time (t). Traditionally, we think of $u(x, t)$ as the value of u at the point (x, t) in space-time—that is, $u(x, t)$ is an element of $\mathbb{R}^n$, where n is the number of components of u. From the dynamical systems perspective, however, time is the "primary" variable, while space plays more of a "secondary" role. To emphasize this hierarchy, we consider u

explicitly as a function of t, $u : t \mapsto u(t)$, and identify $u(x,t)$ with the value of $u(t)$ at x,

$$u : t \mapsto u(t); \quad u(t) : x \mapsto u(t)(x) = u(x,t). \tag{15.1}$$

Suppose u is defined on the (maximal) time interval I. Its value $u(t)$ at a point $t \in I$ is a functions of x, which belongs to a certain function space X. If u is independent of x, $u(t)$ is just a vector of real numbers, so X is finite-dimensional, $X = \mathbb{R}^n$. But in general, X is a space of functions and therefore infinite-dimensional.

The upshot of the reinterpretation (15.1) is that we can consider u as a function which maps a time interval I into X, $u : I \to X$. Partial derivatives with respect to time t become ordinary derivatives with respect to t, partial derivatives with respect to the spatial variable x become operations in the function space X, and the climate model becomes a dynamical system in X. Of course, X cannot be just any function space; it must be defined in such a way that all the terms featuring in the original system of PDEs, including boundary conditions and interface conditions, make sense within X. For example, if X is a space of square-integrable functions and the PDEs involve spatial gradients of the state variables, then the elements of X must be differentiable and their spatial gradients must also be square integrable. These are technical details which need to be worked out on a case-by-case basis; we will simply assume that the function space X meets the requirements.

With these provisos, an *abstract climate model* is a dynamical system for the vector u of all the prognostic variables in a function space X,

$$\frac{d(Du)}{dt} = Au + N(u) + F. \tag{15.2}$$

Here, the symbols D, A, and N represent operators in X, which may involve differentiation with respect to the spatial variables; D is usually a simple linear operator, often diagonal, and in the simplest case D is just the identity I_X in X; A is a *linear* operator with a more complicated structure; and N is a *nonlinear* operator. Both A and N may couple the various components of u. By shifting the origin if necessary we can always achieve the identity $N(0) = 0$. The term F does not involve u and represents external forces, as well as all processes that are not resolved on the scale of the model—for example, turbulent diffusion.

Ignoring the forcing term F, we note that Eq. (15.2) is similar to Eq. (4.18). The origin is a critical point, and the equation describes the dynamics in a neighborhood of the critical point. The terms Au and $N(u)$ represent, respectively, the linear term and all higher-order terms in a Taylor expansion of the right-hand side near $u = 0$.

The analysis of dynamical systems in infinite-dimensional spaces (function spaces) is difficult. To make progress in any particular case, it is necessary to reduce the dimensionality of the abstract climate model (15.2). This is possible if there exists a coordinate system in X where some (presumably a few) coordinates of u dominate the dynamics and the contributions from the remaining coordinates can be ignored. In fact, we have already encountered several instances of this approach in earlier chapters, without explicitly mentioning the functional background. In Fourier analysis, the function space under consideration is a space of periodic functions, where the trigonometric functions (or complex exponentials) form a convenient coordinate system. The components of the spectrum are precisely the coordinates of a periodic function in this coordinate system, so our discussion of the Milankovitch cycles in Section 11.8 rested squarely on this approach. Also, in the zonal EBM of Chapter 12, the temperature depended on latitude θ, the function space was a space of functions of the variable $y = \sin\theta$ on the interval $(-1, 1)$, and the Legendre polynomials provided a convenient coordinate system.

In the following section, we demonstrate how the dimension-reduction procedure is applied to derive the Lorenz equations from the equations of fluid dynamics. The presentation is based on [7, Appendix D].

15.2 ▪ Dimension Reduction: Lorenz Model

In his investigation of GCMs and the feasibility of long-term weather forecasting, Lorenz modeled the Earth's atmosphere as an incompressible fluid situated between two horizontal planes; see Figure 15.1. The planes are a finite distance apart but extend indefinitely in the horizontal directions, so the geometry is that of a horizontal three-dimensional layer of uniform (finite) thickness. The fluid is heated from below and is cooled at the top, like sauce in a pan on a stove. The resulting convective flow is known as *Rayleigh–Bénard flow*. While in practical situations the flow develops in a regular cell-like pattern, we look here for a two-dimensional flow, with one component of the fluid velocity in the horizontal direction and another component in the vertical direction. At the upper and lower boundaries, the flow must be horizontal.

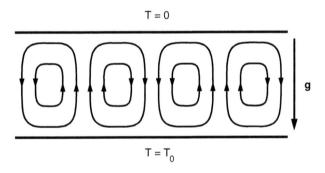

Figure 15.1. *Heat conduction in an incompressible fluid between horizontal planes.*

15.2.1 ▪ Basic Equations

We denote the horizontal direction by x, the vertical direction by y, and the velocity field by $\mathbf{v} = (v_x, v_y)$. The components v_x and v_y are functions of x and y, as well as time t. We assume that the vertical coordinate is scaled so $y = 0$ at the lower boundary (at the Earth's surface) and $y = \pi$ at the upper boundary (the tropopause).

Let $T(x, y, t)$ be the temperature at the position (x, y) at time t. The temperature at the lower boundary is $T_0 > 0$, and the temperature at the upper boundary is 0,

$$T(y = 0) = T_0, \quad T(y = \pi) = 0. \tag{15.3}$$

Heat is transported by convection with the fluid and by conduction from warmer to colder regions within the fluid, so the heat flux is made up of two terms,

$$\mathbf{q} = T\mathbf{v} - \varkappa\nabla T, \tag{15.4}$$

where $\varkappa > 0$ is the thermal conductivity. We assume that the fluid is incompressible, so the velocity field is divergence-free,

$$\nabla \cdot \mathbf{v} = \frac{\partial v_x}{\partial x} + \frac{\partial v_y}{\partial y} = 0. \tag{15.5}$$

With these assumptions, the continuity equation,

$$\frac{\partial T}{\partial t} + \nabla \cdot \mathbf{q} = 0, \tag{15.6}$$

leads to the heat equation,

$$\frac{\partial T}{\partial t} = -\mathbf{v} \cdot \nabla T + \varkappa \Delta T. \tag{15.7}$$

When the fluid is at rest and the only heat transfer is due to conduction, this equation has a solution T^* that is a linear function of just the height y, given by $T^*(y) = (1 - y/\pi)T_0$. We are interested in the deviation $\theta = T - T^*$ of the temperature T from this steady-state profile T^*. The equation governing the evolution of θ is

$$\frac{\partial \theta}{\partial t} = -\mathbf{v} \cdot \nabla \theta + \frac{T_0}{\pi} v_y + \varkappa \Delta \theta, \tag{15.8}$$

where $\Delta = \nabla \cdot \nabla$ is the two-dimensional Laplacian. The boundary conditions are

$$\theta(y = 0) = 0, \quad \theta(y = \pi) = 0. \tag{15.9}$$

Next, we turn our attention to the fluid velocity field $\mathbf{v}$. Recall that the flow is divergence-free; see Eq. (15.5). Furthermore, in the Boussinesq approximation, the density ρ is assumed to be constant (scaled to be equal to 1), except in terms involving gravity; see Eq. (14.40). The fluid velocity is therefore governed by the equation

$$\frac{\partial \mathbf{v}}{\partial t} = -\mathbf{v} \cdot \nabla \mathbf{v} - \nabla p - g' + \nu \Delta \mathbf{v}, \tag{15.10}$$

where p is the pressure, g' the reduced gravity, and ν the kinematic viscosity. The dependence of g' on space and time will be through temperature variations and will be introduced below.

For any divergence-free vector field $\mathbf{v} = (v_x, v_y)$ in two dimensions, there exists a scalar function ψ such that

$$v_x = -\frac{\partial \psi}{\partial y}, \ v_y = \frac{\partial \psi}{\partial x} \quad \text{or} \quad \mathbf{v} = J \nabla \psi, \quad J = \begin{pmatrix} 0 & -1 \\ 1 & 0 \end{pmatrix}. \tag{15.11}$$

The function ψ is known as the *stream function* and is closely related to the vorticity ζ,

$$\zeta = \nabla \times \mathbf{v} = \frac{\partial v_y}{\partial x} - \frac{\partial v_x}{\partial y} = \Delta \psi. \tag{15.12}$$

Because the fluid velocity must be horizontal at the upper and lower boundaries ($v_y = 0$), the stream function must be constant there. If the constant values of ψ at the top and bottom are different, there will be a net gradient of the stream function in the vertical direction and therefore a net flow in the horizontal direction. To exclude this possibility, we assume that ψ is constant and equal to zero at the upper and lower boundaries,

$$\psi(y = 0) = 0, \quad \psi(y = \pi) = 0. \tag{15.13}$$

We now show how to rewrite the equations of motion, Eq. (15.10), in terms of either the stream function or the vorticity. Since the curl of a gradient is zero, we can eliminate the pressure term from Eq. (15.10) by taking the curl of both sides. The resulting equation is the *vorticity equation*,

$$\frac{\partial \zeta}{\partial t} = -\mathbf{v} \cdot \nabla \zeta - \nabla \times g' + \nu \Delta \zeta. \tag{15.14}$$

The reduced gravity g' accounts for any variation of the fluid density; see Eq. (14.40). Assuming a simple linear relation between ρ and the temperature T, we then obtain that $\nabla \times g'$ is proportional to $\partial \theta / \partial x$, so the vorticity equation becomes

$$\frac{\partial \zeta}{\partial t} = -\mathbf{v} \cdot \nabla \zeta + \nu \Delta \zeta + c \frac{\partial \theta}{\partial x}, \tag{15.15}$$

where c is a thermal expansion coefficient. The nonlinearity in both the heat equation and the vorticity equation is due to the terms $\mathbf{v} \cdot \nabla T$ and $\mathbf{v} \cdot \nabla \zeta$, which we rewrite in terms of the stream function ψ as a Jacobian determinant. For any scalar function $f : (x, y) \mapsto f(x, y)$,

$$\mathbf{v} \cdot \nabla f = \frac{\partial \psi}{\partial x} \frac{\partial f}{\partial y} - \frac{\partial \psi}{\partial y} \frac{\partial f}{\partial x} = \frac{\partial(\psi, f)}{\partial(x, y)}.$$

The heat equation (15.8) and vorticity equation (15.15) thus become

$$\begin{aligned}
\frac{\partial \Delta \psi}{\partial t} &= \nu \Delta^2 \psi + c \frac{\partial \theta}{\partial x} - \frac{\partial(\psi, \Delta \psi)}{\partial(x, y)}, \\
\frac{\partial \theta}{\partial t} &= x \Delta \theta + \frac{T_0}{\pi} \frac{\partial \psi}{\partial x} - \frac{\partial(\psi, \theta)}{\partial(x, y)}.
\end{aligned} \tag{15.16}$$

This system of PDEs fits the general framework of Eq. (15.2). The state of the physical system is determined by the functions ψ and θ, which together form a function u,

$$u = \begin{pmatrix} \psi \\ \theta \end{pmatrix} : t \mapsto u(t), \quad t \in I.$$

The values $u(t)$ are functions of the spatial variables x and y, which belong to a function space X, $u(t) \in X$, so $u : I \to X$. The elements of X must be sufficiently differentiable that the right-hand side of Eq. (15.16) defines an element of X. In addition, they must satisfy the boundary conditions (15.9) and (15.13). In this case, there is no forcing term F. The abstract ODE for u is

$$\frac{d(Du)}{dt} = Au + N(u), \tag{15.17}$$

where D, A, and N are operators in X defined by the expressions

$$Du = \begin{pmatrix} \Delta \psi \\ \theta \end{pmatrix}, \quad Au = \begin{pmatrix} \nu \Delta^2 \psi + c \dfrac{\partial \theta}{\partial x} \\ x \Delta \theta + \dfrac{T_0}{\pi} \dfrac{\partial \psi}{\partial x} \end{pmatrix}, \quad N(u) = \begin{pmatrix} -\dfrac{\partial(\psi, \Delta \psi)}{\partial(x, y)} \\ -\dfrac{\partial(\psi, \theta)}{\partial(x, y)} \end{pmatrix}; \quad u = \begin{pmatrix} \psi \\ \theta \end{pmatrix}.$$

15.2.2 ▪ Linearized System

We consider the linearized system,

$$\frac{d(Du)}{dt} = Au, \tag{15.18}$$

and look for solutions of the form

$$u = \begin{pmatrix} \psi \\ \theta \end{pmatrix} = \begin{pmatrix} \xi(t)\,\psi_{a,n} \\ \eta(t)\,\theta_{a,n} \end{pmatrix}, \tag{15.19}$$

where

$$\psi_{a,n} : (x,y) \mapsto \psi_{a,n}(x,y) = \sin(ax)\sin(ny),$$
$$\theta_{a,n} : (x,y) \mapsto \theta_{a,n}(x,y) = \cos(ax)\sin(ny),$$

for some $a > 0$ and $n = 1, 2, \ldots$. The dependence of ξ and η on a and n is suppressed in our notation. Substitution leads to a system of ODEs for ξ and η,

$$\dot{\xi} = -\nu(a^2 + n^2)\xi + \frac{ac}{a^2 + n^2}\,\eta,$$
$$\dot{\eta} = \frac{aT_0}{\pi}\xi - \varkappa(a^2 + n^2)\eta. \tag{15.20}$$

This is a planar dynamical system. The trace of the coefficient matrix is always negative, as long as T_0 is small, the determinant is small, both eigenvalues are negative, and the system is stable. But as T_0 increases, the determinant goes to zero, one of the eigenvalues passes through the origin, and the system loses stability. This is when the onset of the convective flow pattern in Figure 15.1 is expected to occur, and happens when $T_0 = T_{a,n}$,

$$T_{a,n} = \frac{\varkappa\nu\pi(a^2 + n^2)^3}{a^2 c}. \tag{15.21}$$

For fixed a, $T_{a,n}$ is smallest if $n = 1$, so it makes sense to consider only the condition $T_0 > T_{a,1}$.

15.2.3 ▪ Restriction to a Three-Dimensional State Space

We now return to the nonlinear system (15.17) and look for solutions in the subspace spanned by the coordinate vector $u_{a,1}$, keeping a free. A straightforward calculation gives

$$N(u_{a,1}) = \begin{pmatrix} \dfrac{\partial(\psi_{a,1}, \Delta\psi_{a,1})}{\partial(x,y)} \\[2ex] \dfrac{\partial(\psi_{a,1}, \theta_{a,1})}{\partial(x,y)} \end{pmatrix} = \frac{1}{2}a\begin{pmatrix} 0 \\ \sin(2y) \end{pmatrix}.$$

Here, a "new" function, $\sin(2y)$, shows up. However, neither the Laplace operator Δ nor the partial derivative $\partial/\partial x$ applied to this function yields any new function, so it appears promising to include it as a third coordinate function and then truncate the expansion. We therefore look for an approximate solution of Eq. (15.17) of the form

$$u = \begin{pmatrix} \psi \\ \theta \end{pmatrix} = \begin{pmatrix} \xi(t)\,\psi_{a,1} \\ \eta(t)\,\theta_{a,1} \end{pmatrix} - \lambda(t)\begin{pmatrix} 0 \\ \sin(2y) \end{pmatrix}.$$

The only essentially new term is (leaving off the factor $\xi(t)\lambda(t)$)

$$\frac{\partial(\psi_{a,1}, \sin(2y))}{\partial(x,y)} = 2a\cos(ax)\sin(y)\cos(2y)$$

$$= a\cos(ax)(\sin(3y) - \sin(y))$$

$$= a(\cos(ax)\sin(3y) - \theta_{a,1}).$$

The term outside the current coordinate system now is $\cos(ax)\sin(3y)$, and it is this term that we delete to get a projection on a three-dimensional state space. Substitution leads to a system of ODEs for ξ, η, and λ,

$$\dot{\xi} = -\nu(a^2+1)\xi + \frac{ac}{a^2+1}\eta,$$

$$\dot{\eta} = \frac{aT_0}{\pi}\xi - x(a^2+1)\eta - a\xi\lambda, \tag{15.22}$$

$$\dot{\lambda} = -4x\lambda + \tfrac{1}{2}a\xi\eta.$$

15.2.4 ▪ Recovering the Lorenz Equations

We obtain the Lorenz system by rescaling variables. There are seven essentially independent coefficients which can be reduced to only three by rescaling t and the three dependent variables. We rescale time by introducing $t' = t/\tau$, writing again $\dot{f} = df/dt'$, and rescale the dependent variables by putting $\xi = \alpha_1 x$, $\eta = \alpha_2 y$, $\lambda = \alpha_3 z$. (Of course, x and y have nothing to do with the spatial variables in the original problem.) By setting

$$\tau = \frac{1}{x(a^2+1)}, \qquad \alpha_1 = \frac{x(a^2+1)\sqrt{2}}{a},$$

$$\alpha_2 = \frac{x\nu(a^2+1)^3\sqrt{2}}{a^2c}, \qquad \alpha_3 = \frac{x\nu(a^2+1)^3}{a^2c}, \tag{15.23}$$

we obtain the standard form (7.1) of the Lorenz equations,

$$\dot{x} = -\sigma x + \sigma y,$$

$$\dot{y} = \rho x - y - xz, \tag{15.24}$$

$$\dot{z} = -\beta z + xy,$$

with

$$\sigma = \frac{\nu}{x},$$

$$\rho = \frac{aT_0}{x\pi(a^2+1)}\frac{\alpha_1}{\alpha_2} = \frac{a^2cT_0}{x\nu\pi(a^2+1)^3} = \frac{T_0}{T_{a,1}},$$

$$\beta = \frac{4}{a^2+1}.$$

15.3 ▪ Abstract Climate Models

In Chapter 1, we argued for a system-level approach to climate science. We have seen how some of the techniques from the discipline of dynamical systems yield new insight into phenomena like equilibrium states, periodic dynamics, chaotic dynamics, the existence of multiple equilibria, transitions, and bifurcations. A general question is how one should produce intelligent interpretations of these phenomena, either as they are observed in computational and physical experiments, or as they are suggested as possible outcomes of certain forcing scenarios. The results generally indicate that the theory of dynamical systems provides a powerful mathematical language to describe manifestations of collective behavior in deterministic complex physical systems. In the following sections we address some specific questions about this system-level approach.

15.3.1 ▪ Critical Points

In the language of dynamical systems, critical points are either equilibrium points or periodic orbits. How relevant are they in the context of climate science?

An equilibrium point of a GCM would be a climate state which does not change over multiples of the typical time scale (many thousands of years, say). The geological evidence over the last several hundred thousand years suggests that the Earth's climate has not been in an equilibrium state during this time. There have been long periods of constant climate, but they all ended at some point and a transition to a different climate state occurred. A dynamical system for the global climate system or any of its components which admits equilibrium points can therefore only be an approximation. Nevertheless, the approximation is useful, especially when it can be shown that a critical point is surrounded by a *basin of attraction*. Once the system is inside the basin of attraction, either by internal dynamics or by external forcing, it will tend to the equilibrium state at the critical point.

As for periodic orbits, there is substantial evidence for time-periodic patterns in the Earth's climate, with oscillations occurring over periods of several tens of thousands of years. However, as we have argued in Section 11.8, more subtle astronomical patterns are likely to play a role here, although the patterns cannot explain these oscillations completely. More likely, a complex interplay between external (orbital) forcing patterns and internal dynamics of the system lead to the observed oscillations. Of course, there are periodic or approximately periodic phenomena that may be part of the internal dynamics of the climate system itself. For example, there is evidence that the ocean circulation pattern is not unchangeable. Today, the pattern is that of the conveyor belt shown in Figure 3.1. But the evidence points to patterns in the past that were different from, and in some instances the reverse of, what they are today. It is indeed possible that the dynamics of the ocean circulation pattern are internally driven and periodic or approximately periodic, although convincing evidence is missing. Whether the Earth's climate as a whole moves on a periodic orbit is even more debatable. While it is true that the current climate (between two glaciations) will not last forever, we have no idea whether it will return to the current state, let alone whether it will return many more times.

15.3.2 ▪ Stochastic Climate Models

As stated earlier, the theory of dynamical systems provides a mathematical language to describe the time-dependent behavior of *deterministic* physical systems. Indeed, the climate models discussed so far have all been deterministic. They involved differential equations, and the assumption is that unique solutions exist and can be found once initial conditions are specified. Obviously, this assumption is not realistic; there are many sources of uncertainty, which need to be taken into account and introduce stochastic elements into the discussion. For example, a climate model is designed for phenomena on the macroscopic scale; phenomena on the microscopic scale are parameterized, and the values of the parameters are certainly not known with any great accuracy. The initial conditions are another source of uncertainty. They are never completely known and, at best, assembled from incomplete data. Even if data assimilation techniques are applied, as explained in Chapter 20, there is always a residue of uncertainty in the model. Last, but not least, experiments are subject to uncertainty. Deterministic models are by definition unable to capture these uncertainties, so the outcome of a climate model must be considered as a realization from an ensemble of possible outcomes. In other words, a climate model is by its nature *stochastic*.

Dynamical systems theory, to the extent that it has been presented here, is not designed to handle stochasticity. To overcome this limitation, one can extend the theory or one can resort to a computational approach. In computational climate science, stochasticity is simulated by taking an ensemble of initial conditions and generating a range of possible outcomes, which are then assigned weights to yield a probability distribution of outcomes. Following [80], we briefly indicate how this process can be interpreted in the spirit of the abstract climate model (15.2).

Suppose that there is a separation of time scales and that the state vector u which represents the climate system decomposes into a slow component Pu and a fast component Qu. To obtain the governing equations for Pu, one could, in principle, project the system of governing equations onto the slow manifold and ignore the fast variables Qu entirely. Alternatively, one could assume that the system starts from an initial state $((Pu)_0, (Qu)_0)$, where the fast component $(Qu)_0$ is drawn randomly according to some conditional probability measure $\mu(Qu|Pu)$ at $Pu = (Pu)_0$. This conditional probability measure contains all available information on the unresolved values of $(Qu)_0$ if only the $(Pu)_0$ are known. Such a procedure introduces *randomness* into the climate model. The question is what information can be expected from such a stochastic model, and how is this information related to the information gained from a model obtained by simply projecting the original model onto a slow manifold? Furthermore, in what way does this concept lead to a reduced equation of motion for the observables Pu, which obviously will contain some stochastic terms?

15.4 ▪ Exercises

1. Write the zero-dimensional EBM in Eq. (2.10) in the form (15.2). Explain how A and N change if the system is linearized near any of the three equilibrium solutions that can be expected. What is the operator D in this case?

2. Write the zonal EBM (12.8) in the form (15.2). What kind of function space should be employed in this case? What is the operator A? What are possible choices for the operator N? What about the corresponding choices for F? Note that these choices are not unique. If we are interested in the behavior of the abstract dynamical system near the equilibrium temperature given by Eq. (12.41), what should the operator A be?

3. Consider the two-dimensional heat equation, Eq. (15.6), with a constant velocity term $\mathbf{v} = (v_1, v_2)$.

   (i) Verify that, for $x = 1$ and $v_1 = v_2 = 0$, the *heat kernel*

   $$T_K : (x, y, t) \mapsto \frac{1}{4\pi t} e^{-(x^2 + y^2)/(4t)}$$

   is a solution.

   (ii) Construct an explicit solution of Eq. (15.6) from T_K for general $x > 0$, $v_1, v_2 \in \mathbb{R}$.

4. Consider a stationary velocity field in the strip $\mathbb{R} \times [0, \pi]$,

   $$\mathbf{v}(x, y) = (a \cos(nx) \sin(by), \sin(mx) \cos(cy)),$$

   with $m, n = 0, \pm 1, \dots$ and $a, b, c \in \mathbb{R}$.

   (i) Find the vorticity function ζ in the general case.

(ii) Find conditions for the parameters such that the velocity field is divergence-free.

(iii) Find the stream function ψ in the cases where the velocity field is divergence-free. Does the stream function have the same value at the upper and lower boundaries of the domain?

(iv) Verify that $\Delta\psi = \zeta$.

5. Consider the condition $T_0 = T_{a,1}$, where $T_{a,1}$ is defined in Eq. (15.21). This is the smallest temperature at which the zero solution of one of the infinitely many linearized systems loses its stability.

(i) Give a summary of the dependence of $T_{a,1}$ on the physical parameters x, ν, and c.

(ii) Sketch the velocity field $\mathbf{v}$ corresponding to the stream function $\psi_{a,1}$ for a reasonable choice of $a \in [\frac{1}{2}, 1]$.

(iii) The value a for the horizontal length scale of the velocity field $\mathbf{v}$ that will occur "in nature" is still undetermined. It can be found with the following reasoning. If $T_0 < T_{a,n}$ for some a and n, then the corresponding zero solution of Eq. (15.20) is stable, and a nontrivial flow pattern with stream function $\psi_{a,n}$ is not expected. But if $T_0 > T_{a,n}$, then small fluctuations will cause such a nontrivial flow to develop. By a physical version of "Murphy's Law," this will indeed happen at the smallest possible temperature T_0—that is, as soon as $T_0 > T_{a^*,1}$, where a^* minimizes $T_{a,1}$. This value determines the horizontal length scale at which the nontrivial flow pattern in Figure 15.1 will first be observed. Determine a^*.

6. Consider the three-dimensional Lorenz equations in the form given in Eq. (15.22) (before the rescaling). The goal of this exercise is to go through the rescaling steps in more detail.

(i) Rewrite the equations after introducing $t' = t/\tau$, with $\tau = 1/(x(a^2 + 1))$.

(ii) Show that if α_1, α_2, and α_3 satisfy the conditions

$$\frac{\alpha_2 a c}{\alpha_1 x (a^2+1)^2} = \frac{\nu}{x}, \qquad \frac{\alpha_1 \alpha_3 a}{\alpha_2 x (a^2+1)} = 1, \qquad \frac{\alpha_1 \alpha_2 a}{2\alpha_3 x (a^2+1)} = 1,$$

then the system for x, y, and z turns into the standard form (7.1) of the Lorenz equations.

(iii) Show that this set of conditions can be satisfied by choosing α_1, α_2, and α_3 as in Eq. (15.23).

Chapter 16

El Niño–Southern Oscillation

This chapter focuses on El Niño and its atmospheric counterpart, the Southern Oscillation. El Niño–Southern Oscillation (ENSO) is a major driver of the global climate system which can cause significant changes in the global atmospheric circulation. El Niño events occur about every three to seven years and alternate with the opposite phases of La Niña. We present two self-sustaining oscillator models describing the dynamics of ENSO.

Keywords: El Niño–Southern Oscillation (ENSO), recharge-oscillator model, harmonic oscillator, delayed-oscillator model, Kelvin wave, Rossby wave, delay differential equation.

16.1 ▪ El Niño

El Niño or, more precisely El Niño–Southern Oscillation (ENSO), is a quasi-periodic climate pattern that occurs across the equatorial Pacific Ocean roughly every three to seven years. It is characterized by a change in sea surface temperatures (SSTs) in the eastern Pacific off the coast of Peru and accompanying changes in the air pressure difference between the central and western Pacific Ocean (Tahiti and Darwin, Australia). The oceanic phenomenon has been observed for centuries by fishermen in Peru, while the atmospheric oscillation was discovered by the British physicist and statistician SIR GILBERT WALKER in 1923. The connection between the two phenomena was made by the Norwegian-American meteorologist JACOB BJERKNES in 1969.

In normal conditions, the trade winds blow across the tropical Pacific toward the west (see Figure 13.2). Warm surface water piles up in the western Pacific, so that the sea surface is about one-half meter higher in the vicinity of Indonesia than off the coast of Ecuador. Also, the SST off the coast of South America is lower due to an upwelling of cold water from deeper levels. This cold water is nutrient-rich, supporting high levels of primary productivity, diverse marine ecosystems, and major fisheries. The boundary between the warm upper layer of the ocean and the cold deeper part, known as the *thermocline*, in the western Pacific is very deep in such conditions and shallow in the eastern Pacific.

During an El Niño episode, the trade winds in the central and western Pacific relax. As a result, the difference in the sea level between the eastern and western Pacific becomes smaller, and the thermocline becomes more shallow in the west and deeper in the east. The efficiency of upwelling of cold water from deeper levels off the coast of South

America decreases, and the supply of nutrient-rich thermocline water is cut off. Rainfall follows the warm water eastward, with associated flooding in Peru and drought in Indonesia and Australia. The eastward displacement of the atmospheric heat source overlaying the warmest water results in large changes in the global atmospheric circulation, which in turn force changes in weather in regions far removed from the tropical Pacific. This is the phenomenon of *teleconnection* mentioned in Chapter 1. Recent studies suggest that El Niño affected pre-Columbian cultures and that a strong El Niño effect between 1789 and 1793 caused poor crop yields in Europe which, according to some sources, in turn helped touch off the French Revolution [31].

The rise in SST and drastic decline in fish catch during an El Niño episode was well known to Peruvian fishermen. Since the episodes occur primarily around Christmas, the event was named after the Christ Child. Historically, El Niño events occur about every three to seven years and alternate with the opposite phases of below-average temperatures in the eastern tropical Pacific (La Niña). Figure 16.1 shows the SST anomalies under El Niño and La Niña conditions, and Figure 16.2 shows the time series of the SST anomalies in the eastern tropical Pacific since 1875.

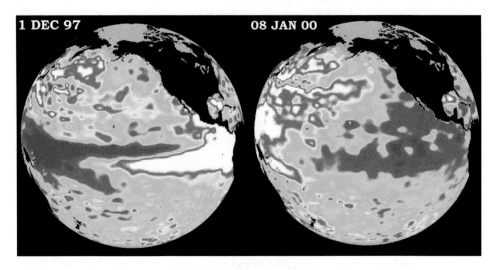

Figure 16.1. *SST anomalies (°C) observed under El Niño conditions (December, 1997; left) and La Niña conditions (January, 2000; right). Reprinted courtesy of NASA/JPL-Caltech.*

ENSO, the combination of these oceanic and atmospheric patterns, actually involves two feedback mechanisms, a positive one which affects the ocean-atmosphere system over the eastern tropical Pacific and leads up to an El Niño event, and a negative one which turns a warm phase into a cold phase and leads to the termination of the event. Since the nature of this negative feedback mechanism is not well understood, several conceptual models have been proposed to describe ENSO. Here we discuss two such models. The first is a simple harmonic oscillator model which is referred to as the *recharge-oscillator* model [45, 46]. It emphasizes the role of a discharge of equatorial heat as the primary agent for negative feedback. The other is a more complicated model, referred to as the *delayed-oscillator* model [4, 5, 99, 106, 115], which emphasizes the role of reflected Kelvin waves at the western boundary of the Pacific as the primary agent for negative feedback.

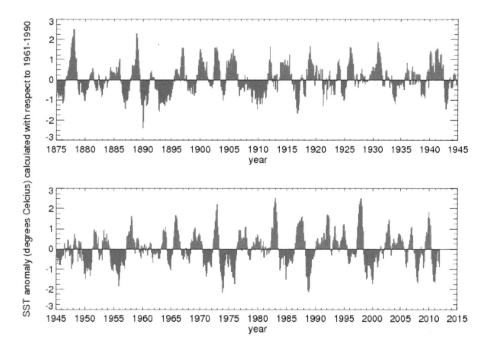

Figure 16.2. *Time series of the SST averaged over the eastern tropical Pacific (region Niño 3) since 1875. The values plotted are monthly mean temperature anomalies (°C) from the mean seasonal cycle. Reprinted with permission from Met Office.*

16.2 ▪ Recharge-Oscillator Model

The ENSO recharge-oscillator model was proposed by Jin in the 1990s [45, 46]. Its main ingredients are the SST anomaly averaged over the central and eastern Pacific, T_E; the thermocline depth anomaly in the equatorial western Pacific, h_W, and eastern Pacific, h_E; and the integrated wind stresses in the equatorial central and western Pacific, S, and eastern Pacific, S_E. All these quantities depend on time t and have suitable dimensions. The wind stress in the central and western Pacific and the two thermocline anomalies are in an approximate equilibrium at all times,

$$h_E = h_W + S. \tag{16.1}$$

During the build up to a "normal state," the thermocline depth in the warm pool in the west adjusts slowly. This process is described by the differential equation

$$\frac{dh_W}{dt} = -r h_W - \alpha S, \tag{16.2}$$

with positive constants r and α. The negative sign in front of α comes from the fact that a large positive wind stress is expected to depress the western thermocline depth anomaly; see Eq. (16.1). The SST anomaly is described by the differential equation

$$\frac{dT_E}{dt} = -c T_E + \gamma h_E + \delta S_E. \tag{16.3}$$

The first term on the right describes the relaxation of the SST due to upwelling and heat exchange between the atmosphere and the ocean. The second term models the thermocline upwelling process. The third term describes an advective feedback process called

Ekman pumping, in which easterly equatorial trade winds result in a net transport of surface water toward the poles and westerly equatorial trade winds result in a net transport in the opposite direction, possibly leading to additional upwelling. This process is caused by opposite Coriolis effects north and south of the equator. The set of equations is now closed by assuming that both wind stress terms S and S_E are proportional to the SST anomaly T_E,

$$S = b\,T_E, \quad S_E = b_E\,T_E, \tag{16.4}$$

and that the feedback effect from Ekman pumping can be ignored, so $\delta\,b_E = 0$. The model then reduces to a harmonic oscillator—a planar system of two coupled linear differential equations for T_E and h_W,

$$\begin{aligned}
\frac{dT_E}{dt} &= -c\,T_E + \gamma(h_W + b\,T_E),\\
\frac{dh_W}{dt} &= -r\,h_W - \alpha\,b\,T_E.
\end{aligned} \tag{16.5}$$

The factor $\gamma b - c$ in the first equation summarizes the tropical ocean-atmosphere interaction, in the spirit of the original Bjerknes hypothesis. The model given in [45, 46] introduces an additional term in the SST equation, which shuts down upwelling for very deep thermoclines—that is, for large h_E. This results in the nonlinear system of equations

$$\begin{aligned}
\frac{dT_E}{dt} &= -c\,T_E + \gamma(h_W + b\,T_E) - \varepsilon(h_W + b\,T_E)^3,\\
\frac{dh_W}{dt} &= -r\,h_W - \alpha\,b\,T_E.
\end{aligned} \tag{16.6}$$

16.2.1 ▪ Dynamical System

The system (16.6) is readily analyzed with the techniques developed in Chapter 4. We introduce reference values h_0 and T_0 for the thermocline depth anomaly and SST anomaly, respectively, and define dimensionless variables and parameters,

$$\begin{aligned}
x &= \frac{T_E}{T_0}, \quad y = \frac{h_W}{h_0}, \quad t' = ct,\\
r' &= \frac{r}{c}, \quad \alpha' = \frac{\alpha}{c}, \quad \gamma' = \frac{h_0}{T_0}\frac{\gamma}{c}, \quad b' = \frac{T_0}{h_0}b, \quad \varepsilon' = \frac{h_0^3}{T_0 c}\varepsilon.
\end{aligned} \tag{16.7}$$

Typical values are $h_0 = 150\,\mathrm{m}$, $T_0 = 7.5\,\mathrm{K}$, and $c = 2\,\text{months}$. The variable t' is just a rescaled version of t; without loss of generality we may drop the prime $'$. We also drop the primes on the parameters. Thus, the system of equations (16.5) becomes

$$\begin{aligned}
\dot{x} &= -x + \gamma(bx + y) - \varepsilon(bx + y)^3,\\
\dot{y} &= -ry - \alpha bx.
\end{aligned} \tag{16.8}$$

This is a planar system of the type discussed in Section 4.6. The coefficient matrix of its linearization about the trivial solution is

$$A = \begin{pmatrix} \gamma b - 1 & \gamma \\ -\alpha b & -r \end{pmatrix}, \tag{16.9}$$

so $\mathrm{trace}(A) = \gamma b - 1 - r$ and $\det(A) = r - (r - \alpha)\gamma b$. These expressions show that it is convenient to combine γ and b into a single parameter, $\lambda = \gamma b$. Then

$$T = \mathrm{trace}(A) = \lambda - 1 - r, \quad D = \det(A) = r - (r - \alpha)\lambda. \tag{16.10}$$

After some algebra we find that

$$T^2 - 4D = -(\lambda - \lambda_1)(\lambda_2 - \lambda), \tag{16.11}$$

where

$$\lambda_1 = \left(\sqrt{\alpha + 1 - r} - \sqrt{\alpha}\right)^2, \quad \lambda_2 = \left(\sqrt{\alpha + 1 - r} + \sqrt{\alpha}\right)^2. \tag{16.12}$$

We are looking for the oscillatory behavior that characterizes the El Niño phenomenon. For oscillatory behavior we must have $D > 0$ and $T^2 - 4D < 0$. This puts two constraints on λ, namely $\lambda < r/(r-\alpha)$ and $\lambda \in (\lambda_1, \lambda_2)$. Then A has a pair of complex conjugate eigenvalues, $\sigma_1 = \xi - i\omega$ and $\sigma_2 = \xi + i\omega$, and the solution of the linearized equation (16.8) is a superposition of oscillatory modes $e^{\xi t}\cos(\omega t)$ and $e^{\xi t}\sin(\omega t)$, where ξ and ω are functions of λ,

$$\xi(\lambda) = \tfrac{1}{2}(\lambda - 1 - r), \quad \omega(\lambda) = \tfrac{1}{2}\sqrt{(\lambda - \lambda_1)(\lambda_2 - \lambda)}. \tag{16.13}$$

The value $\lambda_c = 1 + r$ is a critical value. As long as $\lambda < \lambda_c$, we have $\xi < 0$, and the origin is a stable spiral point. But as λ reaches the value λ_c, the eigenvalues cross the imaginary axis and transit from the left half of the complex plane into the right half. At that point, a Hopf bifurcation occurs in the nonlinear system Eq. (16.8), and the trivial solution loses stability.

At the critical value, the value of ω is

$$\omega_c = \omega(\lambda_c) = \tfrac{1}{2}\sqrt{(\lambda_c - \lambda_1)(\lambda_2 - \lambda_c)} = \sqrt{\alpha(1+r) - r^2},$$

and a nontrivial solution of the linearized system (16.8) is given by

$$\begin{aligned} x(t) &= \cos(\omega_c t), \\ y(t) &= -\frac{\alpha b}{r^2 + \omega^2}\left[r\cos(\omega_c t) + \omega_c \sin(\omega_c t)\right]. \end{aligned} \tag{16.14}$$

All other solutions are obtained by changing the phase and/or amplitude of this solution. The expression for $y(t)$ can be rewritten as

$$y(t) = -\frac{\alpha b}{\sqrt{\alpha(1+r)}}\, x(t - \eta), \quad \eta = \frac{1}{\omega_c}\tan^{-1}\left(\frac{\omega_c}{r}\right), \tag{16.15}$$

which shows that the trajectories of x and y coincide, but y lags behind x with a lag given by η. Thus, this ENSO model predicts that the negative thermocline depth anomaly follows the same oscillatory pattern as the SST anomaly but with a time lag η. An analysis of the full system (16.8) leads to a solution which is similar to Eq. (16.14), with a time lag for the thermocline depth anomaly that is slightly less than η.

In [45], the following values are used for the dimensionless parameters:

$$r = \tfrac{1}{4}, \quad \alpha = \tfrac{1}{8}, \quad \gamma = \tfrac{3}{4}. \tag{16.16}$$

Then $\lambda_1 = 1 - \tfrac{1}{4}\sqrt{7} \approx 0.34$, $\lambda_2 = 1 + \tfrac{1}{4}\sqrt{7} \approx 1.66$, and $\lambda_c = \tfrac{5}{4} = 1.25$, so the constraints on λ are satisfied at λ_c. Furthermore, $\omega_c = \tfrac{1}{8}\sqrt{6} \approx 0.31$. Since the time unit is two months, the predicted period is about 41 months, which is in the range of observed periods. We also obtain $\eta \approx 2.89$, corresponding to a length of about six months between the SST anomaly and the thermocline depth anomaly.

The left panel of Figure 16.3 shows a scaled orbit of the solution (16.14) (thin line) as well as a computed orbit of a periodic solution of Eq. (16.8) (thick line) for $\lambda = 1.5 > \lambda_c$ and $\varepsilon = 0.1$. The right panel shows graphs of the scaled T_E (solid line) and the scaled h_W (dashed line) vs. time for the same values of the parameters λ and ε. The graphs show that the negative depth anomaly h_W lags the SST anomaly T_E by about six months for the nonlinear model as well.

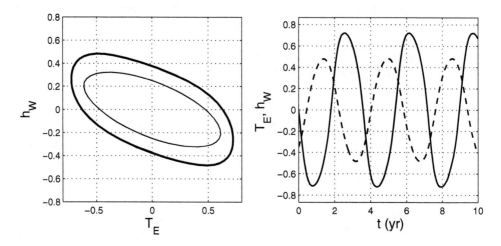

Figure 16.3. *Solutions of Eq. (16.8) and its linearization.*

16.3 ▪ Delayed-Oscillator Model

The delayed-oscillator model is an alternative ENSO model, where Rossby and Kelvin waves play a fundamental role [4, 5, 99, 106, 115]. It is based on the feedback loop illustrated in Figure 16.4. Under normal conditions (top left), the thermocline is much deeper in the western Pacific than in the eastern Pacific. A high-pressure system over the eastern Pacific and a low-pressure system near Indonesia lead to prevailing easterlies which push surface water to the west, leading to higher sea levels in the western Pacific. When these pressure systems and the resulting easterlies weaken due to random fluctuations (top right), the force balance between easterlies and the sloping sea level is destroyed, and several packets of equatorial warm (downwelling) Kelvin waves are created, which propagate eastward and reach the coast of South America in one to two months. The pressure difference and the resulting easterly winds are weakened further, and the eastward surface flow of warm water from the western Pacific leads to a deepening of the thermocline in the east and an increase of the SST there. This starts an El Niño event (bottom left). The Kelvin wave packets are deflected and propagate north and south along the coastlines, changing local wind patterns and leading to warm upwelling of nutrient-poor water (the phenomenon originally observed by the fishermen). The original wind weakening over the central Pacific also creates off-equatorial cold (upwelling) Rossby waves, which are also shown in the bottom left. These waves propagate westward and are reflected from the western boundary as cold Kelvin waves (bottom right), which arrive at the eastern boundary about six months later and terminate the event. The thermocline returns to the asymmetric shape shown in the top left picture, and the atmospheric imbalance shown in the top left is restored. Occasionally, this restoration effect "overshoots," leading to a "La Niña" event with unusually cold surface waters in the eastern Pacific.

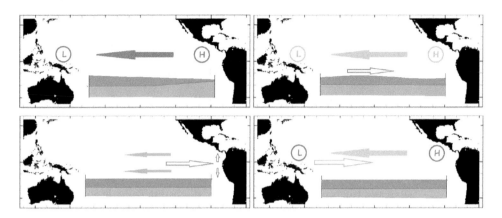

Figure 16.4. *Schematic representation of proposed ENSO oscillation mechanism. Thermocline is depicted schematically in orange and blue, prevailing winds are gray, surface waves are orange (warm) and green (cool), Kelvin waves are hollow arrows, and Rossby waves are solid arrows.*

To model this scenario, we again use the SST anomaly in the eastern equatorial Pacific, T_E, as state variable. We drop the subscript and write T instead of T_E for the remainder of the presentation of the model. The model has two additional auxiliary variables, namely the central equatorial thermocline anomaly h_{c0} and the central off-equatorial thermocline anomaly h_{c1}.

The SST anomaly in the eastern equatorial Pacific T increases because of the arrival of warm equatorial Kelvin waves that originate on the equator in the central Pacific ($h_{c0} > 0$), and decreases because of the arrival of cold equatorial Kelvin waves that originate as westward propagating off-equatorial Rossby waves on and near the central meridian ($h_{c1} < 0$).

Let τ_K and τ_R denote the time needed for Kelvin and Rossby waves, respectively, to cross the entire Pacific basin. The time elapsed between the genesis of these waves and their arrival at the eastern boundary of the Pacific basin is $\frac{1}{2}\tau_K$ (about one month) for the warm Kelvin waves and $\frac{1}{2}\tau_R + \tau_K$ (about six months) for the cold Kelvin waves. Hence, we use the differential equation

$$\dot{T}(t) = -cT(t) + a_0 h_{c0}(t - \tfrac{1}{2}\tau_K) + b_0 h_{c1}(t - (\tfrac{1}{2}\tau_R + \tau_K)) - \varepsilon(T(t))^3, \qquad (16.17)$$

where the dot $\dot{}$ denotes differentiation with respect to t, the constant c describes relaxation effects for the SST anomaly, a_0 and b_0 are positive constants, and the cubic term εT^3 is introduced to stabilize the system.

To close the system, we observe that the thermocline anomaly h_{c0} is a response to the stress of the equatorial central Pacific wind which, in turn, is pretty much a direct response to the eastern Pacific ocean temperature. That is, h_{c0} is proportional to T at all times, with a positive proportionality constant.

The situation for the off-equatorial thermocline depth anomaly in the central Pacific is a bit more complicated. The off-equatorial thermocline depth anomaly h_{c1} is driven by the wind curl off the equator. The latter is *negatively* correlated with the wind stress at the equator which, as just noted, is a more or less direct response to the eastern Pacific SST. Therefore, h_{c1} can also be considered proportional to T, but with a negative proportionality constant.

Thus, we obtain the following equation for the eastern Pacific temperature which includes the effects of Kelvin waves, Rossby waves, and local damping terms.

$$\dot{T}(t) = -cT(t) + aT(t - \tfrac{1}{2}\tau_K) - bT(t - (\tfrac{1}{2}\tau_R + \tau_K)) - \varepsilon(T(t))^3, \tag{16.18}$$

where a, b, c, and ϵ are positive constants. It will be shown in the exercises that if $\varepsilon > 0$, all solutions of this equation remain bounded by some large constant M for all $t > 0$.

Needless to say, Eq. (16.18) is a gross simplification of the real ENSO phenomenon, just like the recharge oscillator model was a simplification. We made some explicit and many more implicit assumptions in the derivation of both models, so we should not even expect to see a reasonable match with observational data. Nevertheless, the equation can serve as a toy model for ENSO. Before discussing some properties of this equation and presenting some numerical results, we give a few more details about Rossby and Kelvin waves.

16.3.1 ▪ Rossby Waves

Oceanic Rossby waves are disturbances in the ocean height that are moving very slowly from east to west. They arise naturally in fluids on a rotating sphere. They also arise in the atmosphere, where they are giant meanders in high-altitude winds. They are named after CARL-GUSTAF ARVID ROSSBY, who first identified them in the atmosphere in 1939.

Oceanic Rossby waves are difficult to detect because of the big difference between the horizontal and vertical scales. Their horizontal scale is of the order of hundreds of km, while the amplitude of the oscillation at the sea surface is just a few centimeters, practically impossible to measure with *in situ* techniques. Oceanic Rossby waves always travel from east to west, following the parallels, and do not go fast—their speed varies with latitude and increases toward the equator but is of the order of just a few cm/s (or a few km/day). This means that at midlatitudes (say, 30 degrees N or S) one such wave may take several months or even years to cross the Pacific. Their detection has become possible through satellite observations. Figure 16.5 shows Rossby waves as small "streaks" with a negative slope in a *Hovmöller diagram* of the sea level anomaly (color-coded) at two different latitudes in the Pacific as a function of longitude (horizontal axis) and time (vertical axis) for the period 1996–2001. The El Niño event that began in the Spring of 1997 is clearly visible in

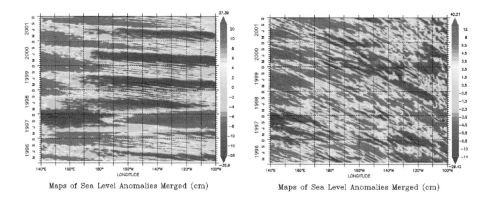

Maps of Sea Level Anomalies Merged (cm)   Maps of Sea Level Anomalies Merged (cm)

Figure 16.5. *Longitude vs. time plot (Hovmöller diagram) of oceanic Rossby waves at two different latitudes (3.9° N, left; 15.1° N, right) in the Pacific Ocean for the period 1996–2001. This image was generated using altimeter products produced by Ssalto/Duacs and distributed by Aviso, with support from Cnes.*

the elevation of sea levels at eastern latitudes and corresponding depression of sea levels at western latitudes. Slanted strips indicate westward moving Rossby waves, the slope being a measure of their phase velocity. The graph shows that the phase velocity of Rossby waves is more or less constant over time. We shall show below that the phase velocity depends only on latitude and that it increases with latitude. The graph also shows that the amplitude tends to decrease with latitude, which can also be predicted mathematically.

The special identifying feature of a Rossby wave is its phase velocity (the velocity of the wave crest), which always has a westward component. The wave's group velocity (associated with the energy flux) can be in any direction; in general, short waves have an eastward group velocity and long waves a westward group velocity.

To derive expressions for the phase and group velocities, we consider the Earth's ocean as a two-dimensional fluid overlaying the Earth's surface. At each point on the Earth's surface we impose a Cartesian coordinate system with the x-axis in the zonal (latitudinal) direction and the y-axis in the meridional (longitudinal) direction.

The velocity of a fluid parcel at any point (x, y) on the surface is represented at any time t by a vector (u, v), where u and v depend on x and y as well as t. We assume that the flow is a perturbation of a mean zonal flow, which is steady and uniform. If U is the (constant) speed of the mean zonal flow, then (u, v) is a perturbation of the mean zonal flow vector $(U, 0)$, so $u = U + u'$ and $v = v'$, where $u', v' \ll U$.

The vorticity of the same fluid parcel is a vector quantity which describes its rotation around a vertical axis. The *planetary vorticity*, f, is associated with the Earth's surface, while the *relative vorticity*, ζ, is associated with the motion of the fluid parcel relative to the Earth's surface. The former is a consequence of the variation of the Coriolis force with latitude, so f depends on y but not on x. The latter is defined in terms of the velocity vector (u, v),

$$\zeta = \frac{\partial v}{\partial x} - \frac{\partial u}{\partial y}. \tag{16.19}$$

If there are no sources or sinks (the fluid is *divergence-free*), we can introduce a scalar *stream function*, ψ, such that

$$u = -\frac{\partial \psi}{\partial y}, \quad v = \frac{\partial \psi}{\partial x}. \tag{16.20}$$

Combining Eqs. (16.19) and (16.20), we obtain the expression relating the relative vorticity and the stream function,

$$\zeta = \Delta \psi. \tag{16.21}$$

Here, Δ is the *Laplace operator*, $\Delta = \partial / \partial x^2 + \partial / \partial y^2$.

Conservation of vorticity is an interplay of several effects and is described by conservation of the quantity $\zeta + f$,

$$\frac{d(\zeta + f)}{dt} = 0. \tag{16.22}$$

Applying the chain rule and retaining only leading-order terms, we have

$$\frac{d(\zeta + f)}{dt} = \frac{\partial \zeta}{\partial t} + U \frac{\partial \zeta}{\partial x} + \frac{\partial f}{\partial y} v.$$

We use the β-plane approximation (14.10), $f \approx f_0 + \beta(y - y_0)$, so $\partial f / \partial y = \beta$. Here, $\beta = (2\Omega \cos \theta)/R$ is the Rossby parameter, θ being the latitude, Ω the angular velocity of the Earth's rotation, and R the mean radius of the Earth.

Replacing ζ by $\Delta\psi$, we obtain a PDE for ψ,

$$\frac{\partial \Delta\psi}{\partial t} + U\frac{\partial \Delta\psi}{\partial x} + \beta\frac{\partial \psi}{\partial x} = 0. \tag{16.23}$$

Since this is a linear equation, any solution is a superposition of elementary solutions. Rossby waves are periodic, both in space and in time, so we look for elementary solutions of the form

$$\psi_{k,\ell}(t,x,y) = e^{i(kx+\ell y-\omega t)},$$

where ω is (2π times) the frequency and k and ℓ are the wave numbers (integers) in the zonal and meridional directions, respectively. These elementary solutions diagonalize the Laplace operator, $\Delta\psi_{k,\ell} = -(k^2+\ell^2)\psi_{k,\ell}$. A necessary condition for the existence of such an elementary solution is that ω, k, and ℓ are related through the *dispersion relation*,

$$\omega = Uk - \beta\frac{k}{k^2+\ell^2}. \tag{16.24}$$

The *phase velocity* c and *group velocity* c_g are then given by

$$c = \frac{\omega}{k} = U - \beta\frac{1}{k^2+\ell^2}, \quad c_g = \frac{\partial\omega}{\partial k} = U - \beta\frac{\ell^2-k^2}{(k^2+\ell^2)^2}. \tag{16.25}$$

This result shows that the phase velocity—that is, the zonal wave speed—is always westward relative to the mean zonal flow. The group velocity can go either way, depending on the relative size of the wave numbers k and ℓ. Short zonal waves (large zonal wave numbers k) enhance the mean zonal flow, and long zonal waves (small zonal wave numbers k) counteract the mean flow. If $\ell = k$, the group speed is the same as the mean zonal flow.

The Rossby parameter is greatest at the equator, so Rossby waves move faster near the equator than at midlatitudes, as can be seen in Figure 16.5.

16.3.2 ▪ Kelvin Waves

A *Kelvin wave* is a wave in the ocean or atmosphere that balances the Earth's Coriolis force against a topographic boundary. The boundary can be a physical boundary like a coastline, in which case the wave is referred to as a *coastal Kelvin wave*. It can also be a virtual boundary like the equator. The equatorial zone acts essentially as a physical boundary for both the Northern and Southern Hemispheres. Disturbances tend to get trapped in a zone around the equator, which acts like a waveguide for *equatorial Kelvin waves*. Figure 16.6 shows an equatorial Kelvin wave in the Pacific Ocean.

A feature of Kelvin waves is that they are nondispersive—that is, the phase speed of the wave crests is equal to the group speed of the wave energy for all frequencies. This means that Kelvin waves retain their shape over time.

When the motion at the equator is to the east, any deviation into the Northern or Southern Hemisphere is brought back toward the equator by the Coriolis force, because the Coriolis force acts to the right of the direction of motion in the Northern Hemisphere and to the left of the direction of motion in the Southern Hemisphere. On the other hand, when the motion at the equator is toward the west, the Coriolis force will not restore a northward or southward deviation back toward the equator. Hence, equatorial Kelvin waves move only toward the east. Both atmospheric and oceanic equatorial Kelvin waves

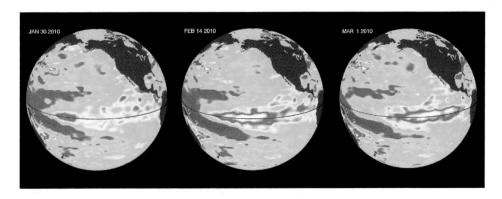

Figure 16.6. *Equatorial Kelvin wave. Reprinted courtesy of NASA/JPL-Caltech.*

play an important role in the dynamics of ENSO by transmitting changes in conditions in the western Pacific to the eastern Pacific.

To find the phase velocity and group velocity of Kelvin waves, we consider a layer of fluid moving over a motionless abyssal layer. We impose a Cartesian (x, y)-coordinate system as in Section 16.3.1. The layer has a finite thickness, h, which may vary with position, and the horizontal motion (relative to the abyssal layer) is described by the velocity vector (u, v). The equations of motion, linearized about a state of rest and a mean thickness H, are

$$\frac{\partial u}{\partial t} - f v + g \frac{\partial h}{\partial x} = 0,$$

$$\frac{\partial v}{\partial t} + f u + g \frac{\partial h}{\partial y} = 0, \tag{16.26}$$

$$\frac{\partial h}{\partial t} + H \left(\frac{\partial u}{\partial x} + \frac{\partial v}{\partial y} \right) = 0.$$

Here, $f(y) = \beta y$ is the *equatorial β-plane approximation* of the Coriolis force. The remaining parameter in Eq. (16.26) is g, the acceleration of gravity (adjusted for the vertical density gradient).

At the equator we can assume that v is small. If we ignore v entirely, Eq. (16.26) reduces to

$$\frac{\partial u}{\partial t} + g \frac{\partial h}{\partial x} = 0,$$

$$f u + g \frac{\partial h}{\partial y} = 0, \tag{16.27}$$

$$\frac{\partial h}{\partial t} + H \frac{\partial u}{\partial x} = 0.$$

We eliminate u by differentiating the third equation with respect to t and the first equation with respect to x, and similarly by differentiating the second equation with respect to t and substituting $\partial u / \partial t$ from the first equation. Thus we obtain a system of two PDEs for h,

$$\frac{\partial^2 h}{\partial t^2} - gH\frac{\partial^2 h}{\partial x^2} = 0,$$

$$\frac{\partial^2 h}{\partial t \partial y} - f\frac{\partial h}{\partial x} = 0. \tag{16.28}$$

The equations are linear in h, so any solution is a superposition of elementary solutions. Equatorial Kelvin waves are traveling waves, propagating in the zonal (x-) direction and decaying in the meridional (y-) direction, so we look for elementary solutions of the form

$$h_k(t,x,y) = \gamma(y)e^{i(kx-\omega t)},$$

where k is the zonal wave number, ω is (2π times) the frequency, and γ is decaying as $|y| \to \infty$. These elementary solutions have the property that $\partial^2 h_k / \partial x^2 = -k^2 h_k$.

Upon substitution, we find the *dispersion relation*

$$\omega^2 = (gH)k^2 \tag{16.29}$$

and an ODE for the function γ,

$$\frac{d\gamma}{dy} = -\frac{\beta k}{\omega}y\gamma(y). \tag{16.30}$$

Since β and ω are both positive, Eq. (16.30) shows that γ decays as $|y| \to \infty$ if and only if k is positive; hence, the only admissible elementary solutions are waves that propagate in the direction of increasing x—that is, eastward. Then it follows from Eq. (16.29) that $\omega = \sqrt{gH}k$, so the *phase velocity* and the *group velocity* coincide and both are equal to

$$c = \sqrt{gH}. \tag{16.31}$$

This implies that equatorial Kelvin waves propagate without dispersion, as if the Earth were a nonrotating planet. A typical phase speed would be about $2.8\,\text{m/s}$, causing an equatorial Kelvin wave to take 2 months to cross the Pacific Ocean between New Guinea and South America.

The equation for γ can be integrated; the solution is $\gamma(y) = e^{-(\beta/2c)y^2}$, so the solution of Eq. (16.28) is a superposition of the elementary solutions

$$h_k(t,x,y) = e^{-(\beta/2c)y^2}e^{ik(x-ct)}. \tag{16.32}$$

A characteristic length scale for decay away from the equator is the *equatorial Rossby radius of deformation*, which is defined as $L_{eq} = \sqrt{c/2\beta}$; its value is of the order of $220\,\text{km}$.

16.4 ▪ Delay Differential Equations

Equation (16.18) is an example of an autonomous *delay differential equation* (DDE). The general form of such equations for a function $x : t \mapsto x(t) \in \mathbb{R}^n$ is

$$\dot{x}(t) = f(x(t), x(t-\tau_1), \ldots, x(t-\tau_p)), \tag{16.33}$$

where $f : (\mathbb{R}^n)^{p+1} \to \mathbb{R}^n$ is a given function. The *delays* $\tau_1, \ldots, \tau_p$ are given; we assume that they are ordered, $0 < \tau_1 < \tau_2 < \cdots < \tau_p$.

A DDE differs fundamentally from an ODE. For example, the solution of the ODE $\dot{x} = f(x)$ is, under sufficient conditions on f, completely determined once the differential equation is supplemented by an initial condition at $t = 0$. By contrast, the solution of an IVP for the DDE (16.33) requires, even in the scalar case ($n = 1$), the specification of $x(t)$ at an "infinite" number of points, namely for all $t \in [-\tau_p, 0]$. One could therefore legitimately ask whether an IVP for a DDE is actually an infinite-dimensional problem, even in the scalar (one-dimensional) case. A discussion of the theory of DDEs would lead us well beyond the scope of this book. We refer the reader therefore to the mathematical literature on the subject; a classic reference is [35].

The basic theorems about existence, uniqueness, and regularity of solutions for equations of the form (16.33) are readily derived from the corresponding theorems for ODEs; see Theorems 4.1, 4.2, and 4.3. One only has to notice that solutions can be found successively on the intervals $[0, \tau_1]$, $[\tau_1, 2\tau_1]$, ... by applying results about ODEs, since in each of these steps the arguments $x(t - \tau_1), \ldots, x(t - \tau_p)$ are already known.

DDEs have much richer dynamics than ODEs. We illustrate this with a discussion of the linearization of Eq. (16.18) about the zero steady state, in the special case where $\tau_K = 0$. Writing $\tau = \frac{1}{2}\tau_R$, $\alpha = a - c$, $\beta = b$, and dropping the cubic term as a result of the linearization, we obtain an equation with a single delay,

$$\dot{T}(t) = \alpha T(t) - \beta T(t - \tau). \tag{16.34}$$

The sign of α is undetermined, and $\beta < 0$. The trial function $t \mapsto e^{\lambda t}$ satisfies Eq. (16.34) if $\lambda \in \mathbb{C}$ satisfies the *characteristic equation*

$$\lambda = \alpha - \beta e^{-\lambda \tau}. \tag{16.35}$$

This transcendental equation has infinitely many complex solutions. For $\beta\tau \leq e^{\alpha\tau - 1}$, there is a real solution, which is given by

$$\lambda^* = \frac{\alpha\tau + W(-\beta\tau e^{-\alpha\tau})}{\tau}. \tag{16.36}$$

Here, $W : [-e^{-1}, \infty) \to [-1, \infty)$ is *Lambert's W function*, which is defined by the equation $y = W(x)$ if and only if $x = ye^y$ for $x \geq -e^{-1}$. This function is implemented in most computer algebra systems.

If $\beta > \alpha$, we have $\lambda^* < 0$. However, this result does not imply that the zero steady-state solution is stable when $\beta > \alpha$. Indeed, for large β additional solutions exist that can grow exponentially (a destabilizing effect of delays). To understand this, we can look for oscillatory solutions. The characteristic equation (16.35) has a purely imaginary solution $\lambda = i\omega$ if $\alpha - \beta\cos(\omega\tau) = 0$ and $\beta\sin(\omega\tau) = \omega$, resulting in the condition

$$\tan(\omega\tau) = \frac{\omega}{\alpha}. \tag{16.37}$$

For given α and τ, this equation has infinitely many solutions ω_k, and thus there are infinitely many $\beta_k = \omega_k / \sin(\omega_k \tau)$ for which there are periodic solutions of the linear equation. If we regard β as a bifurcation parameter for the nonlinear problem (16.18) with the simplification $\tau_K = 0$, we see that there are infinitely many values of β where a pair of complex eigenvalues crosses the imaginary axis and where therefore a Hopf bifurcation can occur. This could never happen for a systems of ODEs, which always have polynomial characteristic equations. Equations with more than one delay, such as the full problem (16.18), are very difficult to discuss with purely analytical techniques.

16.5 • Numerical Investigations

We conclude the chapter with some numerical illustrations. Returning first to the recharge model, Eq. (16.8), we note that the periodic solutions which it produces miss some important qualitative aspects of ENSO dynamics. Specifically, the model fails to predict a variable period between about three and seven years and a variable amplitude. Also, the onset of El Niño, which is in fact often observed around December, does not depend on the season in this model. To improve the model, one can include a seasonal forcing term in the temperature equation and modify Eq. (16.8) as follows,

$$\begin{aligned}
\dot{x} &= -x + \gamma(bx + y) - \varepsilon(bx + y)^3 + M\cos(\tfrac{1}{3}\pi t), \\
\dot{y} &= -ry - \alpha bx.
\end{aligned} \tag{16.38}$$

Here, M indicates the strength of seasonal forcing. (Keep in mind that the time unit in this model corresponds to two months.) Figure 16.7 shows periodic solutions with a period close to three years, which were obtained with the parameter values

$$r = \tfrac{1}{4}, \quad \alpha = \tfrac{1}{8}, \quad b = 2, \quad \gamma = \tfrac{3}{4}, \quad \varepsilon = 0.1, \quad M = 0.25. \tag{16.39}$$

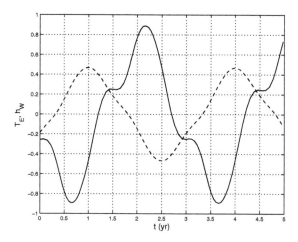

Figure 16.7. *Solutions of Eq.* (16.38).

Next, we turn to the delay equation model (16.18) and investigate whether it has periodic solutions with a period that is close to the observed value. We first consider the simplified case where $\tau_K = 0$, so there is only a single delay, $\tau = \tfrac{1}{2}\tau_R = 0.5$, assuming that time is measured in years. The temperature anomaly satisfies the equation

$$\dot{T}(t) = (a-c)T(t) - b\,T(t-\tau) - \varepsilon(T(t))^3. \tag{16.40}$$

According to the condition (16.37), the linearized equation $\dot{T}(t) = (a-c)T(t) - b\,T(t-\tau)$ has a periodic solution $T(t) = \cos(\omega t)$ if

$$\tan(\omega\tau) = \frac{\omega}{a-c}. \tag{16.41}$$

We expect solutions with $\omega \in (0.8, 2)$, corresponding to an El Niño period of three to seven years. This is possible only if $a - c > 0$. The left panel of Figure 16.8 shows a plot of the period $2\pi/\omega$ for the smallest solution ω of Eq. (16.41) against $a - c$. Given a value for $a - c$ and a corresponding ω, we expect periodic solutions of Eq. (16.40) if $b > \omega_k/\sin(\omega_k\tau)$. Two such solutions are plotted in the right panel of Figure 16.8. They were obtained with $a - c = 1.6$ and $\varepsilon = 0.1$. Then periodic solutions of the linearized equation exist if $b_0 = \omega/\sin(\omega\tau) = 2.5059$, where $\omega = 1.5186$, corresponding to a period of 4.14 years. The solutions plotted in Figure 16.8 were obtained by choosing $b = b_0 + 0.1$ (solid line) and $b = b_0 + 0.2$ (dashed line). They show that an increase in the feedback parameter b leads to a larger amplitude and a shorter period of the oscillation. A solution with an arbitrary nonconstant initial function on $[-\tau, 0]$ was computed on a sufficiently long time interval and was observed to settle on the periodic waveform, indicating that this periodic solution may be orbitally stable.

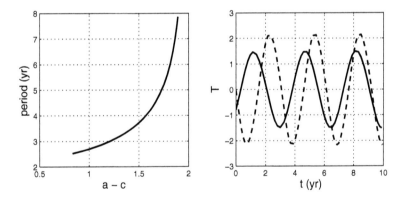

Figure 16.8. *Relation between damping parameter $a - c$ and expected period (left); solutions of Eq. (16.40) for two different values of b (right).*

Numerical simulations indicate that sustained periodic solutions of the full delay equation (16.18) with positive values of τ_K require larger values of the negative feedback parameter b than in the case $\tau_K = 0$. Figure 16.9 shows two periodic solutions, corresponding to basin-crossing times for Kelvin waves, τ_K, of two and three months, respectively. These solutions were obtained with the parameter values $a = 2.6$ and $c = 1$, so $a - c = 1.6$ as in the previous example; $b = b_0 + 0.5$, where $b_0 = 2.5059$ as in the previous example; and $\varepsilon = 0.1$ as before; starting with arbitrary initial functions and letting the solution settle on a periodic one. The model predicts a substantially smaller amplitude and a somewhat longer period if τ_K increases. The solutions depend very sensitively on τ_K. Including a second delay τ_K, even if it is small, therefore makes a significant difference in the solution behavior.

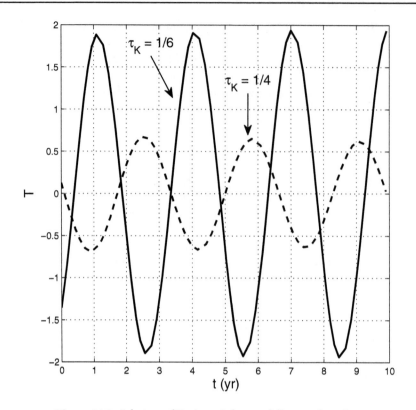

Figure 16.9. *Solutions of Eq. (16.18) for two different values of τ_K.*

16.6 ▪ Exercises

1. Describe the behavior of the solution of the linearized version of Eq. (16.8) if $\lambda = \gamma b < \lambda_1$ or $\lambda > \lambda_2$, where λ_1, λ_2 are given in Eq. (16.12). What is the physical meaning of these conditions?

2. This exercise explains why a cubic term such as the one appearing in Eq. (16.8) can have a stabilizing effect on solutions of ODEs.

   Consider the system of equations

   $$\begin{aligned} \dot{x} &= a_{11}x + a_{12}y - \varepsilon x^3, \\ \dot{y} &= a_{21}x + a_{22}y, \end{aligned} \qquad (16.42)$$

   where the only assumption is that $a_{22} < 0$. Let M and δ be positive constants, to be chosen later. Let (x, y) be a solution of this system on $[0, T]$.

   (i) Find conditions for M and δ such that the following statements are true:
      (a) If $x(T) = M$ and $|y(T)| \leq \delta M^3$, then $\dot{x}(T) < 0$.
      (b) If $x(T) = -M$ and $|y(T)| \leq \delta M^3$, then $\dot{x}(T) > 0$.

   (ii) Find conditions for M and δ such that the following statements are true:
      (a) If $y(T) = \delta M^3$ and $|x(T)| \leq M$, then $\dot{y}(T) < 0$.
      (b) If $x(T) = -\delta M^3$ and $|x(T)| \leq M$, then $\dot{y}(T) > 0$.

(iii) Combine the two conditions and proofs to show that there are $M > 0$ and $\delta > 0$ such that $-M < x(t) < M$ and $-\delta M^3 < y(t) < \delta M^3$ for all $t > 0$ if these inequalities are true for $t = 0$. Hint: Consider the smallest T for which one of these inequalities is violated, and apply the previous results.

(iv) Extend these results to problem (16.8) by rewriting the equations in terms of y and the new dependent variable $z = bx + y$. What is the condition on the coefficients such that the previous argument can be applied?

3. Explain why a "cylinder planet" does not support Rossby waves.

4. A model for coastal Kelvin waves can be derived from Eq. (16.26) by ignoring u, $u = 0$, and assuming that f is constant, $f = f_0$,

$$-f_0 v + g \frac{\partial h}{\partial x} = 0,$$
$$\frac{\partial v}{\partial t} + g \frac{\partial h}{\partial y} = 0, \tag{16.43}$$
$$\frac{\partial h}{\partial t} + H \frac{\partial v}{\partial y} = 0.$$

(i) Eliminate v to obtain a system of PDEs for h.

(ii) Look for elementary solutions of the form $\gamma_\ell(x) e^{i(\ell y - \omega t)}$. Verify that the dispersion relation for coastal and equatorial Kelvin waves are similar and show that the phase velocity $c = \omega/\ell$ and the group velocity $c_g = \partial \omega / \partial \ell$ coincide and are equal to $\pm\sqrt{gH}$.

(iii) Show that a coastal Kelvin wave travels with the coast to its right in the Northern Hemisphere and with the coast to its left in the Southern Hemisphere (see Figure 16.6).

5. Consider the DDE
$$\dot{x}(t) = ax(t) + bx(t - \tau) + f(t), \tag{16.44}$$
where $x : t \mapsto x(t) \in \mathbb{R}$; a, b, and τ are positive constants; and $f \in C(\mathbb{R})$ is a given function.

(i) Verify that Eq. (16.44) has the solution

$$x(t) = e^{at} \varphi(0) + \int_0^t e^{a(t-s)} (b\varphi(s - \tau) + f(s)) ds, \quad t \in [0, \tau]. \tag{16.45}$$

(ii) Explain how Eq. (16.45) can be used to obtain x successively on the intervals $[0, \tau], [\tau, 2\tau], \ldots$.

6. Prove that the solution of the DDE (16.44) that satisfies the condition $x(t) = \varphi(t)$ for all $t \in [-\tau, 0]$ satisfies the estimate

$$|x(t)| \le e^{(|a|+|b|)t} \left(M + \int_0^t |f(s)| ds \right), \quad t \ge 0,$$

where $M = \sup\{|\varphi(t)| : t \in [-\tau, 0]\}$.

7. Find as many explicit solutions of the DDE $\dot{x}(t) = -x(t - \frac{1}{2}\pi)$ as you can.

8. Prove that the real solution λ of Eq. (16.35) is indeed given by formula (16.36).

9. Consider the DDE

$$\dot{x}(t) = -ax(t - \tau), \qquad\qquad (16.46)$$

where $a > 0$. Use the characteristic equation for this equation to prove the following statements:

   (i) If $x(t) = e^{\lambda t}$ is a real-valued solution with $\lambda \in \mathbb{R}$, then $\lambda < 0$.

   (ii) If $x(t) = e^{(\lambda + i\omega)t}$ is a complex-valued solution and $\tau a < \frac{1}{2}\pi$, then $\lambda < 0$. Hint: Write down real the real and imaginary parts of the characteristic equation.

   (iii) If $\tau a = \frac{1}{2}\pi$, then there is a complex-valued solution of the form $x(t) = e^{i\omega t}$. Find ω in this case.

   Taken together, these results suggest that the zero solution of the DDE (16.46) is asymptotically stable if $0 < \tau a < \frac{1}{2}\pi$ and unstable if $\tau a > \frac{1}{2}\pi$.

10. Consider the DDE

$$\dot{x}(t) = \alpha x(t - \tau_1) + \beta x(t - \tau_2) - \varepsilon(x(t))^3 \qquad\qquad (16.47)$$

for $t > 0$, where α and β are real constants, $0 \leq \tau_1 < \tau_2$, and $\varepsilon > 0$.

   Let $m = \max\{|x(t)| : t \in [-\tau_2, 0]\}$ and take $M > \max\{m, (|\alpha| + |\beta|)/\varepsilon)^{1/2}\}$. Prove that $|x(t)| < M$ for all $t > 0$. This result explains the "stabilizing effect" of the cubic term in Eq. (16.18). Hint: Suppose $x(t) = M$ for some $t > 0$. Then there is a smallest $T > 0$ for which this happens. Clearly, then $\dot{x}(T) \geq 0$. Derive a contradiction, using the DDE.

11. Consider the oscillator model (16.18) of ENSO with $\tau_K = 0$ and $a - c > 0$.

   (i) Use a suitable change of variables to reduce the equation to the form

$$\dot{x}(t) = x(t) - \alpha x(t - \tau) - (x(t))^3. \qquad\qquad (16.48)$$

   Then derive conditions for the original parameters such that $\alpha < 1$. Assume for the rest of the problem that $\alpha < 1$.

   (ii) Show that Eq. (16.48) has three critical points,

$$x_1^* = 0, \quad x_2^* = (1 - \alpha)^{1/2}, \quad x_3^* = -(1 - \alpha)^{1/2}.$$

   (iii) Find the perturbation equation in the neighborhood of each critical point by setting $x = x_i^* + y$ ($i = 1, 2, 3$) in Eq. (16.48) and linearizing the resulting equation for y.

   (iv) Derive the characteristic equations of each of the linearizations derived in (iii).

   (v) Show that x_1^* has nonoscillatory exponential growth as $t \to \infty$; hence, x_1^* is always unstable.

   (vi) For x_2^* and x_3^*, find the *neutral stability curves*—that is, the curves in the (α, τ)-plane where $\text{Re}\,\lambda = 0$. Indicate the regions in the (α, τ)-plane where x_2^* and/or x_3^* are linearly stable.

12. In [28], the authors consider the following DDE for the thermocline depth anomaly h in the eastern Pacific:

$$\frac{dh}{dt}(t) = -a\tanh(xh(t-\tau)) + b\cos 2\pi\omega t. \qquad (16.49)$$

The parameter x measures the strength of the ocean-atmosphere coupling; the delay τ results from finite crossing speeds of Kelvin and Rossby waves, as in the other models discussed in this chapter; and the coefficient a measures the strength of the feedback mechanism. The last term in the equation represents an external forcing mechanism with strength b and frequency ω.

(i) Introduce dimensionless variables and parameters such that $a = 1$ and $\omega = 1$.

(ii) Show that the solution depends continuously on all remaining parameters in the problem.

(iii) For $b = 0$, linearize the equation about the zero state and set up the characteristic equation.

(iv) Use a numerical integration scheme to find periodic solutions with period 1 for suitable choices of τ, x, and b. Then show numerically that, for $\tau = 0.56$, $x = 11$, and $b = 1.4$, there are solutions with period 7.

(v) Show numerically that, for $\tau = 0.47$, $x = 10$, and $b = 1$, there are aperiodic solutions.

(vi) The model in [28] was introduced to explain the phenomenon of *phase locking* that has been observed for ENSO, where El Niño and La Niña events tend to occur during the boreal winter. The phase-locking phenomenon should be reflected in solutions that have extrema which occur exclusively within a particular season. The authors of [28] report that all minima of the solutions with period 7 obtained under (iv) above occur in time intervals $[k-\frac{1}{2}, k]$, and all maxima occur in time intervals $[k, k+\frac{1}{2}]$, where k is an integer. Verify this observation with a computation.

13. Consider the general scalar DDE (16.33), where f is a smooth function of all its variables (partial derivatives of arbitrary orders exist). Let $\varphi : [-\tau_p, 0] \to \mathbb{R}^n$ be a given function that is continuous. Consider the unique solution x of the DDE that satisfies $x(t) = \varphi(t)$ for all $t \in [-\tau_p, 0]$. Can you say for which (positive) values of t the second derivative $\ddot{x}(t)$ exists? How about the third derivative? Is this any easier if there are only two delays? What if there is only one delay? The difficulty in locating discontinuities of higher derivatives in advance limits the performance of high-order integration schemes for DDEs. As a result, most numerical schemes for DDEs use only low- (second- or third-) order schemes.

Chapter 17

Cryosphere and Climate

In this chapter we focus on the cryosphere and its relevance for climate. We pay particular attention to sea ice, which mediates the exchange of heat, moisture, and momentum between the ocean and the atmosphere. Sea ice covers a vast range of spatial scales, from microscopic brine inclusions to ice floes covering hundreds of square kilometers. The challenge is to find scaling laws that characterize the behavior of sea ice across multiple scales. We give several examples of such scaling laws.

Keywords: Snow, ice, glaciers, melt ponds, ice-albedo feedback, scaling law.

17.1 ▪ Cryosphere

The term *cryosphere* refers to those portions of the Earth's surface where water is in solid form. It includes sea ice, lake ice, river ice, snow cover, glaciers, ice caps and ice sheets, and frozen ground (which includes permafrost). Thus, there is a wide overlap with the hydrosphere. The cryosphere is the second largest component of the Earth's climate system (after the oceans) in terms of mass and heat capacity and has a significant impact on the surface energy budget through the surface reflectivity (albedo) and the latent heat associated with phase changes. For example, the presence of snow or ice in the polar regions increases meridional temperature differences, which affect winds and ocean currents. Also noteworthy is the fact that, on land, the cryosphere stores about 75% of the world's fresh water.

Presently, ice covers approximately 10% of the land surface, mostly in the form of ice caps and glaciers over Antarctica and Greenland, and approximately 7% of the oceans. In midwinter, snow covers approximately 49% of the land surface in the Northern Hemisphere. Frozen ground has the largest area of any component of the cryosphere; the permanently frozen permafrost region occupies approximately 24% of the land area in the Northern Hemisphere.

Changes in the various components of the cryosphere occur on different time scales, depending on thermodynamic and dynamic characteristics. While it may take more than 10,000 years for surface warming to penetrate an ice sheet and change the temperature at the base, water in crevasses penetrates through a porous ice sheet and affects the temperature at least locally within minutes. Because of the positive ice-albedo feedback, some components of the cryosphere act to amplify both changes and variability, while other components, like glaciers and permafrost, tend to average out short-term variability. In general, all components contribute to short-term changes, but long-term changes, including the ice age cycles, are due mostly to permafrost, ice sheets, and ice shelves.

17.2 · Glaciers, Ice Sheets, and Ice Shelves

Glaciers are found in mountain ranges of every continent and on a few high-latitude oceanic islands. They are slow-moving rivers of ice which drain areas in which snow accumulates, much as rivers drain catchment areas where rain falls (Figure 17.1). Glaciers also flow in the same basic way that rivers do. Although glacier ice is solid, it can deform by the slow creep of dislocations within the lattice of ice crystals which form the fabric of the ice. Thus, glacier ice effectively behaves like a viscous material with an extremely large viscosity. Because glacial mass is affected by long-term changes in precipitation, mean temperature, and cloud cover, glacial mass changes are considered among the most sensitive indicators of climate change.

Figure 17.1. *LeConte Glacier, Alaska.*

Ice sheets are formed mainly from snow that has been compressed into ice over thousands or millions of years. Ice sheets tend to spread under their own weight, transferring mass toward their margins. As an ice sheet reaches a marginal sea or lake, the ice may remain attached to become a floating ice shelf or form narrower ice tongues. The ice shelf loses mass by calving icebergs from the front and by basal melting into the ocean cavity underneath.

The great ice sheets of Greenland and Antarctica are up to two miles thick and hold enough ice to raise sea levels by about 64 m if fully melted. Even a modest change in the ice sheet balance could strongly affect sea levels and the fresh water flux to the oceans. The estimated total contribution to sea level rise from ice melt for the period 1993–2003 is 0.6 to 1.8 mm per year.

The standard technique for modeling ice sheets and ice shelves is based on a simple mass balance, comparing input from snow accumulation with output from ice flow and melt water runoff. These inputs and outputs are difficult to estimate with high accuracy. Mass balance estimates for the Antarctic ice sheet range from growth of 100 gigatons per year (Gt/yr, $1 \text{Gt} = 10^9$ tons) to shrinkage of 200 Gt/yr between 1961 and 2003, with East Antarctic thickening and West Antarctic thinning. Mass balance estimates for the Greenland ice sheet range from growth of 25 Gt/yr to shrinkage of 60 Gt/yr over the same period, with shrinkage of 50 to 100 Gt/yr between 1993 and 2003. In both cases, interannual variability is very large, driven mainly by variability in summer melting and sudden glacier accelerations. In 2003, NASA launched the "Ice, Cloud, and Land Elevation Satellite" (ICESat), using laser altimetry to more accurately measure changes in the Earth's surface elevation. While the Greenland interior is in mass balance, the coastlines are losing ice. Overall, Greenland is losing ice mass at an accelerating rate; from 2002 to 2009, the rate of ice mass loss doubled.

There is general agreement that the ice sheets over Antarctica and Greenland are capable of significantly affecting the sea level, but quantitative estimates vary widely because of inadequate knowledge of the bathymetry and circulation in the largely inaccessible ice shelf cavities.

17.3 ▪ Sea Ice

Sea ice is frozen seawater that floats on the ocean surface. It forms in both the Arctic and the Antarctic in each hemisphere's winter, and it retreats but does not completely disappear in the summer.

Sea ice has a profound influence on the polar physical environment, including ocean circulation, weather, and regional climate. As ice crystals form, they expel salt, which increases the salinity of the underlying ocean waters. This cold, salty water is dense and can sink deep to the ocean floor, where it flows back toward the equator. Sea ice also creates an insulating cap across the ocean surface, which reduces evaporation and prevents heat loss to the atmosphere from the ocean surface. As a result, ice-covered areas are colder and drier than they would be without ice.

Sea water freezes at a temperature of about $-1.8°C$. When seawater begins to freeze, it forms tiny crystals just millimeters wide, called *frazil*. How the crystals coalesce into larger masses of ice depends on whether the sea is calm or rough. In calm seas, the crystals form thin sheets of ice, *nilas*, so smooth they have an oily or greasy appearance. These wafer-thin sheets of ice slide over each other forming *rafts* of thicker ice. In rough seas, ice crystals converge into slushy *pancakes*. These pancakes can slide over each other to form smooth rafts, or they can collide into each other, creating ridges on the surface and keels on the bottom.

Some sea ice holds fast to a coastline or the sea floor (*fast ice*), and some sea ice drifts with winds and currents (*pack ice*). Because pack ice is dynamic, pieces of ice can collide and form much thicker ice. *Leads*—narrow, linear openings in the ice ranging in size from meters to kilometers—continually form and disappear.

As a material, sea ice is quite different from glacial ice. When salt water freezes, the result is a polycrystalline composite of pure ice with inclusions of liquid brine, air pockets, and solid salts. When fluid flows through sea ice, transport is facilitated by brine channels—connected brine structures ranging in scale from a few centimeters for horizontal slices to a meter or more in the vertical direction. Figures 17.2 and 17.3 show the various structures of sea ice and the typical length scale associated with each structure.

Figure 17.2. *Multiscale nature of sea ice—from millimeter to meter. Images reprinted with permission. Top row: left image from CRREL (U.S. Army Cold Regions Research and Engineering Lab) [118], right image from [52]. Middle row: left and center images from Kenneth M. Golden, right image from David Cole. Bottom row: both images from Kenneth M. Golden [29].*

The porosity of sea ice—that is, the volume fraction of brine—increases with temperature and exhibits an interesting and important critical phenomenon [30]. As long as the volume fraction of brine, ϕ, is below a critical value of approximately 5%, sea ice is effectively impermeable to fluid flow. But once ϕ is above the critical value, its permeability increases with ϕ. The critical brine volume fraction $\phi_c \approx 5\%$ corresponds to a critical temperature $T_c \approx -5°C$ for a typical bulk sea ice salinity of 5 psu. In the sea ice research community, this phenomenon is known as the *rule of fives*.

The sea ice pack mediates the exchange of heat, moisture, and momentum between the ocean and the atmosphere. While snow and sea ice reflect most incident sunlight, melt ponds on the surface of sea ice absorb most of it. As melting increases, so does solar absorption, which leads to more melting, and so on. It is believed that this ice–albedo feedback mechanism has contributed significantly to the precipitous losses of Arctic sea ice, which have outpaced the projections of most climate models.

17.3.1 ▪ Modeling Sea Ice

Modeling the reflection, absorption, and transmission of sunlight in sea ice is essential to capture the energy exchange with the atmosphere and ocean, and to estimate the light levels available for photosynthesis by living organisms in the ocean surface layer. But sea ice is a composite material that is difficult to model mathematically, since the size, geometry, and volume fraction of the fluid inclusions of brine depend strongly on the temperature. Generally, the governing equations are conservation laws of mass, momentum, and inter-

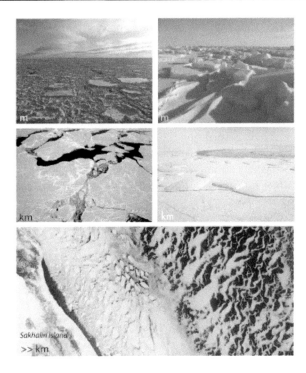

Figure 17.3. *Multiscale nature of sea ice—from meter to hundreds of kilometers. Images reprinted with permission. Top row: both images from Kenneth M. Golden. Middle row: left image from Andreas Klocker, right image courtesy of NSIDC and Tony Worby. Bottom image reprinted courtesy of NASA* [29].

nal energy, which take the form of PDEs. For example, the momentum equation treats sea ice moving at typical speeds as a non-Newtonian fluid with material properties of a theoretical plastic at the breaking point. Gradients in the flow field induce energy dissipation, which can result in sea ice rafting and ridging. These same gradients can cause the breakup of sea ice floes, creating leads. The complexity of the sea ice geometry results in a large range of sea ice thicknesses, which is normally treated in a statistical sense, so that the conservation equations for mass and internal energy involve probability density functions.

The conservation laws contain various parameters (transport coefficients) which characterize bulk properties of sea ice, the most relevant ones being thermal conductivity (how easy it is for heat to flow through the ice), fluid permeability (how easy it is for fluid to flow through the ice), and mechanical strength and fracture characteristics (determining how sea ice breaks up under stress from winds, currents, and collisions with other ice floes). These and other macroscopic parameters must somehow be obtained from information about the microstructure of sea ice. Among the mathematical techniques that have been used for this purpose are homogenization, variational techniques, and percolation theory. In addition, there are various computational techniques to couple mathematical models across a hierarchy of scales.

By way of example, we mention some recent work on fluid permeability for sea ice that made use of the fact that fluid transport in sea ice resembles fluid transport in certain types of crustal sedimentary rock. Both media exhibit critical behavior near a percolation threshold, with comparable values for ϕ_c. Assuming a certain degree of connectivity, particularly on small scales, and self-similarity, using a brine-coated sphere of ice as the basic unit of sea ice microstructure, Golden obtained a simple model for the vertical permeability k of sea ice over a range of brine volume fractions ϕ, namely $k(\phi) =$

$3\phi^3 \times 10^{-8}$ m^2 [29]. The expression appears to match closely the data for the permeability of Arctic sea ice.

17.3.2 ▪ Melt Ponds and Scaling Laws

Melt ponds on the surface of sea ice form a key component of the late spring and summer polar marine environment. Figure 17.4 is an aerial photo taken during the Healy–Oden TRans Arctic EXpedition (HOTRAX) on August 14, 2005. It shows the complex geometry of well-developed Arctic melt ponds and the complexity of their boundaries. The photo suggests an interesting question for mathematics, namely, does melt-pond geometry obey a universal *scaling law*? The question was addressed recently with geometric scaling arguments in [41].

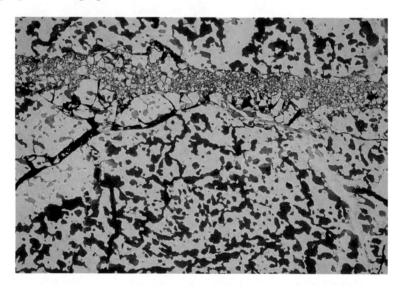

Figure 17.4. *Complex geometry of well-developed Arctic melt ponds. Aerial photo of arctic melt ponds with ship track, taken in August, 2005, on the Healy–Oden Trans Arctic Expedition (HO-TRAX). Reprinted courtesy of Donald Perovich.*

A scaling law is a mathematical relationship between two quantities, usually in the form of a power law. The main attribute of a power law that makes it interesting is its *scale invariance*. Given a relation $f(x) = ax^k$, scaling the argument x by a constant factor c causes only a proportionate scaling of the function itself. Since $f(cx) = a(cx)^k = c^k f(x)$, scaling by a constant c simply multiplies the original power-law relation by the constant c^k. Thus, all power laws with a particular scaling exponent are equivalent up to constant factors, since each is simply a scaled version of the others. In this sense, universal scaling laws capture the essence of seemingly complicated structures and are tremendously useful in characterizing aggregate properties of a complex system. There is evidence that the distributions of a wide variety of physical, biological, and man-made phenomena follow a power law; examples are the magnitude of earthquakes (Gutenberg–Richter law), the distribution of wealth and income (Pareto's law), the size of craters on the moon and of solar flares, the frequencies of words in most languages, the sizes of power outages and wars, and many other quantities.

Define the fractal dimension d of a melt pond in terms of its area A and perimeter P through the scaling relation

$$P \sim A^{d/2}. \tag{17.1}$$

For regular objects like circles and polygons, the perimeter scales like the square root of the area, so $d = 1$. For objects with a more highly ramified structure, such as Koch snowflakes, $d > 1$. As the boundary becomes increasingly complex and starts filling the two-dimensional domain, P scales more and more like A and d approaches its upper bound, which is 2.

Using images of melting Arctic sea ice collected during the 2005 HOTRAX expedition and the MATLAB Image Processing Toolbox, Hohenegger and coworkers computed area-perimeter data for 5,269 melt ponds [41]. The results are shown in Figure 17.5. The top panel shows that there is a "bend" in the scatter plot when the area is approximately $100\,\text{m}^2$. The center panel gives the fractal dimension computed from the data in the top panel. The graph suggests three distinct regimes,

- $A < 10\,\text{m}^2$, simple ponds with Euclidean boundaries and $d = 1$;

- $10\,\text{m}^2 < A < 1000\,\text{m}^2$, transitional ponds, where complexity increases rapidly with size; and

- $A > 1000\,\text{m}^2$, self-similar ponds, where complexity is saturated and $d \approx 2$.

The bottom panel shows melt ponds which are representative of each of the three regimes.

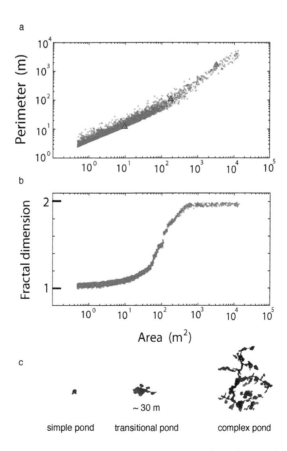

Figure 17.5. *Fractal dimension.* (a) *Area-perimeter data.* (b) *Graph of the fractal dimension d.* (c) *Representative melt ponds* [41].

The results indicate that melt ponds can be divided roughly into three categories, each with its own characteristic length scale, independently of the details of the underlying mechanism of melting and freezing.

17.4 • Exercises

1. (i) Consider a sequence $\{\mathscr{F}_n : n = 0, 1, \dots\}$ of subsets $\mathscr{F}_n$ of the plane, all contained in a sufficiently large square $[0, K] \times [0, K]$, such that $\mathscr{F}_n$ consists of $\lfloor c^n \rfloor$ squares that do not overlap or touch each other, each of side length 2^{-n}, where $0 < c \leq 4$. (The symbol $\lfloor \cdot \rfloor$ denotes the *floor* function of rounding down to the nearest integer.)

   Let A_n be the area of $\mathscr{F}_n$ and let P_n be its perimeter. Find $d > 0$ such that

   $$\lim_{n \to \infty} \frac{P_n}{A_n^{d/2}} = C \tag{17.2}$$

   for some positive constant C.

   (ii) Consider a sequence $\{\mathscr{G}_n : n = 0, 1, \dots\}$ of subsets $\mathscr{G}_n$ of the plane, all contained in a sufficiently large square $[0, K] \times [0, K]$, such that

   $$\mathscr{G}_0 = \emptyset; \quad \mathscr{G}_{n+1} = \mathscr{G}_n \cup \mathscr{F}_n, \quad n = 1, 2, \dots, \tag{17.3}$$

   where $\mathscr{F}_n$ is defined in part (i) with $0 < c < 4$. In addition, all squares in $\mathscr{G}_n$ are to be nonoverlapping and nontouching. Set $\mathscr{G} = \cup_{n=1,2,\dots} \mathscr{G}_n$.

   Explain why one expects to be able to construct such sets $\mathscr{G}_n$ if $c/(4-c) < K^2$. Let A_n be the area of all squares with side length 2^{-n} in $\mathscr{G}$ and let P_n be their perimeter. Find $d > 0$ such that

   $$\lim_{n \to \infty} \frac{P_n}{A_n^{d/2}} \tag{17.4}$$

   exists.

2. Let $\mathscr{G} \subset \mathbb{R}^n$ be an arbitrary bounded set, where $n = 1, 2, 3$. For $\varepsilon > 0$, let $N(\varepsilon)$ be the number of intervals of length ε if $n = 1$, squares of side length ε if $n = 2$, or cubes of side length ε if $n = 3$ that are needed to cover $\mathscr{G}$. If there exist constants $A > 0$ and $d \geq 0$ such that

   $$\lim_{\varepsilon \to 0} N(\varepsilon) \varepsilon^d = A, \tag{17.5}$$

   then d is called the *box count dimension* of $\mathscr{G}$, denoted by $d_B(\mathscr{G})$.

   (i) Find $d_B(\mathscr{G})$ if $\mathscr{G}$ is a finite set of points, a line segment, a smooth surface, and a solid sphere in $\mathbb{R}^3$.

   (ii) Let $\mathscr{G} \subset \mathbb{R}^2$ be given such that $d_B(\mathscr{G}) < 2$. Show that $\mathscr{G}$ must have zero area.

   (iii) Let $\mathscr{G}$ be defined as in Exercise 1. Let $\mathscr{H}$ be the union of all edges of the squares that make up $\mathscr{G}$. Find $d_B(\mathscr{H})$. How is it related to the parameter d that you found in Exercise 1?

3. The box count dimension of a set can be found with image processing software. Look up the documentation of such a software package (for example, the *boxcount* package for MATLAB) and summarize how such methods work and what their limitations are.

4. In the seasonal ice zones (SIZs), various types of ice floes are present, and their sizes range from about one meter to kilometers. Using Landsat imagery and video monitoring of seasonal sea ice in the southern Sea of Okhotsk in February 2003, Toyota, Takatsuji, and Nakayama [111] were able to measure ice floes over three orders of magnitude, from 1 m to 1.5 km.

   Here is a mathematical approach to describing the distribution of ice floe sizes. Such a distribution can be used to estimate the perimeter of all the ice floes in the region, something that would be impossible to do directly. We make the highly simplified assumption that all ice floes are squares of varying side lengths (in meters).

   For an observational area of area $A_0 = \mathcal{O}(10^9)$, let $N(d)$ be the total number of floes with diameter d or less. We treat d like a continuous variable ranging over the interval (d_0, d_1), where $d_0 = \mathcal{O}(1)$ and $d_1 = \mathcal{O}(10^3)$, and assume that $N(d)$ follows a power law distribution,

   $$N(d) = \beta d^{-\alpha}, \quad d \in (d_0, d_1),$$

   for some positive constants α and β.

   (i) Derive the expression

   $$A(d) = \frac{\beta}{2-\alpha}\left(d^{2-\alpha} - d_0^{2-\alpha}\right) \tag{17.6}$$

   for the area of all ice floes with edge length less than or equal to d.

   (ii) Given $\gamma \in (0,1)$, derive approximate conditions for α and β such that a fraction γ of the observational area is covered with ice. Use the parameter values of [111] ($A_0 = 10^9$, $d_0 = 1$, $d_1 = 10^3$, $\alpha = 1.87$) to find β and determine its units.

   (iii) Lateral melting of individual floes is an important contribution to the development of sea ice. Find a formula for the perimeter $P(d)$ of all ice floes with edge less than or equal to d, for a given area of observation, in terms of α and β.

5. The authors of [111] actually reported that the observed ice floe size distribution can be approximately described by two different power laws,

   - $N(d) \approx \beta_1 d^{-\alpha_1}$ if $d > d^*$,
   - $N(d) \approx \beta_2 d^{-\alpha_2}$ if $d \leq d^*$,

   with $\alpha_1 = 1.87$, $\alpha_2 = 1.15$, $d^* = 40$.

   (i) Find conditions on α_i and β_i ($i = 1,2$) such that N is continuous, assuming that d^* is given.

   (ii) Show that if the area γA_0 of ice-covered ocean and the parameters d_0, d^*, d_1 as well as α_1 and α_2 are given, then β_1 and β_2 are uniquely determined.

   (iii) Find a formula for the perimeter of all ice floes in the area of observation in terms of the parameters d_0, d^*, d_1, α_1, α_2 and the observation area A_0.

   The authors of [111] found that the total perimeter of all ice floes in the case of two regimes could be up to 4.8 times less than if there were only one regime, with $\alpha = 1.87$. This result indicates that the size distribution of ice floes smaller than 40 m has a strong effect on the lateral melting process.

Chapter 18

Biogeochemistry

Although the Earth constantly receives energy from the Sun, its chemical composition is essentially fixed, so all the chemical nutrients like carbon, oxygen, nitrogen, phosphorus, calcium, and water must be recycled. The circulation of chemical nutrients through the biological and physical world is known as the biogeochemical cycle. In this chapter we touch upon a few aspects of this cycle: the carbon cycle or, more particularly, the exchange of CO_2 at the ocean-atmosphere interface; the role of plankton in the mixing layer of the Earth's oceans; and algal blooms.

Keywords: Biogeochemical cycle, carbon cycle, nitrogen cycle, partial pressure, ocean-atmosphere CO_2 flux, biological pump, ocean plankton, NPZ model, algal bloom, advection-diffusion equation, self-shadowing, nonlocal PDE, two-point boundary value problem, shooting method.

18.1 ▪ Biosphere and Climate

The *biosphere* is all life on our planet, including all living things as well as the remains of those that have died but have not yet decomposed. It includes life on land and in the oceans—multitudes of plants, animals, fungi, plankton, and bacteria. The biosphere has a great impact on the Earth's climate, mainly (but not exclusively!) through the processes that make up the biogeochemical cycles—primarily the *carbon cycle* and the *nitrogen cycle*.

The carbon cycle is illustrated in Figure 18.1. Carbon in the form of CO_2 is absorbed from the air into the ocean, where it serves as a nutrient for floating plankton doing photosynthesis. This carbon may become part of the plankton's skeleton, or part of the skeleton of a larger animal that feeds on plankton, and then part of a sedimentary rock when the living things die and only skeletons are left behind. Carbon that is a part of rocks may be stored for a long time, but weathering of rocks on land over millions of years eventually adds carbon compounds to surface water, which runs off to the ocean. When fossil fuels are burned, carbon that had been underground is sent into the air as CO_2—a greenhouse gas. Greenhouse gases trap heat, contributing to a warm atmosphere. Without greenhouse gases, the Earth would be a frozen world.

Humans have burned so much fuel that there is approximately 40% more CO_2 in the air today (395 ppm, as of June 2012) than before the Industrial Revolution (280 ppm), about 200 years ago. The atmosphere has not held this much carbon for at least 420,000 years, according to data from ice cores.

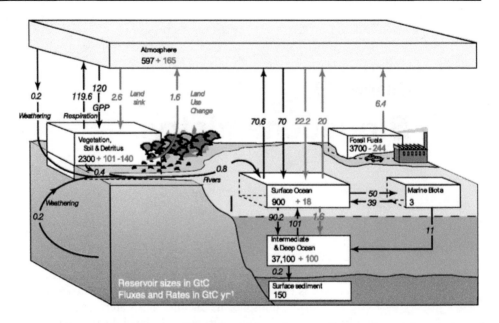

Figure 18.1. *Schematic of the carbon cycle, with estimated figures for stored quantities and fluxes of* CO_2 *in the preindustrial (black) and industrial (red) era. Reprinted with permission from IPCC.*

The story of the *nitrogen cycle* is more complicated. Most of the nitrogen on Earth is in the atmosphere. Approximately 80% of the molecules in the atmosphere are made of two nitrogen atoms bonded together (N_2). All plants and animals need nitrogen to make amino acids, proteins, and DNA, but the nitrogen in the atmosphere is not in a form that they can use. The nitrogen molecules in the atmosphere need to be broken apart to become usable for living organisms, which happens, for example, during lightning strikes or fires, and which can be mediated by certain types of bacteria. When organisms die, their bodies decompose, bringing the nitrogen into soil on land or into ocean water. There it is changed by other types of bacteria into a form that allows it to return to the atmosphere.

18.2 ▪ Carbon Cycle

The ocean plays a central role in the global carbon cycle. It is a huge CO_2 reservoir; it is estimated that in preindustrial times 37,100 Gt of CO_2 was stored in the intermediate and deep ocean and approximately 900 Gt in the surface ocean [103]. The surface layer of the world's oceans takes up a little less than one-half of the CO_2 produced by human activities; however, this is one of the most poorly known figures in climate science, and there is a continuing debate over whether the oceans will continue to be such a large sink of CO_2.

The special role of the ocean in the carbon cycle comes about because of the chemistry. CO_2, like other gases, is soluble in water. However, unlike many other gases (oxygen for instance), CO_2 reacts with water to form carbonic acid (H_2CO_3). The carbonic acid dissociates into a hydrogen ion (H^+) and bicarbonate (HCO_3^-); bicarbonate in turn dissociates into another hydrogen ion and carbonate (CO_3^{2-}). All reactions are reversible,

$$CO_2 + H_2O \Longleftrightarrow H_2CO_3 \Longleftrightarrow H^+ + HCO_3^- \Longleftrightarrow H^+ + H^+ + CO_3^{2-}. \qquad (18.1)$$

Most of the dissolved carbon in the ocean is in the form of bicarbonate (88.6%) and carbonate (10.9%). Neither bicarbonate nor carbonate communicates with the overlying atmosphere, but the conversion of CO_2 gas into bicarbonate and carbonate effectively reduces the dissolved CO_2 in the ocean, thereby allowing more diffusion from the atmosphere.

The exchange of CO_2 between the ocean and atmosphere is controlled by the partial pressures of CO_2 in the surface layer of the ocean, pCO_2^{oc}, and in the atmosphere just above the interface, pCO_2^{at}. If the latter exceeds the former, CO_2 will be taken up by the ocean; otherwise, CO_2 will flow in the opposite direction. A simple formula for the flux of CO_2 at the ocean-atmosphere interface is $F = k(pCO_2^{oc} - pCO_2^{at})$, where the proportionality constant k is the *transfer coefficient*. A negative (positive) flux means CO_2 is being taken up (released) by the ocean.

18.2.1 ▪ Partial Pressure of Atmospheric CO_2

The partial pressure of CO_2 in the atmosphere just above the ocean-atmosphere interface can be computed from the concentration of CO_2 and average total pressure. Just above the ocean-atmosphere interface, the air is a mixture of gases (nitrogen, oxygen, argon, CO_2, etc.). The *partial pressure* of any constituent is the pressure that constituent would exert if it were the only gas present in the volume under consideration. If x_i is the volume fraction of the gas i ($i = 1, 2, \dots$) in the mixture, then the partial pressure is $p_i = x_i p$, where p is the total pressure, $p = \sum_i p_i$. The total pressure at sea level averages 1 atm and the volume fraction of CO_2 at sea level (as of June 2012) is 0.000395 (395 ppm). The partial pressure of CO_2 at sea level is therefore $pCO_2^{at} = 395\ \mu\text{atm}$. The partial pressure is often identified with the volume fraction, in which case the unit is simply ppm.

18.2.2 ▪ Partial Pressure of Oceanic CO_2

The amount of a given gas that dissolves at a given temperature in a given type and volume of liquid is directly proportional to the partial pressure of that gas in equilibrium with that liquid (Henry's law). If c is the concentration of the solute and p the partial pressure of the solute in the gas above the solution, then

$$c = k_H^{-1} p. \tag{18.2}$$

Here, k_H is *Henry's coefficient*; its inverse k_H^{-1} is known as the *solubility*. In pure water at 25°C, the solubility of CO_2 is approximately $0.034\ \text{mol kg}^{-1}\text{atm}^{-1}$.

The content of CO_2 in surface waters is commonly presented as the partial pressure in solution and is computed from Henry's law, Eq. (18.2). Thus,

$$pCO_2^{oc} = k_H [CO_2]. \tag{18.3}$$

(Square brackets indicate concentrations.) In sea water, the solubility of CO_2 depends on the temperature T as well as the salinity S; approximate formulas are given in [98].

To find an expression for $[CO_2]$ in terms of measurable quantities, we consider the carbonate system (18.1) at equilibrium,

$$CO_2 + H_2O \Longleftrightarrow H_2CO_3 \overset{k_1}{\Longleftrightarrow} H^+ + HCO_3^- \overset{k_2}{\Longleftrightarrow} 2H^+ + CO_3^{2-}.$$

Here, k_1 and k_2 are the reaction constants of the dissociation reactions of carbonic acid and bicarbonate, respectively, at equilibrium. They can be expressed in terms of the concentrations of the participating constituents,

$$k_1 = \frac{[HCO_3^-][H^+]}{[CO_2]}, \quad k_2 = \frac{[CO_3^{2-}][H^+]}{[HCO_3^-]}. \tag{18.4}$$

The dissociation constants depend on T and S; approximate formulas are given in [63].

The dissolved forms of bicarbonate, carbonate, and CO_2 are grouped together as *dissolved inorganic carbon* (DIC),

$$DIC = [HCO_3^-] + [CO_3^{2-}] + [CO_2]. \tag{18.5}$$

The concentration of DIC averages $2,248 \, \mu mol \, kg^{-1}$; it is highest in the Antarctic and lowest in the Atlantic where deep water sinks into the ocean depth. The concentration of DIC varies with depth and is highest at depths between 110 and 3,500 m; the vertical gradient is maintained by physico-chemical and biologically mediated processes that transport carbon from the ocean's surface to its interior.

While DIC keeps track of the carbon, the alkalinity keeps track of the charges in the carbonate system. The carbonate alkalinity Alk_C is defined by

$$Alk_C = [HCO_3^-] + 2[CO_3^{2-}]. \tag{18.6}$$

(The carbonate ion is counted twice because it has a double negative charge.) The carbonate alkalinity is part of the total alkalinity (Alk), which includes boron compounds and other minor components. The alkalinity is $2,480 \, \mu mol \, kg^{-1}$ for a typical sample of sea water with salinity $S = 35$, $pH = 8.1$, and temperature $T = 25°C$.

The two equilibrium conditions (18.4), the mass balance for total inorganic carbon, Eq. (18.5), and the charge balance, Eq. (18.6), make up four equations for the six unknowns—$[CO_2]$, $[H_2CO_3]$, $[CO_3^{2-}]$, $[H^+]$, DIC, and Alk. Consequently, when two variables are known, the concentration of each constituent of the carbonate system can be computed. In theory, any two variables can be taken, but in practice only $[CO_2]$, $[H^+]$ (that is, pH), DIC, and Alk can be measured directly. Because the total alkalinity includes boron compounds and other minor components, these species have to be accounted for in the computations. The procedure for computing all components of the carbonate system from any two given quantities, including boron compounds, is described in detail in [126].

The results are illustrated in Figure 18.2. The left panel shows the isobars of oceanic pCO_2 as a function of temperature (°C) and salinity (psu) in the physically relevant regions at a fixed DIC concentration of $2,248 \, \mu mol \, kg^{-1}$ and alkalinity of $2,480 \, \mu mol \, kg^{-1}$.

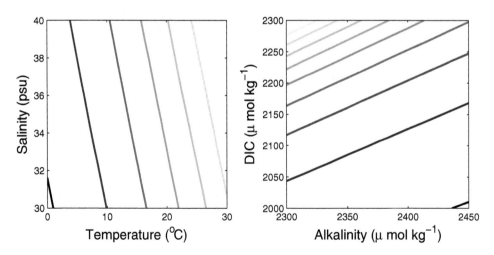

Figure 18.2. *Oceanic pCO_2 isobars as functions of temperature and salinity (left) and as functions of alkalinity and DIC (right). Darker lines correspond to lower values of pCO_2.*

Values range from about 190 μatm in very cold water at low salinity to about 870 μatm in warm and very saline water. The right panel shows the isobars as a function of alkalinity and DIC at a fixed temperature of 20°C and salinity 35 psu. These values range from about 190 μatm for high alkalinity and low DIC to about 2,150 μatm at the other extreme. Both graphs assume a depth of 100 m. The graphs were produced with a MATLAB program developed by Lovenduski [61] based on the equations given in [126].

These diagrams are useful to get an idea of the various competing effects of biological and chemical processes on oceanic pCO_2. For example, the formation of organic matter, which decreases DIC and increases alkalinity (moving diagonally across the right diagram from top left to bottom right), results in a decrease of oceanic pCO_2.

Experimental data on the CO_2 flux at the ocean-atmosphere interface are few and far between, but they have been used by Takahashi et al. [108] to compile a global picture of the annual mean ocean-atmosphere flux of CO_2; see Figure 18.3.

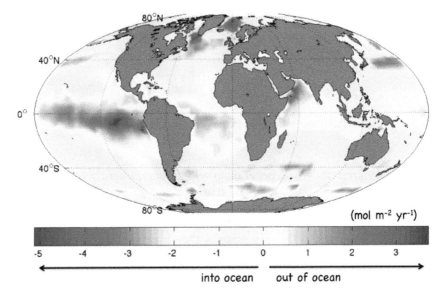

Figure 18.3. *Annual mean ocean-atmosphere CO_2 flux* [108].

18.3 ▪ Carbon Transport into the Deep Ocean

Once CO_2 is in the upper layer of the ocean, there are two mechanisms to transport it into the ocean's interior: the *solubility pump*, which is based on physico-chemical processes, and the *biological pump*, which consists entirely of biologically mediated processes [87].

The solubility pump distributes CO_2 by mixing and ocean currents. The mechanism is especially efficient at higher latitudes, as the uptake of CO_2 as DIC increases at lower temperatures. This process is further enhanced by the THC, which is driven by the formation of deep water at high latitudes where seawater is usually cooler and denser.

The biological pump (Figure 18.4) is a suite of biologically mediated processes that originates with the uptake of CO_2 by phytoplankton, the photosynthesizing organisms in plankton floating or drifting in the upper ocean. About a dozen functional groups of phytoplankton can be distinguished, ranging from unicellular algae to diatoms and cyanobacteria. Phytoplankton needs light and sufficiently high concentrations of nutrients, depending on size, to survive in the ocean's surface layer. Consequently, phyto-

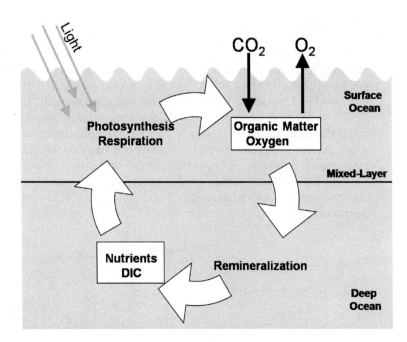

Figure 18.4. *Schematic representation of the biological pump. Reprinted courtesy of Nicholas Cassar.*

plankton tends to do better in upwelling zones, where large bloom concentrations can be observed, subject to seasonal variation.

The organic carbon that forms the biological pump is transported primarily by sinking particulate material, for example dead organisms (including algal mats) or fecal pellets. Carbon reaching the deep ocean by these means is either organic carbon or particulate inorganic carbon (PIC) such as $CaCO_3$. Organic carbon is a component of all organisms, but PIC occurs only in calcifying organisms such as coccolithophores, foraminiferans, and pteropods. In the case of organic material, remineralization (or decomposition) processes such as bacterial respiration return the organic carbon to dissolved CO_2 at a rate which depends upon local carbonate chemistry. As these processes are generally slower than synthesis processes, and because the particulate material is sinking, the biological pump transports material from the surface of the ocean to its depths. A detailed description of the biological pump can be found in [22].

The biological pump plays an extremely important role in the Earth's carbon cycle, and significant effort is being spent by the research community to quantify its strength. However, because it occurs as a result of poorly constrained ecological interactions usually at depth, the processes that make up the biological pump are difficult to measure and efforts to model it have so far not met with notable success.

18.4 ▪ Ocean Plankton

The nutrient-phytoplankton-zooplankton (NPZ) model is a common tool in oceanographic research to describe the dynamics of oceanic plankton. The NPZ model assumes that nitrogen is the factor limiting the growth rate of plankton, that all other resources are in ample supply, and that the state of the plankton system is determined by the nitrogen content of the nutrients, phytoplankton, and zooplankton in the upper ocean. The words "nitrogen content" are usually omitted, and the variables are simply referred to as *nutrients* (N), *phytoplankton* (P), and *zooplankton* (Z).

Nutrients, phytoplankton, and zooplankton drifting in the upper ocean layer interact as part of the food chain and are simultaneously subject to diffusion due to turbulent mixing and advection due to ambient currents. If C denotes the concentration of N, P, or Z, then the change of C in space and time is governed by an *advection-diffusion equation*—that is, a PDE of the type

$$\frac{\partial C}{\partial t} + u\frac{\partial C}{\partial x} + v\frac{\partial C}{\partial y} + (w+w_s)\frac{\partial C}{\partial z} = D_h\left(\frac{\partial^2 C}{\partial x^2} + \frac{\partial^2 C}{\partial y^2}\right) + D_v\frac{\partial^2 C}{\partial z^2} + \left(\frac{\partial C}{\partial t}\right)_{bio}. \quad (18.7)$$

The first term on the left-hand side of the equation gives the explicit change of C with time; the next two terms give the change of C due to advection in the horizontal directions, where u and v are the horizontal components of the fluid velocity; and the last term before the = sign gives the change of C in the vertical direction due to advection and sinking; w is the vertical component of the fluid velocity and w_s is the vertical sinking speed. The terms on the right-hand side of the equation give the change of C due to turbulent diffusion; D_h and D_v are the diffusion coefficients in the horizontal and vertical directions, respectively. The last term gives the change of C due to all biological processes; here, the NPZ model enters the picture.

The NPZ model describes oceanic plankton as a predator-prey system, where phytoplankton is the prey and zooplankton the predator, and nutrients (nitrogen) are consumed and produced by both species. The system is closed, in the sense that there is no external forcing. It is usually given in the form of a "transfer system" of three ODEs for the state variables N, P, and Z, given the amount I of light available,

$$\left(\frac{\partial P}{\partial t}\right)_{bio} = f(I)g(N)P - h(P)Z - i(P)P,$$

$$\left(\frac{\partial Z}{\partial t}\right)_{bio} = \gamma h(P)Z - j(Z)Z, \quad (18.8)$$

$$\left(\frac{\partial N}{\partial t}\right)_{bio} = -f(I)g(N)P + (1-\gamma)h(P)Z + i(P)P + j(Z)Z.$$

The first equation gives the rate of change of the phytoplankton; phytoplankton increases (first term) due to nutrient uptake (g), which is done by photosynthesis in response to the amount of light available (f), and decreases due to zooplankton grazing (second term, h) and death and predation by organisms not included in the model (third term, i). The second equation gives the rate of change of the zooplankton; zooplankton increases (first term) due to grazing (h), but only a fraction γ of the harvest is taken up, and decreases due to death and predation by organisms not included in the model (second term, j). The third equation gives the rate of change of the nutrients; nutrients are lost due to grazing by phytoplankton (first term, f and g) and increased by the left-over fraction $1-\gamma$ of harvested zooplankton (second term, h) and the remains of phytoplankton (third term, i) and zooplankton (fourth term, j). Note that the right-hand sides of Eq. (18.8) sum to zero, so the NPZ model conserves the total amount of nitrogen in the system,

$$(N+P+Z)(t) = (N+P+Z)(0), \quad t \in \mathbb{R}.$$

The most common use of NPZ models is for theoretical investigations, to see how the model behaves if different transfer functions are used. A survey of the merits and limitations of the NPZ model is given in the review article by Franks [25].

In the simplest case, the system of equations (18.7) is solved with a prescribed fluid velocity. But in more realistic cases, the fluid velocity is determined by coupling the system (18.7) to an ocean circulation model and solving the two sets of equations simultaneously. The circulation models may range from a simple one-dimensional model with biological dynamics averaged over the mixing layer to a full three-dimensional model with high-order turbulence-closure submodels.

18.5 ▪ Algal Blooms

Algae are a large and diverse group of simple, typically autotrophic organisms, ranging from unicellular to multicellular forms such as the giant kelps that grow to 65 m in length. They are photosynthetic like plants, but "simple" because their tissues are not organized into the many distinct organs found in land plants. Algae constitute an important component of the biological pump and play a crucial role in the uptake of atmospheric CO_2 by the Earth's oceans and thus in the carbon cycle.

An *algal bloom* is a rapid increase or accumulation in the population of algae in an aquatic system. Algal blooms may occur in fresh water as well as marine environments. Typically, only one or a small number of phytoplankton species are involved, and some blooms may be recognized by discoloration of the water resulting from the high density of pigmented cells. In this section we summarize a model of algal blooms that was first proposed by Klausmeier and Litchman [54]. The model is a reduced NPZ model, where all nutrients, including nitrogen, are assumed to be in ample supply and phytoplankton is the only player in the game. The phytoplankton grows due to photosynthesis, which depends on the availability of energy in the form of light, and is subject to advection by the ambient fluid, diffusion due to turbulent mixing, and removal due to death and scavenging. Once plankton is dead, it sinks to the bottom and is removed from the system.

We begin by making an assumption that will drastically reduce the dimensionality of the problem and enable us to use some new tools of mathematical analysis. We assume that the phytoplankton population is uniformly distributed in the horizontal directions. Of course, this is physically not a realistic assumption, but our goal in this section is to demonstrate that the phenomenon of algal blooming can be triggered by small changes in a few fundamental parameters. As long as the model gives us a way to achieve this goal, we accept its consequences while recognizing its limitations.

The vertical coordinate z measures depth from the ocean surface at $z = 0$ to the bottom at $z = L$. Thus, P is a function of depth z and time t.

The advection-diffusion equation (18.7) for the phytoplankton population reduces to

$$\frac{\partial P}{\partial t} = D\frac{\partial^2 P}{\partial z^2} - V\frac{\partial P}{\partial z} + \left(\frac{\partial P}{\partial t}\right)_{\text{bio}}, \qquad (18.9)$$

where D is the vertical diffusion coefficient and $V = w + w_s$ the downward speed.

The PDE (18.9) must be supplemented by boundary conditions at $z = 0$ and $z = L$ to express the fact that phytoplankton does not leave the ocean, either at the top or at the bottom. By integrating Eq. (18.9) over thin layers $0 < z < \varepsilon$ and $L - \varepsilon < z < L$ and taking the limit $\varepsilon \to 0$, we obtain the boundary conditions

$$D\frac{\partial P}{\partial z} - VP = 0, \quad z = 0, L. \qquad (18.10)$$

The system of biodynamic equations (18.8) reduces to a single equation for P,

$$\left(\frac{\partial P}{\partial t}\right)_{\text{bio}} = gf(I)P - i(P)P. \qquad (18.11)$$

Here we have assumed that the growth rate g of the phytoplankton due to nutrient uptake is constant. Other, more complicated functions are found in the literature, but remember that we try to keep the model as simple as possible.

An expression for f follows from the Lambert–Beer law for light traveling through matter,

$$f(I)(z,t) = I_0 \exp\left(-K_{bg}z - r\int_0^z P(y,t)\,dy\right), \qquad (18.12)$$

where I_0 is the incident light intensity, r is the specific light attenuation coefficient of the phytoplankton, and K_{bg} is the total background attenuation due to all nonphytoplankton components (*turbidity*). Note that this formulation includes light absorption by phytoplankton, so the light gradient changes with a change in the density of the phytoplankton population. This phenomenon is known as *self-shadowing*.

Again, to keep the model as simple as possible, we take the rate of removal of the phytoplankton by death or scavenging constant in time and independent of depth, $i(P) = \ell$.

The complete model for the dynamics of the depth profile of the phytoplankton population thus consists of the PDE

$$\frac{\partial P}{\partial t} = D\frac{\partial^2 P}{\partial z^2} - V\frac{\partial P}{\partial z} + (gf(I) - \ell)P, \quad z \in (0,L), \qquad (18.13)$$

where $f(I)$ is given by Eq. (18.12), together with the boundary conditions (18.10).

Observe that $P(z) = 0$ for $z \in [0,L]$ satisfies the PDE and the boundary conditions. This is the trivial solution, which corresponds to "no bloom."

Eq. (18.13) is *nonlinear*, because of the presence of P in the exponential, and *nonlocal*, because the term f involves values of P at all depths down to z. Clearly, there is no hope that it can be solved analytically.

18.5.1 ▪ Numerical Simulations

Using more or less realistic values of the physical parameters, Huisman et al. [43] solved Eq. (18.13) numerically with a Michaelis–Menten model for the dependence on the light intensity. Varying the turbulent diffusion coefficient D, they found that, under suitable light conditions, the phytoplankton population developed a positive stationary density profile. Some of the steady-state profiles are sketched in Figure 18.5. The figure shows that the stationary depth profile can have a local population density maximum below the surface, provided the phytoplankton growth rate exceeds both the sinking rate and the mixing rate. In these cases, there is "bloom development."

If turbulent diffusion was decreased further below $0.1\ \mathrm{cm^2 s^{-1}}$, the numerical simulations showed the entire phytoplankton population sinking to the bottom of the water column and vanishing in the dark.

The same authors also investigated computationally which conditions were favorable for bloom development. In a systematic study, they varied both the diffusion coefficient and the depth of the water and noted which combinations resulted in bloom development. Their findings are summarized in the diagrams of Figure 18.6. Note the log scales of the axes; the graphs span the entire spectrum from shallow, quiescent lakes to deep, turbulent oceans.

In shallow waters, the eventual depth profile of the phytoplankton may vary depending on the sinking velocity, but there is always sufficient light for bloom development. For bloom development in deep waters, two conditions must be met: growth rates must exceed mixing rates, so that uniform mixing over the entire depth of the water column is

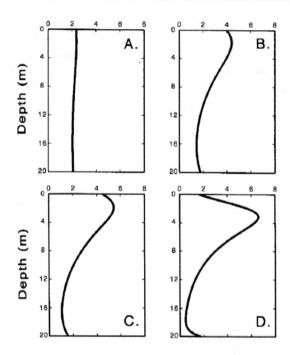

Figure 18.5. *Stationary depth profiles of sinking phytoplankton for four different values of the turbulent diffusion coefficients; (A) $D = 10\,cm^2 s^{-1}$, (B) $D = 1\,cm^2 s^{-1}$, (C) $D = 0.5\,cm^2 s^{-1}$, (D) $D = 0.1\,cm^2 s^{-1}$ [43].*

prevented, and turbulent mixing rates must exceed sinking rates, so that large downward fluxes of phytoplankton are avoided. Only when both conditions are met (middle right of Figure 18.6(A)) can sinking species maintain a population in the euphotic zone.

The no-bloom areas in Figure 18.6(A) are bounded by nearly horizontal and vertical lines. This indicates that the effects of water-column depth and turbulent diffusion on bloom development are essentially independent of one another.

18.5.2 ▪ Analytical Results

The numerical results of the previous section are supported by a recent analysis by Ebert et al. [20]. Following is an outline of their arguments; the details are left as exercises at the end of the chapter.

To begin, the variables are rescaled as follows:

$$t' = K_{bg}^2 D t, \quad z' = K_{bg} z, \quad P'(z', t') = r P(z, t).$$

In terms of these new variables, the PDE (18.13) and the boundary conditions (18.10) take the simpler form (omitting all primes)

$$\frac{\partial P}{\partial t} = \frac{\partial^2 P}{\partial z^2} - C \frac{\partial P}{\partial z} + A(j(P) - B)P, \quad z \in (0, L),$$

$$\frac{\partial P}{\partial z} - CP = 0, \quad z = 0, L,$$

(18.14)

where A, B, and C are positive constants, $A = RI_0/(K_{bg}^2 D)$, $B = \ell/(RI_0)$, $C = V/(K_{bg}D)$,

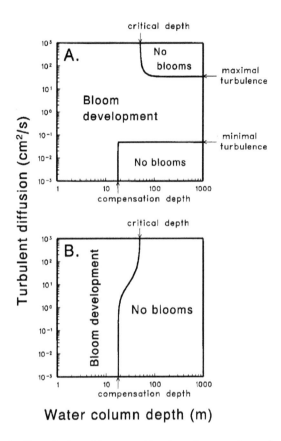

Figure 18.6. *Phytoplankton bloom as a function of water-column depth, L, and turbulent-diffusion coefficient, D. (A) Phytoplankton species with a relatively low sinking velocity, $V = 0.04\,m\,h^{-1}$. (B) Phytoplankton species with a relatively high sinking velocity, $V = 0.40\,m\,h^{-1}$. Each of the graphs is based on a grid of $30 \times 60 = 1,800$ simulations [43].*

and

$$j(P)(z,t) = \exp\left(-z - \int_0^z P(y,t)\,dy\right).$$

Note that B is the ratio of the removal rate, ℓ, and the maximum growth rate at the surface, RI_0, so we may assume that $0 < B < 1$; otherwise, no algal bloom will develop.

We are interested in nontrivial solutions. As a first step, we note that if the initial conditions are such that $P(z,0) > 0$ for all $z \in (0,L)$, then $P(z,t) \ge 0$ for all $z \in (0,L)$ and all $t > 0$. This *maximum principle* is proved in the exercises.

Next, we interpret Eq. (18.14) as an abstract dynamical system for a function $P : t \mapsto P(t)$ with values $P(t)$ in a function space X of functions of the spatial variable $z \in (0,L)$, in the spirit of Section 15.3. Critical points are solutions of the boundary value problem

$$\begin{aligned}
P'' - CP' + A(j(P) - B)P &= 0, \quad z \in (0,L), \\
P' - CP &= 0, \quad z = 0, L,
\end{aligned}$$

(18.15)

(The prime $'$ indicates differentiation with respect to z.) The question is whether this nonlinear boundary value problem has a nontrivial solution.

A standard method to prove the existence of a solution of a two-point boundary value problem is the so-called *shooting method*, where the boundary value problem on $(0, L)$, with one condition at $z = 0$ and another condition at $z = L$, is replaced by an IVP with two conditions at $z = 0$,

$$P'' - CP' + A(j(P) - B)P = 0, \quad z > 0,$$
$$P(0) = P_0, \; P'(0) = CP_0. \tag{18.16}$$

This IVP has a unique solution, which depends continuously on P_0. The solution satisfies the boundary condition of the original boundary value problem at $z = 0$. We follow it, starting at $z = 0$, increasing z until we encounter a value of z where $P'(z) - CP(z) = 0$. If there is no such value of z, then the boundary value problem (18.15) has no other solution besides the trivial solution. If there is such a value, we denote it by L_0; L_0, if it exists, varies continuously with P_0. The value L_0 does not necessarily equal L, but when it does, we have found a solution of the boundary value problem (18.15). The question is whether we can get L_0 to match L by varying P_0.

If P satisfies the differential equation (18.16) and $P' - CP$ vanishes at $z = 0$ and $z = L_0$, integration of the differential equation over the interval $(0, L_0)$ yields a condition on P,

$$\int_0^{L_0} \left(\exp\left(-z - \int_0^z P(y) \, dy \right) - B \right) P(z) \, dz = 0. \tag{18.17}$$

Consider the limit of this condition as $P_0 \to \infty$. After rescaling P,

$$P(z) = P_0 \rho(z), \tag{18.18}$$

the condition (18.17) becomes

$$\int_0^{L_0} \left(\exp\left(-z - P_0 \int_0^z \rho(y) \, dy \right) - B \right) \rho(z) \, dz = 0. \tag{18.19}$$

By assumption, $\rho = \mathcal{O}(1)$ as $P_0 \to \infty$. Recall that L_0 depends on P_0, $L_0 : P_0 \mapsto L_0(P_0)$; we claim that $L_0(P_0) = \mathcal{O}(P_0^{-1})$ as $P_0 \to \infty$. In that case, the main contribution to the integral comes from a small neighborhood of the origin, where $\rho(z) \sim 1$, and the leading term of the exponential is e^{-z}. Thus, in the limit as $P_0 \to \infty$, the condition (18.19) reduces to

$$\int_0^{L_0} \left(e^{-z} - B \right) dz = 0. \tag{18.20}$$

(The above arguments do not constitute a proof in the mathematical sense, but they are the type of arguments that are commonly used in asymptotic analysis to arrive at an order-of-magnitude estimate. The estimate needs to be justified a posteriori.) The integral can be evaluated and yields the condition

$$BL_0 = 1 - e^{-L_0}.$$

This equation can be solved graphically; since $B \in (0, 1)$, the graphs of BL_0 and $1 - e^{-L_0}$ intersect at a unique value L_0. This answers the question of the existence of a solution P of the IVP (18.16) such that $P' - CP$ vanishes at some point L_0 when P_0 is sufficiently large.

Furthermore, if the IVP has such a solution for some P_0, then it also has a solution for any $\tilde{P}_0 < P_0$, and $L_0(\tilde{P}_0) > L_0(P_0)$. Hence, it makes sense to define

$$L^* = \lim_{P_0 \downarrow 0} L_0.$$

Either $L^* = \infty$ or $L^* < \infty$. If $L^* = \infty$, there is a P_0 such that $L_0 = L$ for any given L, so the boundary value problem (18.15) has a nontrivial solution for all L. If $L^* < \infty$, then the condition $L_0 = L$ is satisfied for some P_0 if and only if $L < L^*$; if $L > L^*$, the condition is never met, so the boundary value problem (18.15) has no solution other than the trivial solution.

These results are in agreement with the numerical findings. A nontrivial solution of the boundary value problem (18.15) corresponds to a critical point of the dynamical system (18.14) and represents a pattern of algal blooms. The development of a pattern depends on the depth of the ocean (as well as on the diffusion coefficient and other physical quantities). Of course, the analysis proves only the existence of a pattern, not the type of pattern that may develop.

18.6 ▪ Exercises

1. What does Figure 18.2 tell us about the air-sea CO_2 flux during the Last Glacial Maximum? What about the air-sea CO_2 flux in 2100?

2. What does Figure 18.2 tell us about the air-sea CO_2 exchange in upwelling regions?

3. At location X, the oceanic temperature, salinity, alkalinity, and DIC are 20°C, 35 psu, 2,322 μmol kg^{-1}, and 2,012 μmol kg^{-1}, respectively.

    (i) Use the MATLAB code in [61] to calculate the oceanic pCO_2 at location X.

    (ii) Calculate the change in oceanic pCO_2 due to a 0.5°C increase in surface temperature at location X. (Assume everything else remains unchanged.)

    (iii) Takahashi et al. [107] provide a relationship that summarizes the temperature sensitivity of pCO_2 in a closed system (constant DIC and Alk),

$$\frac{1}{pCO_2} \frac{\partial (pCO_2)}{\partial T} \approx 0.0423(°C)^{-1}.$$

    The sensitivity was determined experimentally. Is it valid at location X?

4. At location Y, the oceanic temperature, salinity, alkalinity, and DIC concentration are 20°C, 35 psu, 2,315 μmol kg^{-1}, and 2,003 μmol kg^{-1}, respectively.

    (i) Due to increased stratification and reduced surface nutrients worldwide, the future is expected to be characterized by a reduction in phytoplankton production. Imagine that this reduces alkalinity to 2,311 μmol kg^{-1} and increases DIC to 2,032 μmol kg^{-1} at location Y. Calculate the change in oceanic pCO_2 from this reduction in phytoplankton production. One proposed mechanism for geoengineering the climate is to artificially supply nutrients to the surface ocean. Based on your calculation, can you justify this geoengineering solution?

(ii) Due to increased emissions of anthropogenic CO_2, the oceanic pCO_2 and DIC at location Y should increase to 439 μatm and 2,050 μmol kg^{-1}, respectively. Calculate the corresponding change in the surface carbonate ion and hydrogen ion concentrations relative to the current value. Discuss the implications of this for calcifying organisms.

5. The trivial solution $P(z) = 0$ for all $z \in [0, L]$, which corresponds to "no bloom," is a solution of the boundary value problem (18.14). Discuss its linear stability properties. (See [125, 124] for a complete analysis.)

6. Prove the maximum principle for Eq. (18.14): *If $P(z, 0) > 0$ for all $z \in (0, L)$, then $P(z, t) \geq 0$ for all $z \in (0, L)$ and $t > 0$.* (Hint: The proof is by contradiction, assuming that, at some time $t^* > 0$, $P(z, t^*)$ has a minimum at $z = z^*$, $P(z^*, t^*) = 0$, and $P(z, t^*) > 0$ for all $z \in (0, L)$, $z \neq z^*$.)

7. Let P be a solution of the IVP (18.16), and let L_0 be the value of z for which $(P' - CP)(z) = 0$. Prove that L_0, if it exists, is unique. (Hint: Integrate the differential equation (18.16) from $z = 0$ to any point $z > 0$. The resulting equation generalizes the condition (18.17). Show that there is a point $z = Z$ with $Z > 0$ such that $j(P) - B$ is positive on $(0, Z)$ and negative on (Z, L).)

8. Consider the two graphs in Figure 18.6. The areas in (A) were obtained for a relatively low sinking velocity and the areas in (B) for a relatively high sinking velocity. Discuss how the two graphs can be connected in a continuous manner by varying the sinking velocity and explain bloom development as a bifurcation phenomenon.

Chapter 19

Extreme Events

Weather extremes capture the public's attention and are often used as arguments in the debate about climate change. While it is generally not possible to attribute a single extreme event to climate change, there are statistical techniques to assess the likelihood of weather extremes in the light of climate change. In this chapter we look in detail at the notion of an extreme event and present a probabilistic framework for their description.

Keywords: Extreme event, temperature anomaly, warm nights, order statistic, distribution-free statistics, exceedance, climate change, tail probability, return period, normal distribution, changing probability distribution, relative risk, hazard function, extreme-value theory.

19.1 ▪ Climate and Weather Extremes

Weather extremes capture the public's attention and are often used as arguments in the debate about climate change. In daily life, the term *extreme event* can refer, for example, to an event whose intensity exceeds expectations, or an event with high impact, or an event that is rare or even unprecedented in the historical record. Some of these notions may hit the mark, but they need to be quantified if we want to make them useful for a rational discussion of climate change. The concern that extreme events may be changing in frequency and intensity as a result of human influences on climate is real, but the notion of extreme events depends to a large degree on the system under consideration, including its vulnerability, resiliency, and capacity for adaptation and mitigation. Since extreme events play such an important role in the current discussion of climate change, it is important that we get their statistics right.

The assessment of extremes must be based on long-term observational records and a statistical model of the particular weather or climate element under consideration. The proper framework for their study is probability theory. The recent special report on managing the risks of extreme events (SREX) [44] prepared by the IPCC describes an extreme event as the "occurrence of a value of a weather or climate variable above (or below) a threshold value near the upper (or lower) ends of the range of observed values of the variable." Normally, the threshold is put at the 10th or 90th percentile of the observed probability distribution function, but other choices of the thresholds may be appropriate given the particular circumstances.

237

To begin with, it is better to use anomalies, rather than absolutes. Anomalies more accurately describe climate variability and give a frame of reference that allows more meaningful comparisons between locations and more accurate calculations of trends. Consider, for example, the global average temperature of the Earth's surface. We have a fairly reliable record going back to 1880, but rather than looking at the temperature itself, we use the *temperature anomaly* to study extremes.

Figure 19.1 shows the global temperature anomaly since 1950 relative to the average of these mean temperatures for the period 1961–1990. We see that the year 2010 tied with 2005 as the warmest year on record. The year 2011 was somewhat cooler, largely because it was a La Niña year; however, it was the warmest La Niña year in recent history. But while the global mean temperature for the year 2010 does not appear much different from previous years, exceeding the 1961–1990 average by only about 0.5°C, the average June temperature exceeded the corresponding 1971–2000 average by up to 5°C in certain regions, and the same year brought heat waves in North America, western Europe, and Russia. Observed temperature anomalies were, in fact, much higher during certain months in certain regions, as illustrated in Figure 19.2, and it is likely that these extremes were even more pronounced at individual weather stations. Since localized extremes cause disruptions of the socioeconomic order, from crop failures to forest fires and excess deaths, they are of considerable interest to the public. To assess the likelihood of their occurrence, we need both access to data and rigorous statistical analysis. Until recently, reliable data have been scarce for many parts of the globe, but they are becoming more widely available, allowing research of extreme events on both global and regional scales. The emphasis shifts thereby from understanding climate models to assessing the likelihood of possibly catastrophic events and predicting their consequences. For what kind of extreme events do we have to prepare? How often do these extremes occur in a stationary climate? What magnitudes can they have? And how does climate change figure in all this?

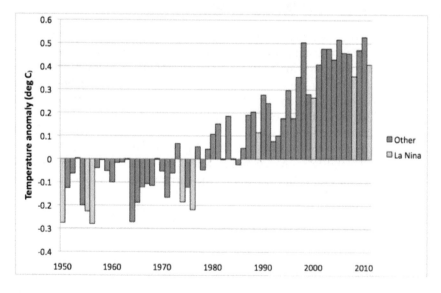

Figure 19.1. *Global temperature anomaly from 1950 to 2011 relative to the base period 1961–1990. Reprinted with permission from WMO* [122].

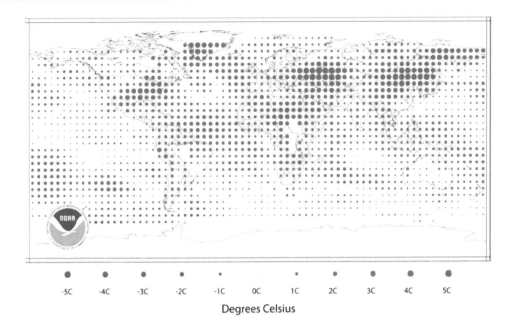

-5C   -4C   -3C   -2C   -1C   0C   1C   2C   3C   4C   5C

Degrees Celsius

Figure 19.2. *Regional temperature anomaly during June* 2010 *relative to the base period* 1971–2000. *Reprinted courtesy of NOAA.*

19.1.1 ▪ Context of Extremes

Obviously, whether a particular weather event is extreme or not depends on the seasonal and location context. A warm day in January in Washington, DC, would be considered a cool day in May at the same location. Therefore, care must be taken when defining extreme events. Here is a good example, which has to do with the number of "warm nights" at a fixed location in the municipality of De Bilt in the Netherlands [55]. By definition, a night is called "warm" if the daily minimum temperature exceeds the 90th percentile of the daily minimum temperature for that particular date during the period 1961–1990. For example, on October 1, the minimum temperature exceeded 10°C in three out of 30 instances (10% of the time) between 1961 and 1990. If, in a given year, the minimum temperature on October 1 exceeds 10°C, then this night is counted as a warm night for that year. Figure 19.3 shows the yearly number of warm nights in De Bilt from 1900 until 2010.

Given the definition, we expect between 30 and 40 warm nights on average in the years between 1961 and 1990, and the graph confirms this. The graph shows a marked increase in the number of warm nights since about 1975. The number of cold nights (not shown) has decreased somewhat during the same time period but not as markedly. While the definition of a "warm night" has been carefully chosen to account for seasonal variation and to avoid subjective definitions, it is not clear whether the recent trend toward more warm nights is part of a larger phenomenon or possibly due to an urban heat island effect, perhaps because of recent development in the area.

19.1.2 ▪ Mathematical Framework

Using a sports analogy, if a baseball player is hitting 20% more home runs during a season in which he is taking performance-enhancing drugs, the drugs are not responsible for every single home run, but they certainly are responsible for the overall change in perfor-

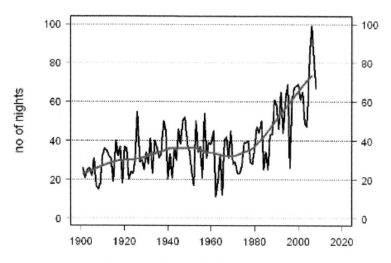

Figure 19.3. *Warm nights registered in De Bilt (Netherlands) from 1900 until 2010* [55].

mance. Thus, attribution can only be done in probabilistic terms, by determining how the likelihood of a certain type of event has changed.

Consider a random variable X which can be observed repeatedly, such as the maximum daily temperature, the mean annual temperature anomaly, or the precipitation in any 24-hour period. Observed values $X_1, X_2, \ldots, X_N$ of this variable are recorded in a *time series* $\{X_i : i = 1, 2, \ldots, N\}$. Reorder the observed values from smallest to largest, breaking ties (if any) randomly, and denote the smallest value by $X_{(1)}$, the second smallest by $X_{(2)}$, and so on. The quantity $X_{(k)}$ is called the kth *order statistic*.

Extremes such as maximum values may be poor indicators of long-term trends if used naively. Let $Y_n = \max\{X_i : i = 1, \ldots, n\}$ be the maximum of the first n observations. It is possible for Y_n to increase because the mean of the X_i increases with i, or because the variability of the X_i increases, or just due to chance. These possibilities are illustrated in Figure 19.4, which shows Y_n as a function of n for three different time series of length $N = 300$. The black curve comes from identically distributed observations, the red curve from observations that have a small positive drift, and the blue curve from data whose variability is slowly increasing. It is impossible to tell by just looking at the graphs which curve comes from which time series.

19.2 ▪ Exceedance

Often, very little is known about the distribution of a random variable which describes a particular climate element, and a detailed climate model may not be available. There are, however, *distribution-free* statistical methods that can be used even in such general situations. Consider again the record of global temperature anomalies for the period 1950–2011 shown in Figure 19.1. The graph shows that between 1950 and 1999, the highest mean global temperature was attained in 1998. This record was then exceeded twice between 2001 and 2011. Could this phenomenon be due to chance alone? More significantly, what kind of information is needed to determine whether this phenomenon is due to chance? We now formalize this question following the presentation in [1].

Assume that we are given N numerical observations, $X_1, \ldots, X_N$, which are all different (no ties) and which are thought to come from the same probability process. Let $X_{(1)} < X_{(2)} < \cdots < X_{(N)}$ be the sequence of order statistics of the sample. A new observa-

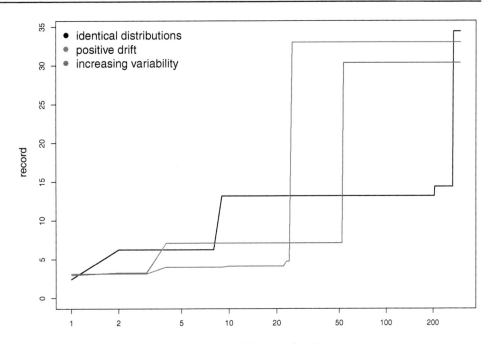

Figure 19.4. *Records of three simulated time series.*

tion X^* is called an *exceedance* of the mth order statistic $X_{(m)}$ from the original record if $X^* > X_{(m)}$. What is the probability that a new observation will exceed $X_{(m)}$? Or, more generally, supposing there are n new observations, what is the probability that there are k exceedances of $X_{(m)}$? The answer is given in the following theorem.

Theorem 19.1. *Let* $\{X_i : i = 1,\dots,N+n\}$ *be a random sample of* $N + n$ *independent identically distributed random variables, satisfying* $X_i \neq X_j$ *with probability 1 whenever* $i \neq j$. *Let* $X_{(1)} < \cdots < X_{(N)}$ *be the order statistics from the first* N *observations in the sample. For any* $k \in \{0,\dots,n\}$, *the probability that there are exactly* k *exceedances of* $X_{(m)}$ *(* $m \in \{1,\dots,,N\}$ *) in the last* n *observations of the sample is*

$$\frac{C(N-m+k;N-m)\,C(m-1+n-k;m-1)}{C(N+n;N)}. \tag{19.1}$$

Here, $C(n;k) = \binom{n}{k} = \frac{n!}{(n-k)!k!}$ *is the binomial coefficient.*

Proof. First, assign the $N+n$ observations *ranks* according to their value, from largest to smallest, by assigning the rank $r_i = \rho$ to X_i if there are exactly $\rho - 1$ observations that are greater than X_i and $N+n-\rho$ observations that are smaller than X_i. With probability 1, there are no ties, so the rank of an observation is a well-defined random variable. Any sequence of ranks $\{r_1,\dots,r_{N+n}\}$ is a (random) permutation of the set $\{1,\dots,N+n\}$. Since the X_i are assumed to be independent and identically distributed, all permutations are equally likely and have probability $1/(N+n)!$. If there are k exceedances of $X_{(m)}$ in the last n observations and $X_{(m)}$ occurs as observation j, then it must be the case that (i) $r_j = N-m+k+1$ for some index $j \in \{1,\dots,N\}$; (ii) $r_i < N-m+k+1$ for exactly $N-m$ indices $i \in \{1,\dots,N\}$ with $i \neq j$; and (iii) $r_i < N-m+k+1$ for exactly k indices $i \in \{N+1,\dots,N+n\}$.

(i) There are N ways to choose an index $j \in \{1,\ldots,N\}$; set $r_j = N - m + k + 1$. (ii) Choose $N - m$ additional values $r_i \le N - m + k$ and assign them to $N - m$ out of the $N - 1$ remaining positions $j \in \{1,\ldots,N\}$. (iii) Choose k positions $i \in \{N+1,\ldots,N+n\}$, assign the other k values $r_i \le N - m + k$ to them, and distribute the remaining $n + m - k - 1$ values $r_i > N - m + k + 1$ to the remaining positions. Basic counting arguments show that these steps can be accomplished in

$$NC(N - m + k; N - m) \frac{(N-1)!}{(m-1)!} \frac{n!}{(n-k)!} (n + m - k - 1)!$$

ways. The desired probability (19.1) follows upon division of this number by $(N + n)!$, after some trivial algebra. $\qquad\square$

Returning to the record displayed in Figure 19.1, we see that during the $n = 11$ years from 2000 to 2010, there were $k = 2$ exceedances of the maximum ($m = 50$) of the $N = 50$ annual mean temperature observations during the period 1950–1999. The expression (19.1) shows that the probability of zero exceedances of a 50-year record during 11 years is approximately 82%, the probability of a single exceedance is approximately 15%, and the probability of two or more exceedances is no more than 3%. Yet, two exceedances were observed, which suggests that the pattern was not just due to chance.

Now consider the number of exceedances of the second highest mean temperature ($m = 49$) during the same period, which was observed in 1997. There were $k = 10$ exceedances during the following 11 years. The expression (19.1) shows that the probability of ten or more exceedances is approximately $1.3 \cdot 10^{-9}$—that is, such an event is virtually impossible, unless the annual global mean temperatures are not stochastically independent or the global climate is changing. This fairly elementary calculation already leads to serious questions about global climate change, independently of any particular climate model.

19.3 ▪ Tail Probabilities and Return Periods

Suppose we define an extreme event as an event where the observed value X_i of a random variable X exceeds some fixed constant M. How long would it take for an extreme event to occur? What is the average value of an extreme event? Both questions can be answered exactly, given the *probability density function* (pdf) f of X.

Start from the *cumulative distribution function* (cdf) F of X,

$$F(x) = \int_{-\infty}^{x} f(x)\,dx = \mathrm{Prob}(X \le x). \tag{19.2}$$

The complement of F is the *tail probability* G,

$$G(x) = 1 - F(x) = \int_{x}^{\infty} f(x)\,dx = \mathrm{Prob}(X > x). \tag{19.3}$$

Since the probability that a single observation exceeds the value M is $G(M)$, the expected number of observations until this extreme event occurs is $1/G(M)$. This quantity is known as the *return period*.

The expected value of an extreme observation X_i is the conditional expected value

$$\mathcal{E}(X_i | X_i \ge M) = \frac{1}{G(M)} \int_{M}^{\infty} x f(x)\,dx = M + \frac{1}{G(M)} \int_{M}^{\infty} G(s)\,ds.$$

We illustrate with two important cases.

Case 1. Algebraically decaying tail probability. Suppose there exist constants $C > 0$ and $\alpha > 1$ such that $G(x) = Cx^{-\alpha}$ for all $x \geq M$. (The restriction $\alpha > 1$ is imposed so that the X_i have finite expected values.) Then the return period is $(1/C)M^{\alpha}$, so the waiting time between two record-setting events is proportional to some power of the record. Furthermore, the expected value of an observation greater than M is $(\alpha/(\alpha-1))M$, so the next record is expected to be a finite multiple of the previous one; successive records will increase approximately like a geometric sequence.

Case 2. Exponentially decaying tail probability. Suppose there exist constants $C > 0$ and $\beta > 0$ such that $G(x) = Ce^{-\beta x}$ for all $x \geq M$. Then the return period is $(1/C)e^{\beta M}$, so the waiting time between two record-setting events is an exponential function of the record and will therefore grow exponentially. The expected value of an extreme observation is $M + 1/\beta$, so the value of a new record is expected to exceed the previous one by a fixed amount; successive records will increase approximately like an arithmetic sequence.

A common way to show return periods graphically is to plot the observed quantity M against the return period. Figure 19.5 shows the results for three normal distributions: $N(0, 1)$, the standard normal distribution; $N(0.5, 1)$, a normal distribution with increased mean and the standard variability; and $N(0, 1.3)$, a normal distribution with the standard mean and increased variability. We see that an event of magnitude $M = 3$ has a return period $T \approx 740$ for $N(0, 1)$, $T \approx 161$ for $N(0.5, 1)$, and $T \approx 95$ for $N(0, 1.3)$, showing that moderate changes in mean and variability can have a dramatic effect on the likelihood of extreme events.

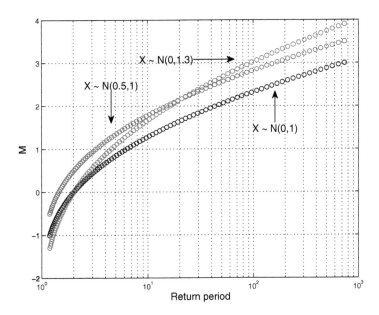

Figure 19.5. *Return periods for the normal distributions $N(0, 1)$, $N(0.5, 1)$, and $N(0, 1.3)$.*

The same graphical techniques can be used to compare the likelihood of extreme weather events under past and current conditions. In 2011, the state of Texas experienced an extraordinary heat wave and drought. The six-month growing season of March–August (MAMJJA) and the three summer months of June–August (JJA) were both, by wide margins, the hottest and driest, breaking a record dating back to 1895. As reported in [82], many simulations of the atmospheric and land surface climate of the decades 1960–1970 and 2000–2010 were performed with a GCM (HadAM3P), using SSTs, sea-ice fractions, and greenhouse gas concentrations from observational data sets. Between 170 and 1400 runs were performed for each year by varying the initial conditions, resulting in a set of plausible realizations of the climate of these decades. Figure 19.6 shows the return periods for the mean temperature in Texas for the MAMJJA period of the years 1964, 1967, 1968, and 2008. (Because simulations under 2011 forcing conditions were not available, 2008 was chosen as a proxy for 2011.) The years from the 1960s were all La Niña years, as was 2008. All model predictions were also compared to actual temperature and precipitation data for 1895–2008, showing good qualitative agreement, although the model computations tended to predict warmer and dryer climates.

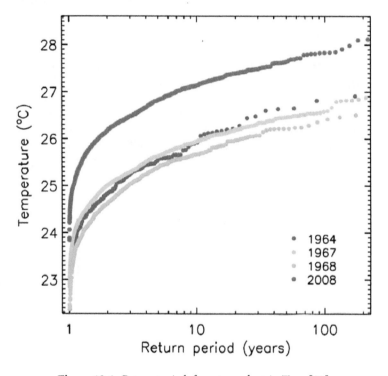

Figure 19.6. *Return periods for extreme heat in Texas* [82].

Figure 19.6 shows that the curve of mean temperatures against return periods shifted up by about 1 degree between the 1960s and 2008, consistent with the observed increase in the global mean temperature of approximately 0.8 degree during the same period. As a consequence, extreme heat events have become much more likely in the last years than they were in the 1960s. For example, the model calculations show that a mean temperature that would have occurred only every 100 years in the 1960s can be expected to occur every 5 to 6 years under 2008 conditions. This does not prove that a given heat wave (like the 2011 heat wave in Texas) is due to climate change, but it shows that a change in the mean temperature which until recently was considered modest can have a significant effect on the likelihood of extreme weather events.

19.4 ▪ Order Statistics, Extreme Value Distribution

We return to the case of a random sample $\{X_i : i = 1,\ldots,N\}$ of size N drawn from a random variable X with cdf F and pdf f. For $k = 1,\ldots,N$, let $X_{(k)}$ be the kth order statistic. The following theorem shows that the pdf of each $X_{(k)}$ can be expressed in terms of F and f.

Theorem 19.2. *The pdf of $X_{(k)}$ is*

$$f_{X_{(k)}}(x) = C(N; k-1)(F(x))^{k-1}(1-F(x))^{N-k} f(x), \tag{19.4}$$

where $C(n; k)$ is the usual binomial coefficient.

Proof. (Sketch) Let $x \in \mathbb{R}$ and let $0 < h \ll 1$. For the kth order statistic $X_{(k)}$ to satisfy the inclusion relation $X_{(k)} \in (x, x+h)$, it must be the case that, at least approximately,

$$X_{(1)} < \cdots < X_{(k-1)} \leq x < X_{(k)} < x+h \leq X_{(k+1)} < \cdots < X_{(N)}. \tag{19.5}$$

(In a rigorous proof one must account for the errors due to this approximation and show that they go to zero faster than h.) The probability that exactly $k-1$ observations in the sample are less than x is $C(N; k-1)(F(x))^{k-1}$, the probability that exactly one of the remaining $N-k+1$ observations is in $(x, x+h)$ is $F(x+h)-F(x)$, and the probability that the remaining $N-k$ observations are all greater than $x+h$ is $(1-F(x+h))^{N-k}$. Since these are independent events, the probability that the sequence of inequalities (19.5) occurs is

$$C(N; k-1)(F(x))^{k-1}(1-F(x+h))^{N-k}(F(x+h)-F(x)).$$

The theorem follows upon division of both sides by h and taking the limit as $h \to 0$. $\square$

Figure 19.7 shows densities for order statistics for a random sample of size $N = 20$, drawn from a t-distribution with two degrees of freedom. Shown are the density of a

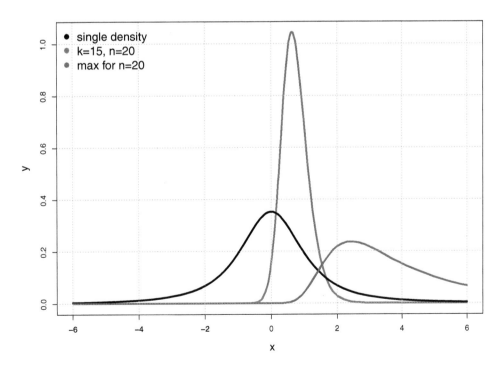

Figure 19.7. *Order statistics for t-distribution.*

single observation (black) and the densities for $X_{(15)}$ (essentially the third quartile) and $X_{(20)}$ (the maximum). The graph shows that the maximum is skewed to the right and has greater variability than the third quartile. The large variability of the maximum makes it an unreliable quantity for statistical computations.

We now consider the case of an infinite random sample and ask what can be said about the probability distribution of $X_{(n)} = \max\{X_i : i = 1, \ldots, n\}$, where n is large. We assume throughout that all tail probabilities $G(x)$ are positive for all $x > 0$. Then arbitrarily large observations are possible, what is possible will eventually happen, and therefore $X_{(n)} \to \infty$ almost surely. Over very long times, the record will exceed all conceivable bounds. One should therefore rescale $X_{(n)}$ (similar to the approach for the central limit theorem) and see whether the rescaled records eventually have a nontrivial distribution. Specifically, we set

$$Y_n = \frac{X_{(n)} - c_n}{a_n},$$

with as yet undetermined sequences $\{a_n\}_n$ and $\{c_n\}_n$, and ask whether, for suitable choices of these sequences, the distributions of the Y_n have a nontrivial limit. The cdf of Y_n is given by

$$F_{Y_n}(x) = \mathrm{Prob}(X_i \leq a_n x + c_n : i = 1, \ldots, n) = (F(a_n x + c_n))^n = (1 - G(a_n x + c_n))^n.$$
(19.6)

Next we recall the well-known calculus result, valid for all convergent sequences $\{z_n\}_n$ and easily proved with L'Hôpital's Rule,

$$\lim_{n \to \infty} z_n = z^* \implies \lim_{n \to \infty} \left(1 + \frac{z_n}{n}\right)^n = e^{z^*}.$$

Thus, if we can choose $\{a_n\}_n$ and $\{c_n\}_n$ such that $\lim_{n \to \infty} nG(a_n x + c_n) = \Phi(x)$ with the same function Φ for all x, then we expect that

$$\lim_{n \to \infty} F_{Y_n}(x) = e^{-\Phi(x)} = F^*(x).$$

That is, the rescaled records have indeed a limiting probability distribution. This problem was solved completely by FRÉCHET, GNEDENKO, GUMBEL, and others in the first half of the 20th century. It turns out that, up to scale changes, there are three possible forms of F^*. Here we present very simple conditions under which two of these limits occur. The third possible limit distribution F^*, not discussed here, concerns the case where $X_i \leq C$ with probability 1 for some $C > 0$, in which case the tail probabilities vanish identically for large x.

Theorem 19.3 (Fréchet distribution). *Assume that there exist constants $C > 0$ and $\alpha > 0$ such that*

$$\lim_{x \to \infty} x^\alpha G(x) = C.$$

Then the rescaled extreme values $Y_n = X_{(n)}/(nC)^{1/\alpha}$ satisfy

$$\lim_{n \to \infty} \mathrm{Prob}(Y_n \leq y) = \begin{cases} \exp(-y^{-\alpha}), & y \in (0, \infty), \\ 0, & y \in (-\infty, 0]. \end{cases}$$
(19.7)

Proof. Let $y > 0$. Define $x_n = nCy^\alpha$ for $n = 1, 2, \ldots$. Then

$$\mathrm{Prob}(Y_n \leq y) = \left(1 - \frac{x_n^\alpha G(x_n)\, y^{-\alpha}}{C} \frac{1}{n}\right)^n \to \exp(-y^{-\alpha}) \text{ as } n \to \infty. \qquad \square$$

Theorem 19.4 (Gumbel distribution). *If there exist constants $C > 0$ and $\alpha > 0$ such that*

$$\lim_{x \to \infty} e^{\alpha x} G(x) = C,$$

then the rescaled extreme values $Y_n = \alpha X_{(n)} - \log C - \log n$ satisfy

$$\lim_{n \to \infty} \text{Prob}(Y_n \leq y) = \exp(-e^{-y}), \quad y \in \mathbb{R}. \tag{19.8}$$

Proof. Let $y \in \mathbb{R}$. Define $x_n = (y + \log n + \log C)/\alpha$ for $n = 1, 2, \ldots$. Then

$$\text{Prob}(Y_n \leq y) = \left(1 - \frac{e^{\alpha x_n} G(x_n)}{C} \frac{e^{-y}}{n}\right)^n \to \exp(-e^{-y}) \text{ as } n \to \infty. \qquad \square$$

An example of a simulation leading to a Fréchet extreme value distribution is given in Figure 19.8.

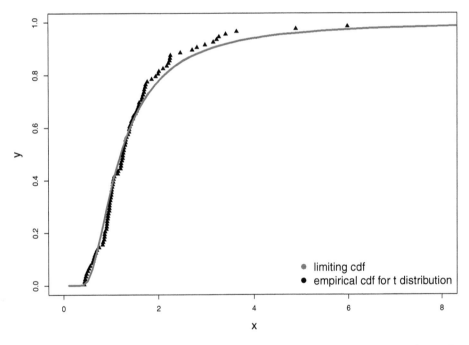

Figure 19.8. *Empirical cdf and limiting distribution function for $X_{(n)}$, for simulations from a t-distribution with two degrees of freedom.*

While the extreme value theory for stationary distributions is very well developed, less is known about extreme values from nonstationary distributions. What is the correct way to scale $X_{(n,\delta)} = \max(X_i + \delta i : i = 1, \ldots, n)$ if the X_i are independent and identically distributed and $\delta > 0$ is small? What if the X_i are correlated? What can one say about the distributions of the corresponding order statistics? Questions like these come up regularly in the assessment of changing climate scenarios.

19.5 ▪ Exercises

1. Here are three possible ways to define a *hot summer*:

   (i) A summer is "hot" if the number of days between June 1 and August 31 with a maximum temperature above a certain level (say 35°C) exceeds a certain number (say 20 days).

   (ii) A summer is "hot" if the average nightly minimum temperature between June 1 and August 31 is higher than that of the previous 10 years.

   (iii) A summer is "hot" if a new record high temperature for that day is observed on at least 10 days between June 1 and August 31.

   Explain why each of these definitions depends on external circumstances and therefore is not universal.

2. In [55], a number of descriptive indices for extremes are proposed. They were developed by the Expert Team on Climate Change Detection and Indices (ETCCDI) under the auspices of CLIVAR (variability and predictability of the ocean-atmosphere system), one of the four core projects of the World Climate Research Programme (WCRP). Among these indices are

   (i) FD, frost days: count of days in a year where the daily minimum temperature is less than 0°C;

   (ii) TR, tropical nights: count of days where the daily minimum temperature is more than 20°C;

   (iii) TN10p, cold nights: count of days where the daily minimum temperature is less than the 10th percentile of the daily minimum temperature, calculated for a five-day window centered on this calendar day for the base period 1961–1990;

   (iv) WSDI, warm spell duration index: count of days in a span of at least six days where the daily maximum temperature is greater than the 90th percentile of daily maximum temperatures calculated for a five-day window centered on this calendar day for the base period 1961–1990;

   (v) CWD, consecutive wet days: maximum length of wet spell (daily precipitation exceeding 1 mm).

   All these indices are based on day counts. An alternative is to use record-setting observations for a given period. For example, instead of using FD, one could record the minimum temperature that was observed in a given year at a given location. Propose corresponding record-based substitutes for the indices TR, TN10p, WSDI, and CWD. Then explain why indices based on day counts are generally considered to be more useful.

3. Refer to the previous exercise. Some of the indices listed there are counts of days where certain physical thresholds are exceeded; others are counts of days where percentiles are exceeded. Percentile-based indices are considered to be more suitable for spatial comparison between different regions. Explain why. Then propose some applications of such indices where day counts based on physical thresholds would be more useful. Think about farming.

4. Let N, n, and k be positive integers satisfying the inequalities $k \leq n \leq N$. Find the probability that the highest k observations out of a sequence of N observations

are seen among the last n observations. Assume that observations are independent and identically distributed random variables. The data in Figure 19.1 show that the highest twelve annual global mean temperatures in the period 1950–2010 were observed between 1997 and 2000. Use your formula to compute the probability that this would happen by chance, assuming that annual global mean temperatures were independent and identically distributed.

5. This and the following exercises show a simple approach to assess the effect of changing probability distributions on the probability of extreme events. It uses related statistical concepts from epidemiology and reliability analysis and is based on [50].

Let X and X' be random variables for the same climatological quantity under different conditions. Examples could be observations of precipitation in the same location in two different decades. The ratio

$$\frac{P(X' \geq M)}{P(X \geq M)} = \frac{G_{X'}(M)}{G_X(M)} \tag{19.9}$$

measures the relative change in probability for observing extreme values of this quantity if climate conditions are changed. In epidemiology, this ratio is known as *relative risk*.

Assume that X and X' have the form

$$X = \sigma Z + \mu, \quad X' = \sigma' Z + \mu', \tag{19.10}$$

where Z is a fixed dimensionless random variable, μ and μ' are location parameters, and σ and σ' are scaling parameters.

Let $f_Z, f_X, f_{X'}$ be the pdfs of Z, X, X'; let $F_Z, F_X, F_{X'}$ be their cdfs, and let $G_Z = 1 - F_Z, G_X = 1 - F_X, G_{X'} = 1 - F_{X'}$ be their tail probabilities.

(i) Express $F_X, F_{X'}, G_X, G_{X'}$ in terms of the corresponding quantities for Z.

(ii) Assume that $\mu' = \mu + h$ and $\sigma' = \sigma$. Show that

$$H(M) = \left[\frac{d}{dh} \frac{G_{X'}(M)}{G_X(M)} \right]_{h=0} = \frac{f_X(M)}{G_X(M)}. \tag{19.11}$$

The function H is known in reliability analysis as the *hazard function* of X.

(iii) Show that

$$P(M \leq X \leq M + \delta \mid X \geq M) \approx \delta H(M). \tag{19.12}$$

(iv) Show that if $\mu' = \mu + h$ and h is small, then the relative risk (19.9) is approximately $1 + hH(M)$.

6. Continue with the notation of the previous exercise. Assume now that $\mu' = \mu$ and $\sigma' = \sigma + h$.

(i) Show that

$$\left[\frac{d}{dh} \frac{G_{X'}(M)}{G_X(M)} \right]_{h=0} = \frac{M - \mu}{\sigma} H(M). \tag{19.13}$$

(ii) Show that if $\sigma' = \sigma + h$ and h is small, then the relative risk (19.9) is approximately $1 + h((M - \mu)/\sigma)H(M)$.

(iii) Compare the magnitude of this change in relative risk to the change that you computed in part (iv) of the previous exercise. Note that usually M is large relative to μ when measured in units of σ. What does this say about the importance of changes in climate variability as opposed to changes in the mean climate, as far as extremes are concerned?

7. The Weibull distribution with shape parameter $k > 0$ and scale parameter $\lambda > 0$ is defined as the random variable with pdf

$$f(x; k, \lambda) = \begin{cases} kx^{k-1}\lambda^{-k}\exp(-(x/\lambda)^k), & x \in (0, \infty), \\ 0, & x \in (-\infty, 0]. \end{cases} \qquad (19.14)$$

This distribution is frequently used in reliability analysis. Show that it is indeed a pdf and find its hazard function.

8. Continue with the Weibull distribution from the previous exercise. Show that its hazard function H is increasing if and only if $k > 1$.

9. Find the hazard function of the Gumbel extreme value distribution (19.8) and show that it approaches a nonzero constant as $x \to \infty$.

10. The standard normal distribution $N(0, 1)$ is known to have tail probabilities with the behavior

$$\lim_{x \to \infty} x\, e^{x^2/2} G(x) = \frac{1}{\sqrt{2\pi}}.$$

Use this fact to prove that, for any $\delta > 0$, the maximum $X_{(n)} = \max\{X_i : i = 1, \ldots, n\}$ of n independent $N(0, 1)$-distributed observations X_i satisfies

$$\lim_{n \to \infty} \mathrm{Prob}\left(\sqrt{2\log n} - \delta \leq X_{(n)} \leq \sqrt{2\log n} + \delta\right) = 1.$$

This result shows that the maxima of normally distributed observations have less variability than an individual observation, in contrast to maxima of observations from heavy-tailed distributions (cf. Figure 19.7).

Chapter 20

Data Assimilation

Data assimilation is a powerful methodology for incorporating data into a mathematical model in a way that is consistent with physical processes. In this chapter we present a Bayesian approach to data assimilation, discuss the Kalman filter algorithm for data assimilation for linear processes and linear data models in the presence of Gaussian noise, and describe a simple version of the ensemble Kalman filter, which relies on stochastic simulation.

Keywords: Process model, data model, probability distribution, uncertainty, Bayesian statistics, reanalysis, filtering, prediction, sequential data assimilation, Markov chain, Kalman filter, extended Kalman filter, ensemble Kalman filter, Lorenz system, predictability.

20.1 ▪ Data Assimilation and Climate

To predict the state of the climate system at any future time or to recreate its past, we must necessarily rely on mathematical models and numerical simulations. Throughout this book, we have encountered several types of models, some of them conceptual and others closer to the "real" world of physics, chemistry, and biology. Most of these models are *process models*—models that are described by systems of equations that determine the state variables (also called *process variables*) and their evolution in time. In *data assimilation*, one links such process models with *data models*—models of observational data of these same state variables, together with their uncertainties. The idea is that, by making it consistent with the available observations, a process model becomes better at predicting future states of the system. Data assimilation is an essential technique in any scientific discipline that is data-rich and for which well-founded predictive mathematical models exist. The technique originated in engineering and has found widespread application in many other disciplines, most notably in weather prediction, where it has extended the ability to predict weather more or less accurately from hours to days. Not only does the technique generate estimates of the state of the weather system, but it also produces an assessment of the uncertainties in the prediction, often in the form of probability distributions for process variables or parameters. For an overview of data assimilation techniques in weather prediction and climate science, we refer the reader to [48, 89].

A typical example of a data model is the record of SSTs obtained with a drifting instrument that periodically reports local measurements. The record contains information about the temperature at the location of the instrument (but not anywhere else). The record may have gaps for times when the instrument is shut down, and there may be uncertainties regarding the drifter's location and due to limited accuracy of the thermometer. Any data model contains an element of randomness.

Data assimilation can be applied to estimate process variables at a certain time using all available observational data, including those made at a later time (*reanalysis* or *smoothing* mode), or to estimate the present state using past and present observations (*analysis* or *filtering* mode), or to estimate process variables that are inaccessible to observations, such as future states or states between measurements (*forecasting* or *predicting* mode). In any of these modes, problems can be approached with a variety of techniques, including optimization methods, which attempt to find the best fit of a parameterized process model to a given set of data; maximum likelihood methods, which work in a similar spirit but have more detailed statistical models for the uncertainties in the process; and Bayesian methods.

20.2 • Example

The following example (adapted from [121]) illustrates the application of data assimilation methodology to reanalysis, filtering, and forecasting.

Consider a process with four real-valued state variables, $\{X_i : i = 1, \ldots, 4\}$. The state variables are related by the process model

$$X_{i+1} = \alpha X_i + \xi_i, \quad i = 1, 2, 3, \tag{20.1}$$

where α is a known positive constant and the random process error terms ξ_i are either identically zero or have a standard normal distribution, $\xi_i \sim N(0, 1)$. (The convention is to use upper-case letters (X, Y, Z, etc.) for random quantities and the corresponding lower-case letters (x, y, z, etc.) for their realizations. Random error terms are indicated by Greek letters.)

The data model consists of two observations, $\{Y_i : i = 2, 3\}$, and is related to the process model through the identities

$$Y_i = X_i + \zeta_i, \quad i = 2, 3, \tag{20.2}$$

where the random observational error terms ζ_i are independent and identically distributed with a standard normal distribution, $\zeta_i \sim N(0, \tau^2)$. The entire model is shown schematically in Figure 20.1. The problem of estimating X_1 from the observations Y_2 and Y_3 is a reanalysis problem, estimating X_2 from Y_2 or X_3 from Y_2 and Y_3 is a filtering problem, and estimating X_4 from Y_2 and Y_3 is a forecasting problem.

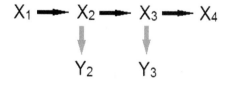

Figure 20.1. *A simple process model (black arrows) together with a data model (gray arrows).*

20.2.1 • Variational Approach

We first demonstrate a variational approach. If the process error terms ξ_i are all zero, we minimize the *cost function J*,

$$J(x_2, x_3; y_2, y_3) = (x_2 - y_2)^2 + (x_3 - y_3)^2. \tag{20.3}$$

Because of the process model relation $x_3 = \alpha x_2$, J reduces to a function J_0 of x_2 alone,

$$J(x_2, \alpha x_2; y_2, y_3) = J_0(x_2; y_2, y_3) = (x_2 - y_2)^2 + (\alpha x_2 - y_3)^2. \qquad (20.4)$$

This function reaches its minimum at $x_2^* = (y_2 + \alpha y_3)/(1 + \alpha^2)$, so x_2^* is the reanalysis value of x_2. The corresponding filtering value of x_3 is $x_3^* = \alpha x_2^*$, the forecast value of x_4 is $x_4^* = \alpha x_3^* = \alpha^2 x_2^*$, and the reanalysis value of x_1 is $x_1^* = \alpha^{-1} x_2^*$. We obtain the same values if we use the process model relation $x_3 = \alpha x_2$ to eliminate x_2 from the cost function (20.3) and minimize with respect to x_3.

These solutions do not give any uncertainty estimates (confidence sets) for the reanalysis and filtering values. In this case, such estimates can be inferred from the data model (20.2) and the explicit form of the solution. However, in the general case, the reanalysis and filtering values are obtained numerically and uncertainty estimates are not immediately available. If the process error terms ξ_i do not vanish, the cost function must be extended with additional terms. We refer the reader to [89] for a more detailed discussion of the variational approach and its relation to the probabilistic approaches discussed in the next sections.

20.2.2 ▪ Maximum Likelihood Approach

Next we demonstrate a maximum likelihood approach to the same example. Assume that the process error variables ξ_i are random and have standard normal distributions. Assume for the moment that X_1 has a fixed but unknown value x_1, which we wish to estimate from the observations Y_2 and Y_3. This is therefore a reanalysis problem. Probability theory shows that

$$Y_2 \sim N(\alpha x_1, 1 + \tau^2), \quad Y_3 \sim N(\alpha^2 x_1, 1 + \tau^2 + \alpha^2), \quad \mathrm{cov}(Y_2, Y_3) = \alpha.$$

The joint distribution of (Y_2, Y_3) is Gaussian with density

$$f_{Y_2, Y_3}(y_2, y_3) \propto \exp\left(-\tfrac{1}{2}(y_2 - \alpha x_1, y_3 - \alpha^2 x_1) \Sigma^{-1}(y_2 - \alpha x_1, y_3 - \alpha^2 x_1)^T\right), \qquad (20.5)$$

where Σ is the covariance matrix,

$$\Sigma = \begin{pmatrix} 1 + \tau^2 & \alpha \\ \alpha & 1 + \tau^2 + \alpha^2 \end{pmatrix}.$$

The proportionality constant implied in Eq. (20.5) does not depend on the unknown parameter x_1. If y_2 and y_3 are actual observational data, the maximization of the expression on the right-hand side of Eq. (20.5) leads to the maximum likelihood estimate

$$\hat{x}_1 = \frac{\alpha(1 + \tau^2)y_2 + \alpha^2 \tau^2 y_3}{\alpha^2 + \alpha^2 \tau^2 + \tau^4}. \qquad (20.6)$$

The reanalysis estimate $\hat{x}_1$ is a linear combination of the observations y_2 and y_3. If $0 < \alpha < 1$, y_2 has the larger weight. If, furthermore, the observation errors ζ_i have very small standard deviations ($\tau \ll 1$), the weight for y_3 is very small, so the reanalysis estimate depends mainly on the observation that was made right after the unknown state. One can show that $\mathscr{E}\hat{x}_1 = x_1$, so the estimate is unbiased, and it is not hard to show that $\hat{x}_1$ has a normal distribution and to find its variance.

Approaches to estimating x_2 (reanalysis) and x_3 (filtering) from observations using the maximum likelihood method are discussed in the exercises. We next introduce a general Bayesian approach to data assimilation and then return to this example.

20.3 ▪ Bayesian Approach

The contemporary approach to analysis and forecasting problems is based on Bayes' rule. This rule was first formulated by the English mathematician and Presbyterian minister THOMAS BAYES (1701–1761). It yields estimates that are asymptotically correct and does not require an appeal to the law of large numbers, which would make little sense in the climate context. To simplify the following presentation, we assume that all state variables and data are finite-dimensional vectors—a reasonable assumption for data but a substantial simplification for state variables.

Suppose we are interested in estimating a vector $\mathbf{X}$ of process variables with a known or assumed pdf f_X. The distribution may come from long-term observations or from another forecast model and is referred to as the *prior distribution* on $\mathbf{X}$. The data model uses a vector $\mathbf{Y}$ of observations of (the components of) $\mathbf{X}$, which may also include other random effects. We assume that for each possible realization $\mathbf{x}$ of $\mathbf{X}$, the conditional distribution $f_{Y|x}$ of $\mathbf{Y}$ given $\mathbf{x}$ is known. This is essentially the data model.

Now, suppose that the observation of $\mathbf{Y}$ at a particular instance results in the value $\mathbf{y}$. The goal is to incorporate this value in the distribution of $\mathbf{X}$ by constructing the conditional distribution $f_{X|y}$ of $\mathbf{X}$ given $\mathbf{y}$. This conditional distribution is referred to as the *posterior distribution* on $\mathbf{X}$. To find its formula, we note that the joint probability density $f_{X,Y}$ of process variables and observations can be expressed in two ways,

$$f_{X,Y}(\mathbf{x},\mathbf{y}) = f_{Y|x}(\mathbf{y})f_X(\mathbf{x}) = f_{X|y}(\mathbf{x})f_Y(\mathbf{y}). \tag{20.7}$$

The prior distribution f_X is assumed to be known, as is the data model $f_{Y|x}$; the posterior distribution $f_{X|y}$ is sought, and the distribution of the observations f_Y is unknown. After dividing both expressions by $f_Y(\mathbf{y})$, we obtain *Bayes' rule*,

$$f_{X|y}(\mathbf{x}) = \frac{f_{Y|x}(\mathbf{y})f_X(\mathbf{x})}{f_Y(\mathbf{y})}. \tag{20.8}$$

Note that both sides depend on $\mathbf{x}$ and $\mathbf{y}$. The quantity $\mathbf{y}$ is given as an observation; therefore the denominator on the right-hand side is fixed, although unknown. The entire equation has the form

$$f_{X|y}(\mathbf{x}) \propto f_{Y|x}(\mathbf{y})f_X(\mathbf{x}), \tag{20.9}$$

where the implied proportionality constant (depending on f_Y) makes the term on the left a pdf. The constant can be found and equality established by computing an integral, either with analytical techniques or with numerical simulations.

In the case of Gaussian random variables, all integrations that would be required in Eq. (20.9) can, however, be replaced by matrix algebra, as the following lemma shows.

Lemma 20.1. *Let $\mathbf{X}$ and $\mathbf{Y}$ be Gaussian random variables such that $\mathbf{X} \sim N(\mu, P)$ and $\mathbf{Y}|\mathbf{X} \sim N(H\mathbf{X}, R)$, where H is a matrix. Then $\mathbf{X}|\mathbf{Y} \sim N(\mu^*, P^*)$ with $\mu^* = \mu + K(\mathbf{Y} - H\mu)$ and $P^* = (I - KH)P$, where K is the* gain matrix,

$$K = PH^T(R + HPH^T)^{-1}. \tag{20.10}$$

Proof. According to Eq. (20.9), the conditional probability density $f_{X|y}$ satisfies

$$f_{X|y}(\mathbf{x}) \propto \exp\left(-\tfrac{1}{2}\left((\mathbf{x} - H\mathbf{y})^T R^{-1}(\mathbf{x} - H\mathbf{y}) - (\mathbf{y} - \mu)^T P^{-1}(\mathbf{y} - \mu)\right)\right). \tag{20.11}$$

Completing the square, we obtain a Gaussian distribution,

$$f_{X|y}(\mathbf{x}) \propto \exp\left(-\tfrac{1}{2}\left((\mathbf{x}-\mu^*)^T(P^*)^{-1}(\mathbf{x}-\mu^*)\right)\right), \tag{20.12}$$

with

$$\mu^* = \mathscr{E}(\mathbf{X}|\mathbf{y}) = (P^{-1}+H^TR^{-1}H)^{-1}(P^{-1}\mu+R^{-1}H^T\mathbf{y}) = \mu + K(\mathbf{y}-H\mu),$$
$$P^* = \mathrm{var}(\mathbf{X}|\mathbf{y}) = (P^{-1}+H^TR^{-1}H)^{-1} = (I-KH)P.$$

We leave the details of these calculations to the reader.   □

The distribution $f_{X|y}$ given in the lemma is the posterior distribution. Its covariance matrix P^* does not depend on the value y. The lemma tells us that $P^* = P - C$, where C is a symmetric positive semidefinite matrix. In this sense, the posterior variance is smaller than the prior variance, and we have reduced uncertainty by using the data. The lemma also shows that the mean of the posterior distribution is the prior mean updated by the gain applied to the difference between the observed $\mathbf{Y}$ and its mean value $H\mu$. Recall that the difference between an observed quantity and its temporal mean is called "anomaly" in climate science, so this concept arises naturally in Bayesian data assimilation.

20.3.1 ▪ Example: Bayesian Approach

We return to the example introduced in Section 20.2. Assume that $X_1 \sim N(\mu_0, \sigma^2)$, where μ_0 and σ^2 are known. Then the column vector $\mathbf{X} = (X_1,\dots,X_4)^T$ of process variables has a multivariate normal distribution $\mathbf{X} \sim N(\mu, \Sigma)$ with mean vector $\mu = (\mu_0, \alpha\mu_0, \alpha^2\mu_0, \alpha^3\mu_0)^T$ and a suitable covariance matrix Σ (see the exercises). This is the prior distribution f_X on $\mathbf{X}$. It does not use any observations. Explicitly,

$$f_X(\mathbf{x}) \propto \exp\left(-\tfrac{1}{2}(\mathbf{x}-\mu)^T\Sigma^{-1}(\mathbf{x}-\mu)\right).$$

The column vector $\mathbf{Y} = (Y_2, Y_3)^T$ of observations and the vector $\mathbf{X}$ of process variables are related by the equation $\mathbf{Y} = H\mathbf{X} + (\zeta_2, \zeta_3)^T$, where $H = \left(\begin{smallmatrix} 0 & 1 & 0 & 0 \\ 0 & 0 & 1 & 0 \end{smallmatrix}\right)$. Given a particular realization x of the process variables, $\mathbf{Y}$ then has a multivariate normal distribution, $\mathbf{Y}|\mathbf{x} \sim N(H\mathbf{x}, \tau^2 I)$, where I is the 2×2 identity matrix. All quantities can be computed from the gain matrix K, which will also be derived in the exercises.

It is instructive to compare the standard deviations of the prior and posterior distributions for X_1 (reanalysis) and for X_3 (filtering). These quantities are computed in Exercise 4. It turns out that $\mathrm{var}(X_1|\mathbf{y}) < \mathrm{var}(X_1) = \sigma^2$ and $\mathrm{var}(X_3|\mathbf{y}) < \mathrm{var}(X_3) = \sigma^2\alpha^4 + \alpha^2 + 1$, as expected. However, $\mathrm{var}(X_1|\mathbf{y})$ cannot be made arbitrarily small, even if τ is small, while $\mathrm{var}(X_3|\mathbf{y}) = O(\tau^2)$. Details are in the exercises.

20.4 ▪ Sequential Data Assimilation

We now focus on reanalysis, filtering, and forecasting for time-dependent processes. The data arrive as a *time series*—a sequence of realizations of a *discrete-time stochastic process*.

Suppose the process variables are $\mathbf{X}(k)$, $k = 0, 1, \dots$, and the observations are $\mathbf{Y}(k)$, $k = 1, 2, \dots$. A starting value $\mathbf{X}(0)$ for the process variables is allowed to incorporate a background state for which no observations are available. We use the notation $\mathbf{X}(0:N)$ for a sequence $\{\mathbf{X}(k) : k = 0, 1, \dots, N\}$ of vector-valued random variables and $\mathbf{x}(0:N)$ for a sequence of its realizations, and similarily for $\mathbf{Y}$. We are interested in estimating $\mathbf{X}(n)$, given $\mathbf{Y}(1:N)$ (reanalysis), or given $\mathbf{Y}(1:n)$ (filtering), or given $\mathbf{Y}(1:n-1)$ (forecasting).

Joint pdfs (also called probability mass functions) and conditional pdfs are identified by suitable indices. For example, $f_{\mathbf{X}(0:n),\mathbf{Y}(1:n)}(\mathbf{x}(0:n),\mathbf{y}(1:n))$ is the joint density function of the process variables $\mathbf{X}(0),\dots,\mathbf{X}(n)$ and observations $\mathbf{Y}(1),\dots,\mathbf{Y}(n)$, and $f_{\mathbf{X}(2:n)|\mathbf{y}(1:n-1)}(\mathbf{x}(2:n))$ is the conditional pdf for $\mathbf{X}(2),\dots,\mathbf{X}(n)$ given observations $\mathbf{y}(1),\dots,$ $\mathbf{y}(n-1)$. The latter is a function of $\mathbf{y}(1:n-1)$ and $\mathbf{x}(2:n)$. Since observations arrive sequentially, one can try to find forecast and filter estimates also sequentially.

20.4.1 ▪ Filtering and Forecasting for Markov Chains

Definition 20.1. *A discrete-time stochastic process* $\mathbf{X}(0:N)$ *has the* Markov property *if its pdfs satisfy*

$$f_{\mathbf{X}(n:N)|\mathbf{x}(0:n-1)}(\mathbf{x}(n:N)) = f_{\mathbf{X}(n:N)|\mathbf{x}(n-1)}(\mathbf{x}(n:N)), \quad n=1,\dots,N, \tag{20.13}$$

for all $\mathbf{x}(n:N)$. *A discrete-time stochastic process that has the Markov property is called a* Markov chain.

Intuitively, the Markov property says that, to predict future observations $\mathbf{X}(n:N)$ of the stochastic process, it is sufficient to know the immediate past $\mathbf{X}(n-1)$. Additional knowledge of the more distant past $\mathbf{X}(0:n-2)$ does not change the predictions of the future. An induction argument shows that the joint distribution of $\mathbf{X}(0:n)$ can then be written as a product of conditional distributions,

$$f_{\mathbf{X}(0:n)}(\mathbf{x}(0:n)) = f_{\mathbf{X}(0)}(\mathbf{x}(0)) \prod_{j=1}^{n} f_{\mathbf{X}(j)|\mathbf{x}(j-1)}(\mathbf{x}(j)), \quad n=1,2,\dots. \tag{20.14}$$

The functions $f_{\mathbf{X}(j)|\mathbf{x}(j-1)}(\mathbf{x}(j))$ are called *transition probabilities* of the Markov chain. We shall also assume throughout that

$$f_{\mathbf{Y}(1:n)|\mathbf{x}(0:n)}(\mathbf{y}(1:n)) = \prod_{j=1}^{n} f_{\mathbf{Y}(j)|\mathbf{x}(j)}(\mathbf{y}(j)), \quad n=1,2,\dots. \tag{20.15}$$

This identity implies that, given the sequence of process variables $\mathbf{X}(0:n)$, the observations $\mathbf{Y}(1:n)$ are independent of one another and their distributions do not depend on $\mathbf{X}(0)$. In particular, it follows that given $\mathbf{x}(i)$, the observation $\mathbf{Y}(i)$ is independent of all other observations $\mathbf{Y}(j)$ $(j \neq i)$.

We can now formulate a basic algorithm for filtering and forecasting of Markov chains.

Algorithm 20.1 (Filtering and prediction). *Given*
(i) *a prior distribution* $f_{\mathbf{X}(0)}$,
(ii) *transition probabilities* $f_{\mathbf{X}(i)|\mathbf{x}(i-1)}$ $(i=1,\dots,N)$ *for the process variables, and*
(iii) *conditional distributions* $f_{\mathbf{Y}(i)|\mathbf{x}(i)}$ $(i=1,\dots,N)$ *for the observations,*
the following algorithm gives the filtering distributions $f_{\mathbf{X}(i)|\mathbf{y}(1:i)}$ *and forecasting distributions* $f_{\mathbf{X}(i)|\mathbf{y}(1:i-1)}$:
  Step 1. *Set*

$$f_{\mathbf{X}(1)}(\mathbf{x}(1)) = \int f_{\mathbf{X}(1)|\mathbf{x}(0)}(\mathbf{x}(1)) f_{\mathbf{X}(0)}(\mathbf{x}(0)) d\mathbf{x}(0),$$

$$f_{\mathbf{X}(1)|\mathbf{y}(1)}(\mathbf{x}(1)) \propto f_{\mathbf{Y}(1)|\mathbf{x}(1)}(\mathbf{y}(1)) f_{\mathbf{X}(1)}(\mathbf{x}(1)),$$

where the proportionality constant is chosen such that a pdf with respect to $\mathbf{x}(1)$ *is generated.*

Step 2. *Suppose $i \in \{2, 3, \ldots, N\}$ and the filtering pdf $f_{\mathbf{X}(i-1)|\mathbf{y}(1:i-1)}$ is given. Set*

$$f_{\mathbf{X}(i)|\mathbf{y}(1:i-1)}(\mathbf{x}(i)) = \int f_{\mathbf{X}(i)|\mathbf{x}(i-1)}(\mathbf{x}(i)) f_{\mathbf{X}(i-1)|\mathbf{y}(1:i-1)}(\mathbf{x}(i-1)) d\mathbf{x}(i-1),$$

$$f_{\mathbf{X}(i)|\mathbf{y}(1:i)}(\mathbf{x}(i)) \propto f_{\mathbf{Y}(i)|\mathbf{x}(i)}(\mathbf{y}(i)) f_{\mathbf{X}(i)|\mathbf{y}(1:i-1)}(\mathbf{x}(i)),$$

where the proportionality constant is chosen such that a pdf with respect to $\mathbf{x}(i)$ is generated.

The correctness of this algorithm can be proved with induction. Step 1 is just the law of total probability (to obtain $f_{\mathbf{X}(1)}$) and Bayes' rule (to obtain $f_{\mathbf{X}(1)|\mathbf{y}(1)}$). For the induction step, the formula for the forecasting distribution is again just the law of total probability, and the filtering distribution can be obtained from

$$f_{\mathbf{X}(i)|\mathbf{y}(1:i)}(\mathbf{x}(i)) = f_{\mathbf{X}(i)|\mathbf{y}(i),\mathbf{y}(1:i-1)}(\mathbf{x}(i))$$

$$\propto f_{\mathbf{Y}(i)|\mathbf{x}(i),\mathbf{y}(1:i-1)}(\mathbf{y}(i)) f_{\mathbf{X}(i)|\mathbf{y}(1:i-1)}(\mathbf{x}(i))$$

$$= f_{\mathbf{Y}(i)|\mathbf{x}(i)}(\mathbf{y}(i)) f_{\mathbf{X}(i)|\mathbf{y}(1:i-1)}(\mathbf{x}(i)).$$

The last equation follows because the observations $\mathbf{Y}(i)$ were assumed to be independent of the other observations conditioned on the $\mathbf{x}(i)$ in Eq. (20.15).

Once the filtering distributions are known, the reanalysis distributions $f_{\mathbf{X}(i)|\mathbf{y}(1:N)}(\mathbf{x}(i))$ can be obtained recursively for $i = N, N-1, \ldots, 0$.

Algorithm 20.2 (Reanalysis). *Given*
(i) *transition probabilities $f_{\mathbf{X}(i)|\mathbf{x}(i-1)}$ ($i = 1, \ldots, N$) for the process variables and*
(ii) *filtering distributions $f_{\mathbf{X}(i)|\mathbf{y}(1:i)}$ ($i = 1, \ldots, N$),*
the following algorithm gives the reanalysis distributions $f_{\mathbf{X}(i)|\mathbf{y}(1:N)}$:
     Step 1. *If $i = N$, the filtering distribution and the reanalysis distribution coincide.*
     Step 2. *Suppose $i \in \{1, 2, \ldots, N-1\}$ and the reanalysis pdf $f_{\mathbf{X}(i+1)|\mathbf{y}(1:N)}$ is given. Set*

$$f_{\mathbf{X}(i)|\mathbf{x}(i+1),\mathbf{y}(1:i)}(\mathbf{x}(i)) \propto f_{\mathbf{X}(i+1)|\mathbf{x}(i)}(\mathbf{x}(i+1)) f_{\mathbf{X}(i)|\mathbf{y}(1:i)}(\mathbf{x}(i)),$$

$$f_{\mathbf{X}(i)|\mathbf{y}(1:N)}(\mathbf{x}(i)) = \int f_{\mathbf{X}(i)|\mathbf{x}(i+1),\mathbf{y}(1:i)}(\mathbf{x}(i)) f_{\mathbf{X}(i+1)|\mathbf{y}(1:N)}(\mathbf{x}(i+1)) d\mathbf{x}(i+1),$$

where the proportionality constant is chosen such that a pdf with respect to $\mathbf{x}(i)$ is generated.

The correctness of this algorithm is proved with backwards induction.

Algorithms 20.1 and 20.2 provide a general framework for reanalysis, filtering, and forecasting. However, they are usually impossible to implement, due to multiple difficulties. For general process models, it is often not possible to obtain all the required transition probabilities, even in very simple cases. In the case of a process described by differential equations, a closed form solution would be required, followed by a complicated change of variables. Even if the transition probabilities were known, each step would require the computation of many integrals, one given explicitly in the algorithm and another to determine the proportionality constants. These integrations are usually impossible to do in closed form, and in the case of five or more state space dimensions they are also difficult to do numerically.

20.5 ▪ Kalman Filtering

If the prior distribution on $\mathbf{X}(0)$ is Gaussian, if the process is described by linear equations, and if the data model is also Gaussian with means that depend linearly on the process variables, then all probability distributions in the reanalysis and filtering algorithms are also Gaussian, and all integrations reduce to matrix manipulations. The result is the famous *Kalman filtering algorithm*, first proposed by the Hungarian-American engineer and mathematician RUDOLF (RUDY) EMIL KÁLMÁN (b. 1930).

Assume that the process variables $\mathbf{X}(i) \in \mathbb{R}^n$ form a linear process model,

$$\mathbf{X}(i) = M_i \mathbf{X}(i-1) + \xi_i, \quad i = 1, \dots, N, \tag{20.16}$$

with $\mathbf{X}(0) \sim N(\mu, \Sigma)$. (The more general process model $\mathbf{X}(i) = M_i \mathbf{X}(i-1) + \mathbf{b}(i) + \xi_i$ can be reduced to Eq. (20.16); see the exercises.) Assume, furthermore, that the data variables $\mathbf{Y}(i) \in \mathbb{R}^m$ are linearly related to the process variables,

$$\mathbf{Y}(i) = H_i \mathbf{X}(i) + \zeta_i, \quad i = 1, \dots, N. \tag{20.17}$$

The M_i and H_i are matrices of suitable dimensions, and the $\xi_i \in \mathbb{R}^n$ and $\zeta_i \in \mathbb{R}^m$ are random variables, independent of each other and of $\mathbf{X}(0)$, distributed as $\xi_i \sim N(0, Q_i)$ and $\zeta_i \sim N(0, R_i)$. The assumptions cover the case where the dimension m_i of the ith observation $\mathbf{Y}(i)$ depends on i or where some of the $\mathbf{Y}(i)$ are absent.

Then the theory of multivariate normal distributions implies that the $\mathbf{X}(i)$ and $\mathbf{Y}(i)$ are also Gaussian, as are all conditional variables. In particular, $\mathbf{X}(1) \sim N(M_1 \mu, Q_1 + M_1 \Sigma M_1^T)$. It is therefore sufficient to describe the means and covariance matrices of these variables. We use the following abbreviations:

$$\mu_{i|i-1} = \mathscr{E}(\mathbf{X}(i)|\mathbf{y}(1:i-1)), \quad \Sigma_{i|i-1} = \operatorname{var}(\mathbf{X}(i)|\mathbf{y}(1:i-1)),$$

$$\mu_{i|i} = \mathscr{E}(\mathbf{X}(i)|\mathbf{y}(1:i)), \quad \Sigma_{i|i} = \operatorname{var}(\mathbf{X}(i)|\mathbf{y}(1:i)),$$

$$\mu_i = \mathscr{E}(\mathbf{X}(i)|\mathbf{y}(1:N)), \quad \Sigma_i = \operatorname{var}(\mathbf{X}(i)|\mathbf{y}(1:N)),$$

with the convention $\mu_{0|0} = \mu$, $\mu_{1|0} = \mathscr{E}\mathbf{X}(1) = M_1 \mu$ and, similarly, $\Sigma_{0|0} = \Sigma$, $\Sigma_{1|0} = Q_1 + M_1 \Sigma M_1^T$. The quantities μ are conditional means and the quantities Σ conditional covariance matrices. The subscript $i|i-1$ refers to a forecasting quantity, the subscript $i|i$ to a filtering quantity (estimate the current state based on current and past observations), and the simple subscript i to a reanalysis quantity (estimate a past state from all available data). The goal is to obtain recursions for all these quantities. Straightforward computations show that

$$\mu_{i|i-1} = M_i \mu_{i-1|i-1}, \quad \Sigma_{i|i-1} = Q_i + M_i \Sigma_{i-1|i-1} M_i^T, \quad i = 1, \dots, N. \tag{20.18}$$

As expected, the forecasting distribution does not depend on new data.

Next, we use induction, applying Lemma 20.1 with $\mathbf{X} = \mathbf{X}(i)|\mathbf{y}(1:i-1)$ and $\mathbf{Y} = \mathbf{Y}(i)|\mathbf{y}(1:i-1)$. The data model has the property that $\mathbf{Y}(i)$ is independent of $\mathbf{Y}(k)$ for $k < i$, so $\mathbf{Y}(i)|\mathbf{y}(1:i-1) = \mathbf{Y}(i)$. According to Lemma 20.1,

$$\mu_{i|i} = \mu_{i|i-1} + K_i \left(\mathbf{y}(i) - H_i \mu_{i|i-1} \right), \tag{20.19}$$

where the *Kalman gain matrix* K_i is given by

$$K_i = \Sigma_{i|i-1} H_i^T \left(H_i^T \Sigma_{i|i-1} H_i + R_i \right)^{-1}, \tag{20.20}$$

provided the matrix inverses all exist. The term $\mathbf{y}(i) - H_i \mu_{i|i-1}$ is called the *innovation* and is conceptually similar to an anomaly in climate science. Also from Lemma 20.1, the filtering covariance matrix is

$$\Sigma_{i|i} = (I - K_i H_i) \Sigma_{i|i-1}. \tag{20.21}$$

The *Kalman filter algorithm* is obtained by alternating the forecasting and filtering step, just as in the general Algorithm 20.1.

Algorithm 20.3 (Kalman filter). *Given*
(i) *a prior distribution* $X(0) \sim N(\mu, \Sigma)$,
(ii) *the process model* (20.16), *and*
(iii) *the linear data model* (20.17),
the following algorithm gives the forecasting distributions $f_{X(i)|y(1:i-1)}$ *and filtering distributions* $f_{X(i)|y(1:i)}$:
   Step 1. *The forecasting and filtering distributions of* $X(1)$ *are given by*

$$X(1) \sim N(\mu_{1|0}, \Sigma_{1|0}),$$
$$X(1)|y(1) \sim N(\mu_{1|1}, \Sigma_{1|1}).$$

   Step 2. *Suppose* $i \in \{2, 3, \ldots, N\}$ *and the filtering distribution at time step* $i-1$ *is known,* $X(i-1)|y(1:i-1) \sim N(\mu_{i-1|i-1}, \Sigma_{i-1|i-1})$. *Then the forecasting and filtering distributions at time step* i *are given by*

$$X(i)|y(1:i-1) \sim N(\mu_{i|i-1}, \Sigma_{i|i-1}),$$
$$X(i)|y(1:i) \sim N(\mu_{i|i}, \Sigma_{i|i}),$$

where $\mu_{i|i-1}$, $\mu_{i|i}$, $\Sigma_{i|i-1}$, *and* $\Sigma_{i|i}$ *are computed from Eqs.* (20.18), (20.19), *and* (20.21).

The data influence only the forecasting and filtering means but not the variance matrices which, in principle, can all be computed in advance. In practical applications, the problem of computing or estimating the covariance matrices becomes important. For those i for which no observations are available, the filtering distribution and the forecasting distribution agree.

The basic reanalysis Algorithm 20.2 can also be rewritten in terms of matrix operations for this situation [121].

20.6 ▪ Numerical Example

The following numerical example illustrates the results of Kalman filtering and reanalysis. The example uses the one-dimensional process model $x_i = \alpha x_{i-1} + \xi_i$ for $i = 1, \ldots, 30$, where $\alpha = 0.8$ and

$$\xi_i \sim N(0, q^2), \quad q = 0.4.$$

The value x_0 is drawn from a standard normal distribution. A typical sequence is plotted in black in Figure 20.2. The data model is $y_i = h_i x_i + \zeta_i$, where

$$\zeta_i \sim N(0, r^2), \quad r = 0.1; \quad h_i = \begin{cases} 0.1, & i = 11, \ldots, 20, \\ 1, & i = 1, \ldots, 10, 21, \ldots, 30. \end{cases}$$

The data model is set up so that for $i = 11, \ldots, 20$, there is a period of "low observability." Figure 20.2 shows a single realization of the true process, the filtering estimates, and the reanalysis estimates. It is clear that these estimates are not the same. As expected, all estimates are much closer to the true values where $h_i = 1$.

Figure 20.3 shows computed standard deviations (bold lines) for the forecasting estimates $x_{i|i-1}$ (black), filtering estimates $x_{i|i}$ (blue), and reanalysis estimates $x_{i|N}$ (red), together with their negative values. Forecasting standard deviations are always larger than

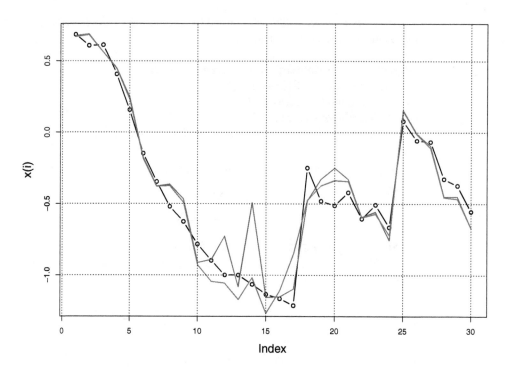

Figure 20.2. *Simulated Kalman filter example; true process (black), filtering estimate (blue), and reanalysis estimate (red).*

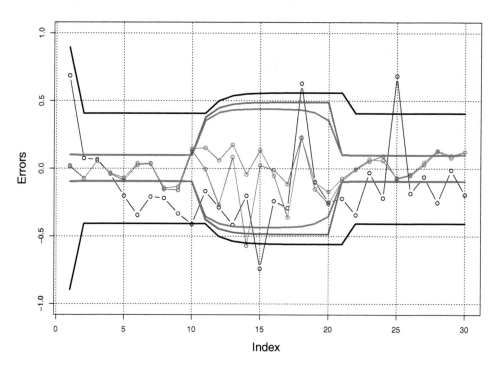

Figure 20.3. *Errors and error standard deviations for Kalman filter example; forecasting error (black), filtering error (blue), and reanalysis error (red).*

filtering standard deviations which, in turn, are larger than reanalysis standard deviations. During the interval of low observability ($i = 11, \ldots, 20$), all these standard deviations are larger. Also plotted in the same figure (thin lines) are the forecasting errors $x_i - x_{i|i-1}$ (black), filtering errors $x_i - x_{i|i}$ (blue), and reanalysis errors $x_i - x_{i|N}$ (red) for a single realization.

20.7 ▪ Extensions

The Kalman filter and reanalysis algorithms have theoretical and practical limitations if the error distributions are not Gaussian or if the process model is nonlinear. In the latter case, even Gaussian errors typically become immediately non-Gaussian, and biases (systematic errors) appear. Another practical difficulty arises because in weather forecasting or climate science temporal and spatial features may lead to state variables with millions of components, with error covariance matrices with 10^{12} or more entries.

20.7.1 ▪ Extended Kalman Filter

Nonlinear process and data models are often handled by linearization. The resulting algorithm is known as the *extended Kalman filter*. In essence, it uses the nonlinear process model (without noise terms) to compute the forecasting estimate and uses linearization about the most recent filtering estimate to compute the forecasting covariance and the gain matrix. The filtering estimate and its covariance are then computed from these quantities more or less in the same way as in Algorithm 20.3. This approach works well in engineering applications with a modest number of process variables but quickly becomes infeasible in high-dimensional situations. Variational methods for data assimilation that were mentioned earlier can avoid these difficulties but do not readily produce an assessment of errors.

20.7.2 ▪ Ensemble Kalman Filter

The *ensemble Kalman filter* (EnKf), introduced in the 1990s, uses stochastic simulation techniques known as *Monte Carlo* methods; [21] is a review article by one of the inventors of EnKf. The archetypical use for a Monte Carlo approach is the computation of an expected value $\mathcal{E}F(\mathbf{X})$ of a random variable $\mathbf{X}$ with density $f_X(\mathbf{x})$. Formally, the definition is $\mathcal{E}F(\mathbf{X}) = \int F(\mathbf{x})f_X(\mathbf{x})\,d\mathbf{x}$, where the integral is over the space in which $\mathbf{X}$ takes its values. Numerical computation of the integral is essentially impossible if the dimension of $\mathbf{X}$ exceeds 10 or so. In a Monte Carlo approach, one draws m independent random samples $\mathbf{x}(1:m)$ from the distribution of $\mathbf{X}$ and approximates the expected value,

$$\mathcal{E}F(\mathbf{X}) \approx \frac{1}{m}\sum_{i=1}^{m} F(\mathbf{x}(i)). \tag{20.22}$$

By the law of large numbers, the right-hand side converges almost surely to the correct expected value as $m \to \infty$, and the speed of convergence can be determined (it is always slow).

For a simple version of EnKf, we assume a nonlinear process model,

$$\mathbf{X}(i) = \mathcal{M}_i(\mathbf{X}(i-1)) + \xi_i, \tag{20.23}$$

where the $\mathcal{M}_i$ are functions from the range of the $\mathbf{X}(i)$ to itself. All other assumptions are left unchanged—the model errors ξ_i are Gaussian, there is a linear data model (20.17), and the background state is Gaussian, $\mathbf{X}(0) \sim N(\mu, \Sigma)$. Suppose we are given an estimate $\hat{\mathbf{x}}_{i-1|i-1}$ of the filtering mean at $i-1$, an estimate $\hat{\Sigma}_{i-1|i-1}$ of the filtering covariance matrix at this time step, and an estimate $N(\hat{\mathbf{x}}_{i-1|i-1}, \hat{\Sigma}_{i-1|i-1})$ of the distribution of $\mathbf{X}(i-1)|\mathbf{y}(1 : i-1)$. For the simulation, draw m independent samples $\mathbf{x}^j_{i-1|i-1}$ $(j = 1, \ldots, m)$ from this distribution, propagate them forward with the process, and add simulated model errors $\eta_j \sim N(0, Q_i)$ that have the same distribution as the model errors ξ_i at this time step. The result is an *ensemble* of simulated forecasts,

$$\mathbf{x}^j_{i|i-1} = \mathcal{M}_i(\mathbf{x}^j_{i|i-1}) + \eta_j, \quad j = 1, \ldots, m. \tag{20.24}$$

The forecasting estimate is now computed as the sample mean of this ensemble,

$$\hat{\mathbf{x}}_{i|i-1} = \frac{1}{m} \sum_{j=1}^{m} \mathbf{x}^j_{i|i-1}, \tag{20.25}$$

and the sample covariance matrix provides an estimate $\hat{\Sigma}_{j|j-1}$ of the forecasting covariance matrix. Next, compute the gain matrix, just as in (20.20) but with the estimated covariance,

$$\hat{K}_j = \hat{\Sigma}_{j|j-1} H_j^T \left(H_j^T \hat{\Sigma}_{j|j-1} H_j + R_j \right)^{-1}. \tag{20.26}$$

To compute a filtering estimate, the ensemble of forecasts is adjusted using an innovation term and gain matrix as in Eq. (20.19). There is only one observation $\mathbf{y}(i)$ available, but it turns out that using it unchanged for all innovations in the ensemble tends to underestimate the variability of the filtering distribution. Hence, the innovation term $\mathbf{y}(i) - H_i \mu_{i|i-1}$ in the ordinary Kalman filter is replaced by an innovation ensemble $\mathbf{y}(i) + e_j - H_i \mathbf{x}^j_{i|i}$, where the perturbations e_j are simulated observation errors with the same distribution as the errors in the data model (20.17). One can therefore compute an ensemble of simulated filtering states,

$$\mathbf{x}^j_{i|i} = \mathbf{x}^j_{i|i-1} + \hat{K}_i \left(\mathbf{y}(i) + e_j - H_i \mathbf{x}^j_{i|i-1} \right). \tag{20.27}$$

This ensemble is used to produce estimates $\hat{\mathbf{x}}_{i|i}$ of the filtering mean and $\hat{\Sigma}_{i|i}$ of the filtering covariance matrix. One then uses $N(\hat{\mathbf{x}}_{i|i}, \hat{\Sigma}_{i|i})$ as an estimate of the filtering distribution at time step i, and the algorithm has completed a step. There is also a reanalysis version of this method.

If the process model is actually linear, such that Eq. (20.23) reduces to Eq. (20.16), then the ensemble Kalman forecasting and filtering estimates converge in the limit of large ensemble size to those of the ordinary Kalman filter algorithm. However, if the $\mathcal{M}_i$ are nonlinear, then the forecasting and filtering distributions become non-Gaussian, and there will be biases from the ensemble approach that do not disappear with large ensemble size. These biases must be assessed or possibly corrected separately.

If the space of process variables is high-dimensional (10^6 or 10^8 is not uncommon), the ensemble approach successfully avoids the problem of high-dimensional integration and the manipulation of huge covariance matrices. However, typically the ensemble size is much smaller, perhaps $m = \mathcal{O}(10^2)$. Then the sample covariance matrices have rank at most m and cannot possibly give all covariances correctly. On the other hand, in such situations the process vector $\mathbf{x}(i)$ may describe physical quantities at different locations across a region or around the globe. Then one often multiplies $\hat{\Sigma}_{i|i-1}$ or $\hat{\Sigma}_{i|i}$ elementwise with a "cut-off" matrix C whose entries are small far from the diagonal (for component pairs that have little to do with each other). This trick eliminates spurious large correlations at distant locations and at the same time tends to restore full rank to the estimated covariance matrices. Care must be taken to avoid destroying teleconnections.

The EnKf is now widely used. Many versions have been developed, and the literature has grown quite large; [49] is a recent overview by one of the leading experts.

20.8 ▪ Data Assimilation for the Lorenz System

To illustrate the EnKf technique, we apply the algorithm to the Lorenz model (7.1) with the fixed parameter values $\sigma = 10$, $\beta = \frac{8}{3}$, and $\rho = 28$. The attractor is shown in Figure 7.2.

Let $\mathbf{x} = (x, y, z) : t \mapsto \mathbf{x}(t) = (x(t), y(t), z(t))$ be the solution of the Lorenz system (7.1) which satisfies the initial data $\mathbf{x}(0) = \mathbf{x}_0 = (x_0, y_0, z_0) \in \mathbb{R}^3$. The process model associated with the Lorenz equations is the map $\mathcal{M} : \mathbb{R}^3 \to \mathbb{R}^3$ defined by

$$\mathcal{M}(\mathbf{x}_0) = \mathbf{x}(1) \in \mathbb{R}^3, \quad \mathbf{x}_0 \in \mathbb{R}^3. \tag{20.28}$$

There is no closed formula for $\mathcal{M}$, so $\mathcal{M}$ must be computed numerically by solving the system of differential equations for each $\mathbf{x}_0$.

Figure 20.4 shows the x-component of a trajectory for $0 \le t \le 6$ for initial data $x_0 = -7.3$, $y_0 = -11.5$, $z_0 = 17.8$. This trajectory switches from one "sheet" of the attractor to the other near $t = 2$. Also shown in Figure 20.4 are the x-components of ten trajectories from a solution ensemble with initial data $\mathbf{x}_0' = \mathbf{x}_0 + (\xi_1, \xi_2, \xi_3)$, where the ξ_i are $N(0, 1)$ random variables, as well as the ensemble mean computed by averaging 1000 trajectories of this ensemble. For $t > 1.5$ or so, the ensemble mean is seen to differ dramatically from the true solution. The specific reason in this case is that trajectories from the ensemble switch from one leaf of the attractor to the other at times that can be very different from $t = 2$. Essentially, if the initial data are known only up to random errors that have standard normal distributions, then the true trajectory becomes unpredictable for $t > 1.5$ or so. Averaging over many ensemble trajectories does not eliminate this bias. This situation is typical for nonlinear systems of differential equations; for linear systems, the ensemble mean always equals the true trajectory.

Figure 20.5 again shows the x-component of the true trajectory, together with the corresponding results of the EnKf, for a ensemble of size 50. The data model is given by Eq. (20.17), with $H = \left(\begin{smallmatrix} 1 & 1 & 0 \\ 0 & 1 & 1 \end{smallmatrix}\right)$. Therefore, only a two-dimensional projection of the true trajectory, corrupted by random noise, is observed at each time. The error terms ζ_i are standard normal $N(0, 1)$. Each assimilated trajectory starts at a filtering state at $t = i$ (red circle) and ends at a forecasting state at $t = i + 1$ (green circle). Evidently, the assimilation processes is only partially successful in approximating the correct trajectory. For example, on the intervals $1.5 < t < 2$ and $4.5 < t < 5$, the assimilated trajectories are not even qualitatively correct. Nevertheless, it is remarkable that a significant portion of the true trajectory is recreated correctly over the entire interval of interest.

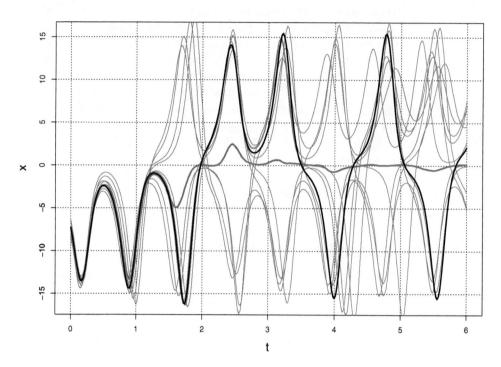

Figure 20.4. *Lorenz system: x-component without data assimilation; exact trajectory (black), some trajectories from the solution ensemble (red), and ensemble mean (blue).*

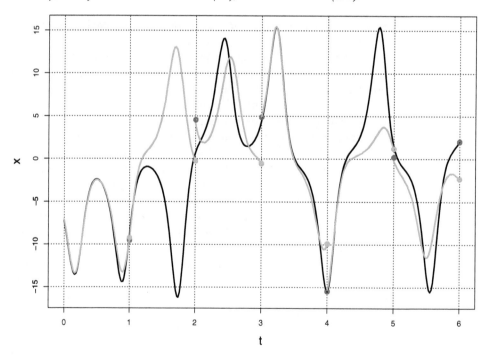

Figure 20.5. *Lorenz system: x-component with data assimilation; exact trajectory (black) and assimilated trajectories (green), starting with filtering states (red circles) and ending in forecasting states (green circles).*

20.9 • Concluding Remarks

Data assimilation has been very important for weather prediction, where it has extended both the range and reliability of forecasts. In the best-case scenario, there is a correct dynamical model with sufficiently high resolution and a well-designed observational network that can provide accurate data. In the next-best-case scenario, there may be a somewhat deficient dynamical model, but the observational network is still good enough to provide adequate data. Then it is still possible to obtain estimates in agreement with nature, as long as dynamical model errors are recognized and incorporated adequately.

In climate science, data assimilation has been used since the late 1980s. It is having an increasingly significant impact, since it allows improved and faster estimation of both internal and external parameters. Parameter estimation is particularly promising in the biogeochemistry of the climate system, where it is often difficult to measure rates directly *in situ*. Realistic dynamical models and feasible measurements of quantities such as concentration fields of plankton, in combination with advanced data assimilation techniques, can lead to surprisingly realistic estimates of these rates.

Data assimilation techniques have been applied to create reliable uniform background data for the past (reanalysis) and to obtain information about unobservable quantities such as ocean upwelling and chemical reaction rates in the ocean. A typical example is described in [11], where an EnKf approach is used to reconstruct the ocean climate for the period 1958–2001.

Interestingly, there are specific problems with heterogeneous data sources associated with all reanalysis exercises for ocean circulation for the 20th century. Since about 1980, satellites with circumpolar orbits have provided reliable records of planet-wide uniformly distributed ocean measurements. But prior to about 1980, observations came mainly from shipboard observations that were concentrated along major shipping routes. There are also other more subtle trends in the data collection efforts that are known to introduce spurious trends during data assimilation. For example, the typical height at which shipboard anemometers are mounted has increased since the middle of the last century, leading to biased results for wind strengths and patterns. In the past, when data assimilation was used mainly for weather prediction, such slow trends did not matter.

Ridgwell et al. [88] describe a similar project for biogeochemistry data. Here, an EnKf is employed in an iterative fashion to obtain steady-state information about geochemical parameters such as uptake rates and concentrations of phosphate and calcium carbonate in the ocean at preindustrial times.

The EnKf is known to run into problems when the underlying process model is highly nonlinear and has many unstable equilibria. Other methods have been proposed for such situations. For example, Apte, Jones, and Stuart [2] develop a Bayesian approach to address data assimilation questions coming from drifting buoys, which are known to have trajectories that are very sensitive to changes in the initial data.

20.10 • Exercises

1. Consider a physical process in which real-valued state variables x_i, $i = 1, 2, \ldots$, are generated according to the rule $x_{i+1} = f(x_i)$, where f is a given real-valued function. The states are observed according to the data model $y_i = x_i + \xi_i$, where the ξ_i are independent random variables with an $N(0, \sigma^2)$-distribution.

   (i) Assume that observations $y_1, y_2, \ldots, y_N$ are available and that you want to estimate x_1. Describe in general terms how a cost function should be set up whose minimum is expected to give an estimate for x_1.

(ii) Consider the special case $f(x) = 4x(1-x)$. Begin with the case $N = 3$, $\sigma = 0.1$, $x_1 = 0.2$. Construct this cost function for several realizations of the y_i and plot it, using a computer. Is the minimum of the cost function where you expect it to be?

(iii) Repeat part (ii) with a larger σ, for example $\sigma = 0.5$. Repeat with the previous σ and a larger N, for example $N = 6$. Describe your observations. Does it help to have more observations available?

The cost function may have multiple local minima, which can lead to serious numerical difficulties.

2. Consider the process model (20.1) together with the data model (20.2). Derive the joint distribution of Y_2, Y_3 as a function of the unknown parameter x_2 and use it to obtain the maximum likelihood estimate $\hat{x}_2$. Interpret your result in the case where $\tau \ll 1$.

3. Consider again the process model (20.1) together with the data model (20.2). Derive the joint distribution of Y_2, Y_3 as a function of the unknown parameter x_3 and use it to obtain the maximum likelihood estimate $\hat{x}_3$. Interpret your result in the case where $\tau \ll 1$. Use $x_2 = \alpha^{-1}(x_3 - \xi_2)$.

4. Consider the process model (20.1) and assume that $X_1 \sim N(\mu_0, \sigma^2)$.

   (i) Compute the covariance matrix Σ of the random vector $\mathbf{X} = (X_1, \ldots, X_4)^T$.

   (ii) Compute the gain matrix K from the definition (20.10).

   (iii) Show that

$$\mathrm{var}(X_1|\mathbf{y}) = \frac{\sigma^2\left(\tau^4 + \left(\alpha^2 + 2\right)\tau^2 + 1\right)}{\tau^4 + ((\alpha^2 + 1)\sigma^2\alpha^2 + \alpha^2 + 2)\tau^2 + \alpha^2\sigma^2 + 1}$$

and

$$\mathrm{var}(X_3|\mathbf{y}) = \frac{\tau^2\left(\alpha^2\sigma^2 + \left(\sigma^2\alpha^4 + \alpha^2 + 1\right)\tau^2 + 1\right)}{\tau^4 + ((\alpha^2 + 1)\sigma^2\alpha^2 + \alpha^2 + 2)\tau^2 + \alpha^2\sigma^2 + 1}.$$

5. Consider the general process model $\mathbf{X}(i) = M_i\mathbf{X}(i-1) + \mathbf{b}(i) + \xi_i$, $i = 1, \ldots, N$, where $\mathbf{X}(i), \mathbf{b}(i), \xi_i \in \mathbb{R}^n$ and M_i are $n \times n$ matrices. Define $\overline{\mathbf{X}}(i)$ by

$$\overline{\mathbf{X}}(0) = 0; \quad \overline{\mathbf{X}}(i) = M_i\overline{\mathbf{X}}(i-1) + \mathbf{b}(i), \quad i = 1, \ldots, N, \qquad (20.29)$$

and set $\tilde{\mathbf{X}}(i) = \mathbf{X}(i) - \overline{\mathbf{X}}(i)$. Show that the $\tilde{\mathbf{X}}(i)$ satisfy the recursion (20.16).

6. Use the MATLAB code for the Lorenz equations given in Section C.1 to explore the behavior of the ensemble mean of the Lorenz equations for other initial data $\mathbf{x}_0$. Can you find initial data such that the ensemble mean stays close to the correct solution for an interval of length $T = 5$?

7. Consider the linear process model (20.16) together with the data model (20.17), and assume that all matrices are independent of i, so $M_i = M$, $H_i = H$, $Q_i = Q$, and $R_i = R$ for all i. Assume that the forecasting covariates matrices $\Sigma_{i|i-1}$ converge to a limit Σ_0. Derive the equation

$$\Sigma_0 = Q + M\left(\Sigma_0 - \Sigma_0 H^T(H\Sigma_0 H^T + R)^{-1}H\Sigma_0\right)M^T,$$

and derive similar equations for the limits of the matrices $\Sigma_{i|i}$ and K_i. The above equation is known as a matrix RICCATI equation. It reduces to an ordinary quadratic equation if the dimension of the state space is 1.

Appendix A

Units and Symbols

We follow the International System of Units. The System specifies seven *base units* or *SI units*. The five SI relevant for the present text are
- meter (m) for length,
- kilogram (kg) for mass,
- second (s) for time,
- kelvin (K) for temperature,
- mole (mol) for the amount of substance.

The additional SI units are ampere (A) for electric current and candela (cd) for luminous intensity. All other units of measurement are formed by products of the powers of base units; these other units are called *derived SI units*. For example, the derived unit of area is square meter (m^2) and of density is kilograms per cubic meter (kg/m^3).

The joule (J) and watt (W) are derived SI units of energy, $J = kg\,m^2\,s^{-2}$, $W = J\,s^{-1}$. 1 joule is approximately equal to the energy required to lift a small apple (a mass of about $102\,g = 1/9.81\,kg$) one meter straight up [120].

Temperatures are measured in kelvins or degrees Celsius. The Kelvin and Celsius scales are related by a shift transformation: the zero point on the Kelvin scale is absolute zero, the zero point on the Celsius scale is the temperature at which water freezes, $0°C = 273.15\,K$, and the magnitude of a kelvin on the Kelvin scale is the same as a degree on the Celsius scale. In 2010, the global mean temperature was $9.46°C = 282.61\,K$ for the Earth's land surface, $16.59°C = 289.74\,K$ for the Earth's sea surface, and $14.52°C = 287.67\,K$ for the Earth's combined land and sea surface [74].

The following symbols are used in the text. Some symbols have multiple meanings, depending on the context in which they appear.

α	Vector of regression parameters
α	Albedo
α	Thermal contraction coefficient, $\alpha = 1.5 \cdot 10^{-4}\ \mathrm{deg}^{-1}$
β	Aspect ratio, parameter in the Lorenz model
β	Rossby parameter in β-plane approximation of Coriolis effect
β	Saline contraction coefficient, $\beta = 8 \cdot 10^{-4}\ \mathrm{psu}^{-1}$
γ	Orbit
ξ, ζ	Random errors
ζ	Vorticity
θ	Latitude, $y = \sin\theta$
θ	Potential temperature

$\varkappa$	Thermal conductivity
λ	Bifurcation parameter
ρ	Mass density
ρ	Rayleigh number, parameter in Lorenz model
σ	Prandtl number, parameter in Lorenz model
σ	Stefan–Boltzmann constant, $\sigma = 5.67 \cdot 10^{-8}\ \mathrm{Wm^{-2}K^{-4}}$
φ	Solution of a differential equation
ϕ_t	Dynamical system
ψ	Stream function
Δ	Laplace operator
$\boldsymbol{\Omega}$	Earth's rotation vector
Ω	Angular frequency of rotating Earth
$\mathbf{a}_C$	Coriolis acceleration
c_p	Specific heat at constant pressure
f	Coriolis parameter
$\mathbf{g}$	Acceleration due to gravity
$\mathbf{n}$	Unit normal vector
p	Hydrostatic pressure, units of bar, $1\,\mathrm{bar} = 10^5\ \mathrm{Nm^{-2}}$
ppm(v)	Parts per million (by volume)
psu	Practical salinity unit
$\mathbf{q}$	Heat flux
q	Flow in box models
r	Correlation coefficient
s	Distribution of incoming solar radiation as function of latitude
$\mathbf{v}$	Velocity vector
A, B	Coefficients in Budyko's model $A + BT$ of outgoing radiation, best estimate for Northern Hemisphere $A = 203.3\ \mathrm{Wm^{-2}}$, $B = 2.09\ \mathrm{Wm^{-2}(deg\ C)^{-1}}$
Alk_C	Carbonate alkalinity, $2,480\ \mu\mathrm{mol\,kg^{-1}}$ for seawater with $S = 35\,\mathrm{psu}$, $\mathrm{pH} = 8.1$, $T = 25^\circ\mathrm{C}$
C	Species concentration
C	Heat capacity, units of joule per kelvin $(\mathrm{JK^{-1}})$
$C(n; k)$	Binomial coefficient
D	Determinant of a matrix
D	Diffusion coefficient $(\mathrm{Wm^{-2}deg^{-1}})$
DIC	Dissolved inorganic carbon
E	Energy, units of watts per square meter $(\mathrm{Wm^{-2}})$
E	Eccentricity in Milankovitch's theory of glacial cycles
$\mathbb{E}$	Rate-of-strain tensor
$\mathscr{E}$	Expected value
F	Fourier matrix
$\mathbf{F}_C$	Coriolis force
H	Virtual salt flux
J	Rotation matrix for plane rotation by $\frac{1}{2}\pi$
K	Gain matrix in data assimilation
Kyr	Thousands of years
L	Environmental lapse rate, value in the troposphere 6 to 7 deg/km
Myr	Millions of years
N	Brunt–Väisälä or buoyancy frequency

N	Nutrient concentration
P	Phytoplankton concentration
P	Precession in Milankovitch's theory of glacial cycles
P_n	Legendre polynomial of degree n
Q	Fraction of the solar constant, $Q = \frac{1}{4}S_0$
Q^{65}	Average daily insolation at 65° North latitude
R	Gas constant, $R = 8.314\,\mathrm{JK^{-1}mol^{-1}}$
R	Radius of the Earth
R^2	Coefficient of determination
$\mathbb{R}$	Set of real numbers
S	Model skill
S	Salinity (psu). Salinity of sea water is approximately 35 psu ($\approx$ 3.5% salt by weight)
S_0	Solar constant, $1{,}368\,\mathrm{Wm^{-2}}$
$\mathbb{S}$	Stress deviator tensor
Sv	Sverdrup, measure of volume transport (not an SI unit), $1\,\mathrm{Sv} = 10^6\,\mathrm{m^3 s^{-1}}$
T	Minimal period of a periodic solution of a differential equation
T	Temperature
T	Tilt (obliquity) in Milankovitch's theory of glacial cycles
T	Trace of a matrix
$\mathbb{T}$	Cauchy stress tensor
Z	Zooplankton concentration

Appendix B

Glossary

Many of the following definitions are taken directly from [103].

Abyssal zone The abyssopelagic layer or pelagic zone that contains the very deep benthic communities near the bottom of the ocean. At depths of 4,000 to 6,000 meters, this zone remains in perpetual darkness and never receives daylight.

Advection A transport mechanism of some property of the atmosphere or ocean, such as heat, humidity, or salinity due to bulk flow. In climate science, advection describes the predominantly horizontal large-scale motions of the ocean and atmosphere. The term advection is sometimes used as a synonym for convection; technically, convection is the sum of transport by diffusion and advection.

Advection-diffusion equation A mathematical equation (partial differential equation) describing the physical phenomena where particles, energy, or other physical quantities are transferred inside a physical system due to two processes: diffusion and advection (convection).

Aerosols Airborne solid or liquid particles that reside in the atmosphere for at least several hours and influence climate, for example by scattering and absorbing radiation or by acting as cloud condensation nuclei.

Albedo Fraction of solar radiation reflected back by a surface or object.

Anomaly The departure from a reference value or long-term average. A positive temperature anomaly indicates that the observed temperature was warmer than the reference value, while a negative anomaly indicates that the observed temperature was cooler than the reference value.

Atmosphere The gaseous envelope surrounding the Earth.

Autonomous differential equation An ordinary differential equation for a function of one independent variable (time) where the forcing function does not depend explicitly on time.

Bayesian method A method by which a statistical analysis of an unknown or uncertain quantity is carried out in two steps. First, a prior probability distribution is formulated on the basis of existing knowledge. In the second step, newly acquired data are introduced using Bayes' rule to update the prior distribution into a posterior distribution.

Bifurcation theory The mathematical theory of changes in the qualitative or topological structure of a family of functions, such as the integral curves of a family of vector fields and a family of solutions of a differential equation. Most commonly applied to the mathematical study of dynamical systems, where a small change in one or more of the parameters (bifurcation parameters) of the system causes a sudden qualitative or topological change in its behavior.

Biosphere The component of the Earth's climate system comprising all ecosystems and living organisms in the atmosphere, on land, or in the ocean, including derived dead organic matter.

Box model A discretization of an ocean circulation model consisting of a finite number of boxes, each filled with sea water and maintaining a uniform temperature and salinity. The boxes exchange heat and salt with one another through connecting capillary pipes at the bottom and corresponding overflow mechanisms at the top, and with their environment.

Carbon cycle The flow of carbon in various forms (for example, CO_2) through the climate system.

Chaotic dynamics Irregular behavior of the orbits shown by a (deterministic) dynamical system. The term is often used to indicate that the system shows sensitive dependence on initial conditions.

Climate The statistical description in terms of the mean and variability of relevant quantities (surface temperature, precipitation, wind, etc.) over a period of time ranging from months to thousands or millions of years.

Climate change A change of the state of the climate system that can be identified by changes in the mean and/or variability of its properties and that persists for an extended period, typically decades or longer.

Climate model A representation of the climate system based on the physical, chemical, and biological properties of its components, their interactions, and feedback processes, and accounting for all or some of its known properties. The representation can be in terms of mathematical equations (mathematical model), mathematical algorithms (computational model), or computer software (computer model). The climate system can be represented by models of varying complexity—that is, for any component or combination of components a hierarchy of models can be identified, differing in such aspects as the number of spatial dimensions, the extent to which physical, chemical, and biological processes are explicitly represented, or the level at which empirical parameterizations are involved.

Climate system A complex system consisting of five major components: the atmosphere, the hydrosphere, the cryosphere, the lithosphere, and the biosphere, and the interactions between them.

Climate variability Variations beyond the mean state and other statistics of the climate on all spatial and temporal scales.

Continuity equation A partial differential equation that describes a conserved quantity such as mass or energy during transport processes.

Convection The predominantly horizontal locally induced motions of fluid regardless of cause. It includes fluid movement both by bulk motion (advection) and by the motion of individual particles (diffusion).

Coriolis effect A virtual acceleration or force effect that arises during motions on a rotating sphere.

Cryosphere The component of the Earth's climate system comprising all snow, ice, and frozen ground (permafrost) on and beneath the surface of the Earth and ocean.

Data assimilation The process by which observational data are incorporated into a computational model of a physical or biological system.

Differential equation A mathematical equation for a function of one or more independent variables involving the function and its derivatives.

Diffusion A mechanism that results in mixing or mass transport without requiring bulk motion. Together with advection, this is convection.

Dipole A pair of regions such that certain variables of interest move together within each region but move in opposite ways when compared between regions.

Dynamical system A process whose evolution in time is described by deterministic equations, usually systems of differential equations.

El Niño–Southern Oscillation (ENSO) A coupled ocean-atmosphere phenomenon, with a preferred time scale of three to seven years, in the equatorial Pacific Ocean. It is often measured by the surface pressure anomaly difference between Darwin, Australia, and Tahiti and the sea surface temperature in the central and eastern equatorial Pacific.

Energy balance The difference between the total incoming and the total outgoing energy. If this balance is positive, warming occurs; if it is negative, cooling occurs. Averaged over the globe and over long periods of time, the balance must be zero. A perturbation of this balance is called radiative forcing.

Equation of state A thermodynamic equation describing the state of matter (fluids, solids) under a given set of physical conditions. It is a constitutive equation which provides a mathematical relationship between two or more state variables associated with the matter, such as its temperature, pressure, volume, or internal energy.

Extreme event An event that is rare at a particular place and time. Normally, an event is considered "rare" if it falls in the 10th or 90th percentile of the observed probability density function.

Fast Fourier transform (FFT) An algorithm to compute rapidly all periodic components of a discrete signal.

Forcing The action of an agent outside the climate system (volcanic eruption, solar variation, anthropogenic action, etc.) causing a change in the climate system.

Fourier analysis The representation or approximation of functions by sums of trigonometric functions or, more conveniently, complex exponential functions.

General circulation model (GCM) A mathematical model for the behavior of planetary atmospheres and/or oceans, based on partial differential equations such as continuity equations and the Navier–Stokes equation.

Glacial cycle A cycle consisting of a planet-wide glaciation period and a warm period, as observed repeatedly during the past 800 Kyr.

Global mean surface temperature An estimate of the global mean air temperature near the surface of the Earth.

Greenhouse effect The trapping of heat by greenhouse gases (GHGs) within the surface-troposphere system. An increase in the concentration of GHGs leads to an increased opacity for infrared radiation of the atmosphere.

Greenhouse gases (GHGs) Gaseous constituents of the atmosphere that absorb and emit radiation of specific wavelengths within the spectrum of infrared thermal radiation emitted by the Earth's surface, the atmosphere itself, and by clouds. Water vapor (H_2O), carbon dioxide (CO_2), nitrous oxide (N_2O), methane (CH_4), and ozone (O_3) are the primary GHGs.

Heat capacity The measurable physical quantity that characterizes the amount of heat required to change a substance's temperature by a given amount.

Hydrosphere The component of the Earth's climate system comprising liquid surface and subterranean water.

Insolation The amount of solar radiation reaching the Earth by location and by season.

Keeling curve A graph of the CO_2 concentration in the Earth's atmosphere. It is based on continuous measurements taken at the Mauna Loa Observatory in Hawaii since 1958. The Keeling curve provided the first significant evidence of rapidly increasing CO_2 levels in the atmosphere.

Kelvin wave A wave in the ocean or in the atmosphere that is caused by the interaction of the Coriolis effect and a topographic boundary (a coastline) or a waveguide (the equator). Kelvin waves retain their shape. Equatorial Kelvin waves always move eastward.

Lithosphere The upper layer of the solid Earth, both on land and in the ocean, which comprises all crustal rocks and the cold, mainly elastic part of the uppermost mantle.

Lorenz system A system of differential equations that arises from a model truncation of the Navier–Stokes equation. Its solutions exhibit chaotic dynamics.

Mixing layer A layer near the surface of the ocean which is homogenized by turbulent mixing generated by winds, cooling, or other processes such as evaporation and sea ice formation.

Model skill A dimensionless numerical measure for the prediction capability of a model, relative to a reference model.

Navier–Stokes equation The fundamental partial differential equation of hydrodynamics governing the velocity field in a viscous fluid.

Paleoclimate Climate during periods prior to the development of measuring instruments.

Parameterization The technique of representing processes that cannot be explicitly resolved at the spatial or temporal scale of a climate model by means of simplified relations.

Photosynthesis The process by which plants take CO_2 from the air or bicarbonate in water to build carbohydrates, releasing oxygen in the process.

Plankton Microorganisms living in the upper layer of aquatic systems. A distinction is made between phytoplankton, which depend on photosynthesis, and zooplankton, which feed on phytoplankton.

Power spectrum The set of magnitudes of all periodic components of a given signal. It can be found using the fast Fourier transform.

Proxy An indicator that is interpreted using physical and biophysical principles to make inferences about paleoclimate variations.

Reanalysis The result of processing past weather and climate data using state-of-the-art forecasting models and data assimilation techniques.

Regression analysis A statistical technique for estimating the relationships among variables for which observations are available.

Rossby wave In oceans, variations in the sea surface height or in the thermocline that travel at speeds depending on latitude.

Scaling law A mathematical relationship between two quantities, usually in the form of a power law. Such relationships can be transformed to simpler form by introducing dimensionless variables and parameters.

Shallow water equations A simplified set of differential equations for fluid motion in shallow basins, obtained from the Navier–Stokes equation.

Sea ice Ice found at sea that has originated from the freezing of seawater.

Sea surface temperature (SST) The subsurface bulk temperature in the top few meters of the ocean.

Solar radiation Electromagnetic radiation emitted by the Sun. Solar radiation has a distinctive spectrum determined by the temperature of the Sun, with a maximum in the range of visible wavelengths.

Spectral method A mathematical technique for transforming a partial differential equation into an infinite system of ordinary differential equations or an ordinary differential equation into a system of algebraic equations.

Stratosphere The region of the atmosphere above the troposphere, extending from about 10 km (9 km at high latitudes, 16 km in the tropics) to about 50 km in altitude.

Teleconnection An observed connection between climate variations that occur over widely separated parts of the globe and that may also be separated in time.

Thermocline The layer of maximum vertical temperature gradient in the ocean lying between the mixing layer and the abyssal ocean.

Thermohaline circulation (THC) Large-scale circulation in the ocean, driven by density differences and external forces (winds, tides, etc.).

Tropopause Boundary between the troposphere and the stratosphere.

Troposphere The lowest part of the Earth's atmosphere, from the surface to about 10 km in altitude at midlatitudes (9 km at high latitudes, 16 km in the tropics).

Urban heat island An urban area that is warmer than the surrounding rural area due to human activity.

Appendix C

MATLAB Codes

C.1 ▪ MATLAB Code for Lorenz Equations

1. Define an anonymous function, or save it as a separate function file.

```
mylorenz = @(x,sigma,rho,beta)...
[ sigma * (x(2) - x(1)); ...
rho * x(1)-x(1) * x(3) - x(2); ...
x(1) * x(2) - beta * x(3) ];
```

2. Specify the parameters and define an anonymous function that can be used by an ODE solver.

```
sigma = 10; beta = 8/3; rho = 8;
lorenzxdot = @(t,x) mylorenz(x,sigma,rho,beta)
```

These two lines need to be run each time any of the parameters is changed.

3. Choose initial values and a time interval for the solution and compute an approximate solution using a built-in Runge–Kutta solver.

```
x0 = [3;4;1]
Tmax = 40;
[time,solvec] = ode45(lorenzxdot, [0,Tmax],x0);
```

The solutions can then be plotted next to each other (see Figure 7.1) or in a three-dimensional plot (see Figure 7.2).

In order to simulate the Lorenz equations for an ensemble of initial data, define the functions mylorenz and lorenzxdot as above, together with suitable choices for the parameters σ, β, ρ. Let $T_{\max}$ be the length of the time interval, and let N be the ensemble size (the number of simulations). The following code computes the means of N simulations with initial data $\mathbf{x}_0 + \xi$, where $\xi \sim N(0, \gamma^2)$ is a normal random vector.

```
N = 100;
gamma = 1;
Tmax = 40;
x0 = [3;15;1];
tt = linspace(0,Tmax,10*Tmax);
xEns = 0*tt;
yEns = 0*tt;
zEns = 0*tt;
for j = 1:N
[time,solvec] = ode45(lorenzxdot, ...
     [0,Tmax],x0 + gamma*randn(3,1));
t = time;
xEns = xEns + spline(t,solvec(:,1),tt);
yEns = yEns + spline(t,solvec(:,2),tt);
zEns = zEns + spline(t,solvec(:,3),tt);
end
xEns = xEns/N;
yEns = yEns/N;
zEns = zEns/N;
```

The ensemble mean can then be plotted together with the "true" solution or with individual ensemble solutions.

C.2 • MATLAB Code for Regression Analysis

Assume that (i) the response data are given in a column vector $\mathbf{y}$ of length n, and (ii) the p predictors are given in a matrix $\mathbf{X}$ of size $n \times p$, as in Eq. (9.7). In the special case of simple linear regression, where only a single column vector $\mathbf{x}$ of predictors is given, the matrix $\mathbf{X}$ must be generated, for example with the command X = [ones(n,1),x];

The commands to compute the regression parameters α_j, the fits (predictions) $\hat{y}_i$, the residuals $r_i = y_i - \hat{y}_i$, and the coefficient of determination R^2 are

```
alpha2 = X\y;
yhat = X*alpha2;
residual = y-yhat;
Rsquared = (norm(yhat-mean(y))/norm(y-mean(y)))^ 2;
```

The commands to plot the residuals (in observation order), the residuals vs. the fits, a histogram of the residuals, and a normal probability of the residuals are

```
plot(residual)
plot(yhat,residual)
hist(residual)
normplot(residual)
```

C.3 • MATLAB Code for Delay Differential Equations

MATLAB has a built-in low-order solution method for DDEs called dde23. It can solve systems of equations of the general form

$$\dot{\mathbf{x}}(t) = f(t, \mathbf{x}(t), \mathbf{x}(t - \tau_1), \dots, \mathbf{x}(t - \tau_p)). \qquad \text{(C.1)}$$

To solve this problem numerically, the function $(t, \mathbf{x}, \mathbf{z}) \mapsto f(t, \mathbf{x}, \mathbf{z})$ must be specified, where $\mathbf{z}$ is the vector of all delayed arguments. In addition, the delays, the interval $[T_0, T_1]$ on which the solution is desired, and the initial function on $[T_0 - \tau_p, T_0]$ must be specified.

1. To solve the DDE

$$\dot{x}(t) = c x(t) - \epsilon x^3(t) + a x(t - \tau_1) + b x(t - \tau_2) + M \cos \omega t \qquad \text{(C.2)}$$

for $0 < t < T$, define an anonymous function, or save it as a separate function file.

```
myENSOdde = @(t,x,Z,c,a,b,d,M,omega) ...
c*x - d*x^3 + a*Z(1) + b*Z(2) + M*cos(omega * t);
```

2. Specify the parameters and define an anonymous function that can be used by the DDE solver, for example like this.

```
c = -1.; a = 2.6; b = -2.7; d = 0.1; ...
tau0 = 1/24; tau1 = .5; M = .5; omega = 2*pi; ...
mydde = @(t,x,z) myENSOdde(t,x,Z,c,a,b,d,M,omega)
```

These lines need to be run each time the parameters are changed.

3. Choose an initial function, e.g., the function $\varphi(t) = \sin 3t$. Also choose a time interval for the solution. Then compute an approximate solution using the built-in DDE solver.

```
mystart = @(x) sin(3*x);
myDDEsol = dde23(dde1,[tau1 tau2], mystart ,[Tstart
Tend]);
```

The output is an object `myDDEsol` that contains the time coordinates of the computed solutions in the vector `myDDEsol.x` and all computed components of the solution in the vector `myDDEsol.y`. For the example given here, the solution can then be plotted with the command

```
plot(myDDEsol.x, myDDEsol.y)
```

Bibliography

[1] H. ABARBANEL, S. KOONIN, H. LEVINE, G. MACDONALD, AND O. ROTHAUS, *Statistics of Extreme Events with Applications to Climate*, Tech. Report, The MITRE Corporation, McLean, VA, January 1992. http://dodreports.com/ada247342. (Cited on p. 240)

[2] A. APTE, C. K. R. T. JONES, AND A. M. STUART, *A Bayesian approach to Lagrangian data assimilation*, Tellus A, 60 (2008), pp. 336–347. (Cited on p. 265)

[3] V. I. ARNOLD, *Ordinary Differential Equations*, MIT Press, Cambridge, Massachusetts, USA, 1973. Translated and edited by Richard A. Silverman. (Cited on p. 43)

[4] D. S. BATTISTI, *Dynamics and thermodynamics of a warming event in a coupled tropical atmosphere-ocean model*, Journal of the Atmospheric Sciences, 45 (1988), pp. 2889–2919. (Cited on pp. 194, 198)

[5] D. S. BATTISTI AND A. C. HIRST, *Interannual variability in a tropical atmosphere-ocean model. Influence of the basic state, ocean geometry and nonlinearity*, Journal of the Atmospheric Sciences, 46 (1989), pp. 1687–1712. (Cited on pp. 194, 198)

[6] W. BROECKER, *The great ocean conveyor*, Natural History Magazine, 97 (1987), pp. 47–82. (Cited on p. 29)

[7] H. BROER AND F. TAKENS, *Dynamical Systems and Chaos*, Springer-Verlag, New York, Berlin, Heidelberg, London, Paris, Tokyo, Hong Kong, 2010. (Cited on pp. 43, 48, 185)

[8] P. BROHAN, J. J. KENNEDY, I. HARRIS, S. F. B. TETT, AND P. D. JONES, *Uncertainty estimates in regional and global observed temperature changes: A new dataset from 1850*, Journal of Geophysical Research, 111, D12106 (2006). (Cited on p. 97)

[9] F. BRYAN, *High-latitude salinity effects and interhemispheric thermohaline circulations*, Nature, 323 (1986), pp. 301–304. (Cited on p. 83)

[10] M. I. BUDYKO, *The effect of solar radiation variations on the climate of the Earth*, Tellus, 21 (1969), pp. 611–619. (Cited on pp. 19, 141, 146, 156)

[11] J. A. CARTON AND B. S. GIESE, *A reanalysis of ocean climate using Simple Ocean Data Assimilation (SODA)*, Monthly Weather Review, 136 (2008), pp. 2999–3017. (Cited on p. 265)

[12] I. CHUINE, P. YIOU, N. VIOVY, B. SEGUIN, V. DAUX, AND E. LE ROY LADURIE, *Grape ripening as a past climate indicator*, Nature, 432 (2004), pp. 289–290. Reconstructed data available at ftp://ftp.ncdc.noaa.gov/pub/data/paleo/historical/france/burgundy2004.txt. (Cited on p. 100)

[13] E. A. CODDINGTON AND N. LEVINSON, *Theory of Ordinary Differential Equations*, McGraw–Hill, Inc., New York, Toronto, London, 1955. (Cited on p. 43)

283

[14] J. E. COLE, R. B. DUNBAR, T. R. MCCLANAHAN, AND N. MUTHIGA, *Malindi, Kenya Coral Oxygen Isotope Data, IGBP PAGES/World Data Center—A for Paleoclimatology Data Contribution Series # 2000-050.* http://www.ncdc.noaa.gov/paleo/corals.html. (Cited on p. 114)

[15] G. COMPO, J. WHITAKER, AND P. SARDESHMUKH, *Bridging the gap between climate and weather,* SciDAC Review, Spring (2008). http://www.scidacreview.org/0801/pdf/climate.pdf. (Cited on pp. xiii, 102)

[16] J. W. COOLEY AND J. W. TUKEY, *An algorithm for the machine calculation of complex Fourier series,* Mathematics of Computation, 19 (1965), pp. 297–301. (Cited on p. 127)

[17] R. COURANT AND D. HILBERT, *Methods of Mathematical Physics,* Vol. 1, Interscience Publishers, Inc., New York, London, 1953. (Cited on p. 146)

[18] G. C. DANIELSON AND C. LANCZOS, *Some improvements in practical Fourier analysis and their application to x-ray scattering from liquids,* Journal of the Franklin Institute, 233 (1942), p. 365. (Cited on p. 127)

[19] H. DYM AND H. P. MCKEAN, *Fourier Series and Integrals,* Academic Press, New York, San Francisco, London, 1972. (Cited on p. 131)

[20] U. EBERT, M. ARRAYAS, N. TEMME, B. SOMMEIJER, AND J. HUISMAN, *Critical conditions for phytoplankton blooms,* Bulletin of Mathematical Biology, 63 (2001), pp. 1095–1124. (Cited on p. 232)

[21] G. EVENSEN, *The ensemble Kalman filter: Theoretical formulation and practical implementation,* Ocean Dynamics, 53 (2003), pp. 343–367. (Cited on p. 261)

[22] P. G. FALKOWSKI AND M. J. OLIVER, *Mix and match: How climate selects phytoplankton,* Nature Reviews Microbiology, 5 (2007), pp. 813–819. Review. Erratum in: Nature Reviews Microbiology (2007)5(12):966. (Cited on p. 228)

[23] J. B. J. FOURIER, *Théorie analytique de chaleur,* Firmin Didot, père et fils, Paris, 1822. (Cited on p. 123)

[24] ———, *The Analytical Theory of Heat,* Cambridge University Press, Cambridge, UK, and New York, NY, USA, 1878. Translated by A. Freeman. (Cited on p. 123)

[25] P. J. S. FRANKS, *NPZ models of plankton dynamics: Their construction, coupling to physics, and application,* Journal of Oceanography, 58 (2002), pp. 379–387. (Cited on p. 229)

[26] F. R. GANTMACHER, *The Theory of Matrices,* AMS Chelsea Publishing, American Mathematical Society, Providence, Rhode Island, USA, 1959. Translated from Russian, first AMS printing 2000. (Cited on p. 90)

[27] GEWEX, *Global Precipitation Climatology Project (GPCP),* Global Energy and Water Cycle Experiment. http://www.gewex.org/gpcpdata.htm. (Cited on p. 96)

[28] M. GHIL, I. ZALIAPIN, AND S. THOMPSON, *A delay differential model of ENSO variability: Parametric instability and the distribution of extremes,* Nonlinear Processes in Geophysics, 15 (2008), pp. 417–433. (Cited on p. 211)

[29] K. M. GOLDEN, *Climate change and the mathematics of transport in sea ice,* Notices of the American Mathematical Society, 56 (2009), pp. 562–584. (Cited on pp. xv, 216, 217, 218)

[30] K. M. GOLDEN, S. F. ACKLEY, AND V. I. LYTLE, *The percolation phase transition in sea ice,* Science, 282 (1998), pp. 2238–2241. (Cited on p. 216)

[31] R. H. GROVE, *Global impact of the 1789–93 El Niño*, Nature, 393 (1998), pp. 318–319. (Cited on p. 194)

[32] J. GUCKENHEIMER AND P. HOLMES, *Nonlinear Oscillations, Dynamical Systems, and Bifurcations of Vector Fields*, Springer-Verlag, New York, Berlin, Heidelberg, London, Paris, Tokyo, Hong Kong, third printing, revised and corrected ed., 1985. (Cited on pp. 43, 48)

[33] J. HALE, *Ordinary Differential Equations*, Wiley-Interscience, New York, NY, USA, 1969. (Cited on p. 43)

[34] J. HALE AND H. KOÇAK, *Dynamics and Bifurcations*, Springer-Verlag, New York, Berlin, Heidelberg, London, Paris, Tokyo, Hong Kong, 1991. (Cited on p. 43)

[35] J. HALE AND S. M. VERDUYN-LUNEL, *Introduction to Functional Differential Equations*, Springer-Verlag, New York, Berlin, Heidelberg, London, Paris, Tokyo, Hong Kong, 1993. (Cited on p. 205)

[36] J. HANSEN, L. NAZARENKO, R. RUEDY, M. SATO, J. WILLIS, A. DEL GENIO, D. KOCH, A. LACIS, K. LO, S. MENON, T. NOVAKOV, J. PERLWITZ, G. RUSSELL, G. A. SCHMIDT, AND N. TAUSNEV, *Earth's energy imbalance: Confirmation and implications*, Science, 308 (2005), pp. 1431–1435. (Cited on p. 15)

[37] P. HARTMAN, *Ordinary Differential Equations*, Birkhäuser Verlag, Basel, Switzerland, second ed., 1982. (Cited on p. 43)

[38] J. D. HAYES, J. IMBRIE, AND N. J. SHACKLETON, *Variations in the Earth's orbit: Pacemaker of the ice ages*, Science, 194 (1976), pp. 1121–1132. (Cited on p. 134)

[39] D. HENRY, *Geometric Theory of Semilinear Parabolic Equations*, Lecture Notes in Mathematics, Vol. 840, Springer-Verlag, New York, Berlin, Heidelberg, London, Paris, Tokyo, Hong Kong, 1981. (Cited on p. 183)

[40] P. F. HOFFMAN AND D. P. SCHRAG, *Snowball Earth*, Scientific American, 282 (2000), pp. 68–75. (Cited on p. 21)

[41] C. HOHENEGGER, B. ALALI, K. R. STEFFEN, D. K. PEROVICH, AND K. M. GOLDEN, *Transition in the fractal geometry of Arctic melt ponds*, The Cryosphere Discussions, 6 (2012), pp. 2161–2177. (Cited on pp. xv, 218, 219)

[42] E. HOPF, *Abzweigungen einer periodischen Lösung von einer stationären Lösung eines Differentialsystems*, Berichte Math.-Phys. Kl. Sächs. Akad. Wiss. Leipzig, 94 (1942), pp. 1–22. (Cited on p. 73)

[43] J. HUISMAN, M. ARRAYÁS, U. EBERT, AND B. SOMMEIJER, *How do sinking phytoplankton species manage to persist?*, The American Naturalist, 159 (2002), pp. 245–254. (Cited on pp. xv, 231, 232, 233)

[44] IPCC/WG I & II, *Special Report on Managing the Risks of Extreme Events and Disasters to Advance Climate Change Adaptation (SREX)*, 2011. http://www.ipcc-wg2.gov/SREX/. (Cited on p. 237)

[45] F.-F. JIN, *An equatorial ocean recharge paradigm for ENSO. Part I: Conceptual model*, Journal of the Atmospheric Sciences, 54 (1997), pp. 811–829. (Cited on pp. 194, 195, 196, 197)

[46] ———, *An equatorial ocean recharge paradigm for ENSO. Part II: A stripped-down coupled model*, Journal of the Atmospheric Sciences, 54 (1997), pp. 830–847. (Cited on pp. 194, 195, 196)

[47] P. D. JONES, M. NEW, D. E. PARKER, S. MARTIN, AND I. G. RIGOR, *Surface air temperature and its changes over the past 150 years*, Reviews of Geophysics, 37 (1999), pp. 173–199. (Cited on pp. 152, 153)

[48] E. KALNAY, *Atmospheric Modeling, Data Assimilation and Predictability*, Cambridge University Press, Cambridge, UK, and New York, NY, USA, 2003. (Cited on p. 251)

[49] ———, *Ensemble Kalman filter: Current status and potential*, in Data Assimilation: Making Sense of Observations, W. A. Lahoz, B. Khattatov, and R. Ménard, eds., Springer-Verlag, New York, Berlin, Heidelberg, London, Paris, Tokyo, Hong Kong, 2010. (Cited on p. 263)

[50] R. KATZ AND B. BROWN, *Extreme events in a changing climate: Variability is more important than averages*, Climate Change, 21 (1992), pp. 289–302. (Cited on p. 249)

[51] J. KAWALE, S. LIESS, A. KUMAR, A. GANGULY, M. STEINBACH, N. SAMATOVAAND, F. SEMAZZI, P. SNYDER, AND V. KUMAR, *Data guided discovery of dynamic climate dipoles*, 2011. NASA Conference on Intelligent Data Understanding. (Cited on pp. xi, 8)

[52] Y. KAWANO AND T. OHASHI, *A mesoscopic numerical study of sea ice crystal growth and texture development*, Cold Regions Science and Technology, 57 (2009), pp. 39–48. (Cited on pp. xv, 216)

[53] C. D. KEELING, R. B. BACASTOW, A. E. BAINBRIDGE, C. A. EKDAHL, P. R. GUENTHER, AND L. S. WATERMAN, *Atmospheric carbon dioxide variations at Mauna Loa Observatory, Hawaii*, Tellus, 28 (1976), pp. 538–551. Data are available at http://www.esrl.noaa.gov/gmd/ccgg/trends/. (Cited on p. 117)

[54] C. A. KLAUSMEIER AND E. LITCHMAN, *Algal games: The vertical distribution of phytoplankton in poorly mixed water columns*, Limnology and Oceanography, 46 (2001), pp. 1998–2007. (Cited on p. 230)

[55] A. M. G. KLEIN-TANK, F. W. ZWIERS, AND X. ZHANG, *Guidelines on Analysis of Extremes in a Changing Climate in Support of Informed Decisions for Adaptation*, Tech. Report WCDMP-No. 72, World Meteorological Organization (WMO), 2009. (Cited on pp. xv, 239, 240, 248)

[56] J. LASKAR, *Astronomical Solutions for Earth Paleoclimates*. http://www.imcce.fr/Equipes/ASD/insola/earth/earth.html. (Cited on pp. 132, 137, 138)

[57] J. LASKAR, A. FIENGA, M. GASTINEAU, AND H. MANCHE, *La2010: A new orbital solution for the long-term motion of the Earth*, Astronomy & Astrophysics, 532 (2011), p. A89. (Cited on pp. 132, 137)

[58] J. LASKAR, F. JOUTEL, AND F. BOUDIN, *La93: Orbital, precessional and insolation quantities for the Earth from −20 Myr to +10 Myr*, Astronomy & Astrophysics, 270 (1993), p. 522. (Cited on p. 132)

[59] J. LASKAR, P. ROBUTEL, F. JOUTEL, M. GASTINEAU, A. C. M. CORREIA, AND B. LEVRARD, *A long-term numerical solution for the insolation quantities of the Earth*, Astronomy & Astrophysics, 428 (2004), pp. 261–285. (Cited on p. 132)

[60] E. N. LORENZ, *Deterministic nonperiodic flow*, Journal of the Atmospheric Sciences, 20 (1963), pp. 130–141. (Cited on p. 87)

[61] N. LOVENDUSKI, *Air-sea carbon dioxide exchange*, MATLAB code distributed at Workshop on "Ocean Ecologies and Their Physical Habitats in a Changing Climate" at the Mathematical Biosciences Institute, 2011. http://mbi.osu.edu/2010/ws6materials/co3eq.m. (Cited on pp. 227, 235)

[62] V. LUCARINI AND P. H. STONE, *Thermohaline circulation stability: A box model study. Part I: Uncoupled model*, Journal of Climate, 18 (2005), pp. 501–529. (Cited on p. 83)

[63] T. J. LUEKER, A. G. DICKSON, AND C. D. KEELING, *Ocean pCO_2 calculated from dissolved inorganic carbon, alkalinity, and equations for k_1 and k_2; Validation based on laboratory measurements of CO_2 in gas and seawater at equilibrium*, Marine Chemistry, 70 (2000), pp. 105–119. (Cited on p. 225)

[64] J. E. MARSDEN AND M. MCCRACKEN, *The Hopf Bifurcation and Its Applications*, Springer-Verlag, New York, Berlin, Heidelberg, London, Paris, Tokyo, Hong Kong, 1976. (Cited on pp. 73, 74)

[65] R. MCGEHEE AND C. LEHMAN, *A paleoclimate model of ice-albedo feedback forced by variations in Earth's orbit*, SIAM Journal on Applied Dynamical Systems, 11 (2012), pp. 684–707. (Cited on pp. xiii, xiv, 133, 134, 135, 136, 138, 143)

[66] MCRN, MATHEMATICS AND CLIMATE RESEARCH NETWORK. http://www.mathclimate.org/. (Cited on p. xx)

[67] N. MEIER, *Grape Harvest Records as a Proxy for Swiss April to August Temperature Reconstructions*, Master's Thesis, Universität Bern, Switzerland, 2007. (Cited on pp. xiii, 101)

[68] M. MILANKOVIČ, *Kanon der Erdbestrahlung und seine Anwendung auf das Eiszeitenproblem*, University of Belgrade, 1941. (Cited on p. 131)

[69] MPE2013, MATHEMATICS OF PLANET EARTH 2013. http://mpe2013.org/. (Cited on p. xvii)

[70] NCAR, *Southern Oscillation Index (SOI)*. http://www.cgd.ucar.edu/cas/catalog/climind/soi.html. (Cited on p. 10)

[71] E. NISBET, *Cinderella science*, Nature, 450 (2007), pp. 789–790. (Cited on p. 117)

[72] NOAA, *20th Century Reanalysis*. http://www.esrl.noaa.gov/psd/data/20thC_Rean/. (Cited on p. 101)

[73] ——, *Global Precipitation Climatology Project (GPCP)*. http://www.ncdc.noaa.gov/oa/wmo/wdcamet-ncdc.html. (Cited on p. 96)

[74] ——, *Global Surface Temperature Anomalies* (2011). http://www.ncdc.noaa.gov/cmb-faq/anomalies.php#mean. (Cited on p. 269)

[75] ——, *NCEP/NCAR Reanalysis* 1. http://www.esrl.noaa.gov/psd/data/gridded/data.ncep.reanalysis.html. (Cited on p. 9)

[76] ——, *South Pole, Antarctica, United States [SPO]*. http://www.esrl.noaa.gov/gmd/obop/spo/summary.html. (Cited on p. 113)

[77] G. R. NORTH, *Theory of energy-balance climate models*, Journal of the Atmospheric Sciences, 32 (1975), pp. 2033–2043. (Cited on pp. 138, 141, 144)

[78] G. R. NORTH, R. F. CAHALAN, AND J. A. COAKLEY, JR., *Energy balance climate models*, Reviews of Geophysics and Space Physics, 19 (1981), pp. 91–121. (Cited on pp. 141, 144, 152)

[79] G. R. NORTH AND J. A. COAKLEY, *Differences between seasonal and mean annual energy balance model calculations of climate and climate sensitivity*, Journal of the Atmospheric Sciences, 36 (1979), pp. 1189–1204. (Cited on p. 20)

[80] D. OLBERS, *A gallery of simple models from climate physics*, in Stochastic Climate Models, Birkhäuser Verlag, Basel, Switzerland, 2001, pp. 3–63. (Cited on p. 191)

[81] J. PEDLOSKY, *Geophysical Fluid Dynamics*, Springer-Verlag, New York, Berlin, Heidelberg, London, Paris, Tokyo, Hong Kong, 1986. (Cited on p. 178)

[82] T. C. PETERSON, P. A. STOTT, AND S. HERRING, *Explaining extreme events of 2011 from a climate perspective*, Bulletin of the American Meteorological Society, 93 (2012), pp. 1041–1067. (Cited on pp. xvi, 243, 244)

[83] J. R. PETIT, J. JOUZEL, D. RAYNAUD, N. I. BARKOV, J.-M. BARNOLA, I. BASILE, M. BENDERS, J. CHAPPELLAZ, M. DAVIS, G. DELAYQUE, M. DELMOTTE, V. M. KOTLYAKOV, M. LEGRAND, V. Y. LIPENKOV, C. LORIUS, L. PÉPIN, C. RITZ, E. SALTZMAN, AND M. STIEVENARD, *Climate and atmospheric history of the past 420,000 years from the Vostok ice core, Antarctica*, Nature, 399 (1999), pp. 429–436. (Cited on p. 135)

[84] S. RAHMSTORF, *Bifurcations of the Atlantic thermohaline circulation in response to changes in the hydrological cycle*, Nature, 378 (1995), pp. 145–149. (Cited on p. 83)

[85] ——, *On the freshwater forcing and transport of the Atlantic thermohaline circulation*, Climate Dynamics, 12 (1996), pp. 799–811. (Cited on p. 83)

[86] ——, *Ocean circulation and climate during the past 120,000 years*, Nature, 419 (2002), pp. 207–214. (Cited on pp. xi, 29, 30)

[87] J. A. RAVEN AND P. G. FALKOWSKI, *Oceanic sinks for atmospheric CO_2*, Plant, Cell and Environment, 22 (1999), pp. 741–755. (Cited on p. 227)

[88] A. RIDGWELL, J. C. HARGREAVES, N. R. EDWARDS, J. D. ANNAN, T. M. LENTON, R. MARSH, A. YOOL, AND A. WATSON, *Marine geochemical data assimilation in an efficient Earth System Model of global biogeochemical cycling*, Biogeosciences, 4 (2007), pp. 87–104. (Cited on p. 265)

[89] R. B. ROOD AND M. G. BOSILOVICH, *Reanalysis: Data assimilation for scientific investigation of climate*, in Data Assimilation: Making Sense of Observations, W. A. Lahoz, B. Khattatov, and R. Ménard, eds., Springer-Verlag, New York, Berlin, Heidelberg, London, Paris, Tokyo, Hong Kong, 2010. (Cited on pp. 251, 253)

[90] C. ROOTH, *Hydrology and ocean circulation*, Progress in Oceanography, 11 (1982), pp. 131–149. (Cited on pp. 83, 84)

[91] O. E. RÖSSLER, *An equation for continuous chaos*, Physics Letters, 57A (1976), pp. 397–398. (Cited on p. 94)

[92] ——, *An equation for hyperchaos*, Physics Letters, 71A (1979), pp. 155–157. (Cited on p. 94)

[93] D. RUELLE AND F. TAKENS, *On the nature of turbulence*, I, Communications in Mathematical Physics, 20 (1971), pp. 167–192. (Cited on p. 48)

[94] ——, *On the nature of turbulence*, II, Communications in Mathematical Physics, 23 (1971), pp. 343–344. (Cited on p. 48)

[95] C. RUNGE AND H. KÖNIG, *Vorlesungen über numerisches Rechnen*, Springer, Berlin, 1924. (Cited on p. 127)

[96] T. RUTISHAUSER, *Cherry Tree Phenology. Interdisciplinary Analyses of Phenological Observations of the Cherry Tree in the Extended Swiss Plateau Region and Their Relation to Climate Change*, Master's Thesis, Universität Bern, Switzerland, 2003. Reconstructed data available at ftp://ftp.ncdc.noaa.gov/pub/data/paleo/historical/europe/switzerland/swiss-cherry-phenology2003.txt. (Cited on p. 100)

[97] B. SALTZMAN, *Finite amplitude free convection as an initial value problem–I*, Journal of the Atmospheric Sciences, 19 (1962), pp. 329–341. (Cited on p. 87)

[98] J. L. SARMIENTO AND N. GRUBER, *Ocean Biogeochemical Dynamics*, Princeton University Press, Princeton, New Jersey, USA, 2006. (Cited on p. 225)

[99] P. S. SCHOPF AND M. J. SUAREZ, *Vacillations in a coupled ocean-atmosphere model*, Journal of the Atmospheric Sciences, 45 (1987), pp. 549–566. (Cited on pp. 194, 198)

[100] J. R. SCOTT, J. MAROTZKE, AND P. H. STONE, *Interhemispheric thermohaline circulation in a coupled box model*, Journal of Physical Oceanography, 29 (1999), pp. 351–365. (Cited on p. 83)

[101] SCRIPPS, *La Jolla Pier, California*. http://scrippsco2.ucsd.edu/data/ljo.html. (Cited on p. 122)

[102] W. D. SELLERS, *A global climate model based on the energy balance of the Earth-atmosphere system*, Journal of Applied Meteorology, 8 (1969), pp. 392–400. (Cited on pp. 141, 146, 156)

[103] S. SOLOMON, D. QIN, M. MANNING, Z. CHEN, M. MARQUIS, K. B. AVERYT, M. TIGNOR, AND H. L. MILLER, EDS., *IPPC, 2007: Climate Change 2007: The Physical Science Basis. Contributions of Working Group I to the Fourth Assessment Report of the Intergovernmental Panel on Climate Change*, Cambridge University Press, Cambridge, UK, and New York, NY, USA, 2007. (Cited on pp. 1, 224, 273)

[104] H. STOMMEL, *Thermohaline convection with two stable regimes of flow*, Tellus, 13 (1961), pp. 224–230. (Cited on pp. xii, 36, 37, 77, 78, 79, 81, 82, 83)

[105] S. H. STROGATZ, *Nonlinear Dynamics and Chaos: With Applications to Physics, Biology, Chemistry, and Engineering*, Perseus Books Publ., L.L.C., Reading, Massachusetts, USA, 1994. (Cited on pp. xii, 57)

[106] M. J. SUAREZ AND P. S. SCHOPF, *A delayed action oscillator for ENSO*, Journal of the Atmospheric Sciences, 45 (1988), pp. 3283–3287. (Cited on pp. 194, 198)

[107] T. TAKAHASHI, J. OLAFSSON, J. GODDARD, D. W. CHIPMAN, AND S. C. SUTHERLAND, *Seasonal variation of CO_2 and nutrients in the high-latitude surface oceans: A comparative study*, Global Biogeochemical Cycles, 7 (1993), pp. 843–878. (Cited on p. 235)

[108] T. TAKAHASHI, S. C. SUTHERLAND, R. WANNINKHOF, C. SWEENEY, R. A. FEELY, D. W. CHIPMAN, B. HALES, G. FRIEDERICH, F. CHAVEZ, A. WATSON, D. C. E. BAKKER, U. SCHUSTER, N. METZL, H. YOSHIKAWA-INOUE, M. ISHII, T. MIDORIKAWA, Y. NOJIRI, C. SABINE, J. OLAFSSON, TH. S. ARNARSON, B. TILBROOK, T. JOHANNESSEN, A. OLSEN, R. BELLERBY, A. KÖRTZINGER, T. STEINHOFF, M. HOPPEMA, H. J. W. DE BAAR, C. S. WONG, B. DELILLE, AND N. R. BATES, *Climatological mean and decadal changes in surface ocean pCO_2, and net sea–air CO_2 flux over the global oceans*, II, Deep-Sea Research II, 56 (2009), pp. 554–577. (Cited on pp. xv, 227)

[109] F. W. TAYLOR, *Elementary Climate Physics*, Oxford University Press, Oxford, UK, and New York, NY, USA, 2005. (Cited on pp. 17, 37)

[110] E. C. TITCHMARSH, *Introduction to the Theory of Fourier Integrals*, Oxford University Press, Oxford, UK, and New York, NY, USA, second ed., 1937. (Cited on p. 131)

[111] T. TOYOTA, S. TAKATSUJI, AND M. NAKAYAMA, *Characteristics of sea ice floe size distribution in the seasonal ice zone*, Geophysical Research Letters, 33 (2006), p. L02616. (Cited on p. 221)

[112] K. E. TRENBERTH, J. T. FASULLO, AND J. KIEHL, *Earth's global energy budget*, Bulletin of the American Meteorological Society, 90 (2009), pp. 311–323. (Cited on pp. xi, 15)

[113] K. K. TUNG, *Simple climate modeling*, Discrete and Continuous Dynamical Systems – Series B, 7 (2007), pp. 651–660. (Cited on p. 27)

[114] ———, *Topics in Mathematical Modeling*, Princeton University Press, Princeton, New Jersey, USA, 2007. (Cited on p. 146)

[115] E. TZIPERMAN, *Lecture 2: ENSO toy models*, in Conceptual Models of the Climate. Lectures on ENSO, the Thermohaline Circulation, Glacial Cycles and Climate Basics, Neil J. Balmforth, ed., Woods Hole Oceanographic Institution, Woods Hole, MA, USA, 2001. (Cited on pp. 194, 198)

[116] F. VERHULST, *Nonlinear Differential Equations and Dynamical Systems*, Springer-Verlag, New York, Berlin, Heidelberg, London, Paris, Tokyo, Hong Kong, 1990. (Cited on p. 43)

[117] H. VON STORCH AND F. W. ZWIERS, *Statistical Analysis in Climate Research*, Cambridge University Press, Cambridge, UK, and New York, NY, USA, 1999. (Cited on pp. xiii, 97, 107)

[118] W. F. WEEKS AND A. ASUR, *The Mechanical Properties of Sea Ice*, Cold Regions Science and Engineering, II-C3 (1967). (Cited on pp. xv, 216)

[119] E. R. WIDIASIH, *Dynamics of Budyko's energy balance model*. Submitted for publication, 2012. (Cited on p. 146)

[120] WIKIPEDIA, *Joule*. http://en.wikipedia.org/wiki/Joule. (Cited on p. 269)

[121] C. K. WIKLE AND L. M. BERLINER, *A Bayesian tutorial for data assimilation*, Physica D, 230 (2007), pp. 1–16. (Cited on pp. 252, 259)

[122] WMO, WORLD METEOROLOGICAL ORGANIZATION, *Provisional Statement on the Status of the Global Climate*, 2011. (Cited on pp. xv, 238)

[123] J. ZACHOS, M. PAGANI, L. SLOAN, E. THOMAS, AND K. BILLUPS, *Trends, rhythms, and aberrations in global climate 65 Ma to present*, Science, 292 (2001), pp. 686–693. ftp://ftp.ncdc.noaa.gov/pub/data/paleo/contributions_by_author/zachos2001/zachos2001.txt. (Cited on p. 135)

[124] A. ZAGARIS AND A. J. DOELMAN, *Emergence of steady and oscillatory localized structures in a phytoplankton–nutrient model*, Nonlinearity, 24 (2011), pp. 3437–3486. (Cited on p. 236)

[125] A. ZAGARIS, A. DOELMAN, N. N. PHAM THI, AND B. P. SOMMEIJER, *Blooming in a nonlocal, coupled phytoplankton-nutrient model*, SIAM Journal on Applied Mathematics, 69 (2009), pp. 1174–1204. (Cited on p. 236)

[126] R. E. ZEEBE AND D. WOLF-GLADROW, CO_2 *in Sea Water: Equilibrium, Kinetics, Isotopes*, Elsevier, Ltd., Amsterdam, The Netherlands; San Diego, CA, USA; Oxford, UK; London, UK, 2001. (Cited on pp. 226, 227)

[127] F. W. ZWIERS AND H. VON STORCH, *On the role of statistics in climate research*, International Journal of Climatology, 24 (2004), pp. 665–680. (Cited on p. 95)

Index